THE BOOK OF NUMBERS

COMPILED BY THE EDITORS OF HERON HOUSE

PELHAM BOOKS
London

First published in Great Britain in 1979 by
Pelham Books Ltd.
52 Bedford Square
London WC1B 3EF
By arrangement with Heron House Publishing Limited

ISBN: 0 7207 1119 3

Printed in Great Britain by
Hollen Street Press Ltd at Slough, Berkshire
and bound by Dorstel Press Ltd at Harlow, Essex

Editorial Director: Peter Verstappen
Senior Editor: Tessa Clark
Research Director: Esther Piette Runnalls
Art Directors: Tim Fitzgerald
                        Oliver Hickey

Executive Editor: Michael Brook
Statistical Consultant: Geoff Dicks
Senior Writer: Paul Strathern
Illustrator: Nigel Paige

Writers: Penny Naylor, Junius Adams with: John Abineri,
                Robin Cooper, John Cunningham, Carole Hay,
                Nigel Hinton, Rose Lockwood, Patricia Lyons,
                Dennis Maddison, Kevin Rose,
                Kevin O'Sullivan, Marguerite Tarrant

Copy Editors: Alison Fell, Lynn Greenwood, James McCarter,
                De Robinson

Editorial   Meg Dorman, Kerry Green, Becky Llewellen,
Assistants: Maureen Ramm, Maggi Renshaw, Brian Scott

Research   Jill Benson, Liz Cashman, Carole Dufrechou,
Assistants: Ellen Frank, Joe Friedman, Ingela Hedlund,
                Christine Hoover, Beatrice Kernan, Adam Low,
                Fleur Osmanson, Judi Rosenthal, Carey Schofield,
                Susan Sickenger, Jamie Somerville, Hilary Whyte

TABLE OF CONTENTS

INTRODUCTION

Many of us are terrified by numbers. Yet, paradoxically, all of us use numbers every day.

- Are my salary increases keeping me ahead of the game?
- How many other people share my views?
- There are 32 children in my child's class and only one teacher. Is that a decent ratio?
- How does my lifestyle compare with others in the U.K. and in other developed countries?
- Is my sex life below average, average or simply spectacular?

We constantly use numbers to compare ourselves and our lives with the lot of others. It's a great competitive sport and we've always been a sporting country.

The Book of Numbers gives you the ammunition to play this game to your heart's content. Back in the 17th century, Sir Thomas Browne said that numbers have "a secret magic" all their own. In the pages which follow we've attempted to take out the mystery while leaving that magic.

Sex, taxes, making money and trying to hold on to it, holidays, crime, the weather and whether or not we are all going to survive – these topics form the very essence of our daily life. They also cover just some of the topics you'll find in the pages of this book.

Whatever the aspect of man in his environment we've attempted to give you all the available numbers. For numbers are facts and facts are what we all require to draw conclusions.

We've drawn some general conclusions. However, you'll find many facts that are particularly relevant to your own situation ranging from your economic status compared to the rest of the world to how many holidays you get; from how you like to spend your leisure to how often you're taken to bed. That's the inherent fun of *The Book of Numbers*.

"How do I love thee – let me count the ways".
Elizabeth Barrett Browning, *Portuguese Sonnets*

Even in romance, numbers play a part. How many affairs we've had, how old we were when we lost our virginity, how old the person was with whom we first slept. All of these are based on numbers. Where possible, we've now attempted to tell you how your experience relates to the world at large.

The Numbers Jungle

Numbers frighten most of us, yet they govern our lives from before birth until after death. Numbers can tell you what the odds are of having a child and the odds of it surviving. By the same token, numbers can be used to predict when you are apt to die (it varies widely from country to country) and to indicate what will happen to the estate you leave.

We understand how daunting numbers can seem. Therefore, we've employed a talented team of writers who understand both numbers and people. Each is expert in the fields he covers. All have been told to think of the reader as someone who may not understand numbers and/or is usually frightened by them.

All the Facts Between Two Covers
Now, for the first time, we've assembled the world's most fascinating facts in one book and presented them in a uniform, compact and (we hope) easy-to-understand way. The results will let you compare your own life with other Britons and with citizens of other countries.

Numbers do have one beauty: they represent an absolute and thereby allow us to compare like with like. This is exactly what we've attempted to do in the pages which follow. As a result, you can compare practical matters (how much rent you pay for your flat – or a steak – compared with a housewife in Paris) and enjoy whimsical facts (the West Germans really *do* drink more beer than anybody else), as well as more weighty matters (for example, how the government spends your money).

How This Book Was Created
The world is not without its share of books on numbers. The United Nations publishes literally hundreds of volumes. So does the British Government. In fact, most governments are weighed down with numbers. They have to be. You can't make a major governmental decision without them. Private concerns use numbers to make business decisions. Social action groups and lobbyists use them to put forward their case, unions to bargain for their members and to lobby for improved conditions. So the world is full of numbers. With diligent research, good research contacts, and lots of money there is no figure which you could not determine. So why have we bothered to do this book?

The answer is quite simple. This is the first book that takes most of the world's available statistics and puts them on a roughly comparable basis as a single resource. So you can quickly find the facts you want to make interesting comparisons.

You can make these comparisons because our researchers have quite literally circled the globe to bring back answers for you. Our people have waded through statistics at the World Bank in Washington to get financial facts.

We've visited the Danish Institute on Sex for information on private lives and called on Montreal's International Civil Aviation Organisation to find out all about airlines. We've talked to experts to uncover the world's worst airports and Interpol in Paris, Scotland Yard in London and the F.B.I. in Washington to ferret out criminal statistics. These and more than 500 other sources have been tapped, including a wide number of government and United Nations agencies, banks, special research units and other experts in a vast variety of fields.

"A witty statesman said you might prove anything with figures"
Thomas Carlyle, *Chartism*

We've tried to make the figures in this book as accurate as possible but there are some problems. In the first place, many countries simply don't have adequate statistics. This is particularly true in the less developed world where, in many cases, such rudimentary facts as census statistics are almost non-existent. Rulers often don't even know how many people they rule. Other countries have statistics, but withhold them. For these reasons, we've confined our basic tables to the major industrial countries which produce accurate, roughly comparable facts. These are:

Australia	Japan
Austria	The Netherlands
Belgium	Norway
Canada	Denmark
France	Sweden
West Germany	Spain
Ireland	Switzerland
Italy	United Kingdom
	United States

To a degree, virtually all governments, all over the world, use figures to suit themselves. They do so for a variety of reasons. First, because a politician's primary concern is his own survival, so, when elections loom, figures have an unnatural – albeit predictable – propensity to look good. The increase in the cost of living goes down. The balance of trade looks better. Taxes can be held or cut.

Even where figures do exist, there can be disconcerting problems. These, we've discovered, come in a variety of forms. The first is *when* the figures were compiled. It is patently regrettable, but understandable, that all countries don't conduct the same kind of research at the same point in time. So we've had to make a difficult choice between using older comparable statistics and more up-to-date facts. Please do remember similar facts for different years are not precisely comparable.

Numbers Versus Words
Without going into great detail, there's the problem of research base. Exactly how many people were surveyed? What precisely were they asked? What are the criteria applied to each statistic? We live in a world of numbers but research is conducted on the basis of words. Many responses are subject to widely different interpretations.

Some Troublesome Topics
Some topics have given us far more trouble than others. Three immediately come to mind: personal consumption, sex, and crime.

Quite maddeningly, we know that research exists on how often people change their underwear, yet it's confidential and we haven't been able to get those figures. Some commercial firms have been extraordinarily forthcoming. Others have decided that there's nothing to be gained and much to be lost by releasing their figures. Understandable, perhaps. After all, if you were a pet food company would you want it readily known that the denizens of some countries feed their pets a better protein diet than many citizens in the same country receive? But then, if you were the government concerned, would you want that known?

This explains why we've had trouble coming up with certain consumption facts for various countries.

Sex has presented similar problems. Since Kinsey did his pioneer study back in 1948, an incredible amount has been published on this subject, dear to the hearts of (virtually) one and all. Surprisingly, however, to the best of our knowledge, there are no uniform world-wide statistics. Furthermore, the studies of individual countries show wild disparities in what they set out to do and how accurate they are in going about it. What's more, sex is a private matter. So we are dealing with what people report, not what they do.

This applies equally to other personal matters, including abortion, mental health, and the use of drugs. The world abounds in estimates of illegal abortions, yet no accurate statistics can (or could) exist. It isn't difficult to determine the total number of admissions to mental institutions. It is impossible to determine how many different people have been admitted to mental institutions. Then there's the broader question: what exactly is mental health or illness? By the same token, drug use and drug addiction are complicated matters. Marijuana is relatively socially acceptable in the U.K. today. In Spain, it used to command a 30-year prison sentence. On a multi-national basis, no one knows how many people have used drugs; so reliable statistics are few and far between.

Finally, and – to our mind – quite incredibly, uniform crime statistics are a major problem. You'd think that to determine how many people were shot in a country would be a simple matter. Nothing could be further from the truth. Selected countries, like the U.K., have these figures. Other countries, like Italy, may not even list them. For example, in Italy, before 1977, if you had shot your wife to defend your honour because she had been sleeping with someone else, you wouldn't have committed an indictable offence.

"Round numbers are always false"
Dr. Samuel Johnson to Boswell

Perhaps, to a degree, this is true. On the other hand, statisticians will tell you that all figures are rounded. Even a number with three decimal points is rounded from one with five … or ten (or more). So, yes, our figures are rounded up or down to make them more readily understandable to you, the reader. Whether we've used absolute numbers, or decimal points, we've made sure that the figures in each table are directly comparable.

What's more, the reader should be warned that it's temptingly easy to make over simplifications on the basis of abstract figures. For example, in Japan in 1976 they had 0.01 fatal accidents for every million man hours worked in their factories. This means a person would have to work eight hours a day for over 30,000 years before he could expect to be killed at his work. A likely tale! Nor does this convey how safety regulations vary in different countries.

What People Say Versus What People Do
In many cases, figures not only have a "secret magic" but a

13

mystery about them. Italian men say they have extremely advanced ideas about women's liberation, but those attitudes don't translate into action when it comes to helping around the house, as you'll see when you refer to the final chapter. This difference between what people say and do is one of the many paradoxes of *The Book of Numbers*.

Our Credentials

Our work draws on a world-wide data bank including information collected by, amongst others, Gallup International, MORI, NOP and the International Labour Organisation.

In addition, we've been helped by a wide variety of multi-national banks and other business concerns, who have been good enough (normally for the first time) to open up their research secrets to us. Much of the book's most interesting data derives from these sources: where you'll find the world's bustiest women, where the use of anti-perspirants is an exception rather than a rule, which countries are car-craziest – and much more.

We hope you'll use *The Book of Numbers* in a variety of ways. First, it will give you answers to many of the world's most interesting questions on a roughly comparable basis. If nothing else, we hope the book will help to settle (or set off) many an argument over an after-dinner drink.

Second, insofar as possible, it lays out all the figures. If you want to link figures on alcoholism to lack of capitalist incentives, be our guest. If you want to link the degree of socialism to economic welfare, the figures are all here.

Third, we hope that, over time, *The Book of Numbers* will become a standard reference work of interest to those who need to obtain accurate facts on a multi-national basis.

The Funny Side of Numbers

On a lighter note, if you want to know where to go to find sexually liberated women or avoid getting robbed or murdered, we've tried to give you the answers.

We doubt that many readers will read *The Book of Numbers* at one sitting. Instead, we've tried to make it a mine of the world's most illuminating and unusual facts.

People, the land, our environment and how we use (and abuse) it, money, taxes, work, housing, what we do in our hours off, what we eat, our sexual attitudes and habits, what we buy and what we think. We've tried to provide answers in abundance to whatever piques your curiosity.

We're sure we don't have all the answers. We'd like your help in these and other fields. Towards that end, we've included a questionnaire at the back of the book.

This is the first edition of what we hope will be a hardy perennial. We need your help to make this so. In the meantime, we hope that *The Book of Numbers* will be a continuing source of amusing answers.

EXPLANATORY NOTES

We have emphasised the extremes on the tables. The highest figure is outlined by a bold box, the lowest by a dotted box.

We applied average exchange rates prevailing at the time of the data.

COUNTRY GLOSSARY

Benelux: Belgium, the Netherlands, Luxembourg.

E.E.C.:   European Economic Community. Member countries: Belgium; Denmark; France; Italy; Luxembourg; the Netherlands; the Republic of Ireland; West Germany; the United Kingdom.

Great Britain: England; Scotland; Wales. Excludes Northern Ireland.

Ireland: The Republic of Ireland.

Scandinavia: Denmark; Finland; Norway; Sweden. Iceland is also part of Scandinavia but was not included in the surveys used in this book.

United Arab Emirates: Abu Dhabi; Ajman; Dubai; Fujairah; Sharjah; Ummai Qaiwan; Rasal Khaimah.

United Kingdom: England; Scotland; Wales; Northern Ireland.

15

People

People – where would we be without them? If you're interested in learning more about the world's population, this section is the one for you.

With the first table, we start right at the beginning, with facts and figures about birth rates. You'll be able to see at a glance just where the most productive countries are when it comes to babies, and which are the countries where the birth rate is much lower. The birth rates of some countries are more than double those of others. Can you guess which are which? The table doesn't just stop here. Because when you're born your problems are only just beginning. We've got figures which compare your chances of surviving the first year of life depending on where you happen to be born.

But if you're reading this, then you've already survived. You'll probably find yourself looking with more interest at the other figures in the table. Life expectancy, for example. The figures show just how long you can expect to live in each of the countries on the table. If your own country comes near the bottom of the list, perhaps there's still time to think of moving. By the way, we've got separate figures for men and women. It's common knowledge that women generally outlive men. But by how much? And is this the same everywhere? Our table gives you the answers.

As well as giving the facts on birth rates, this table also shows the death rates for different countries. By comparing these two, you'll be able to get some idea of the rate of population growth. You may be surprised to see that in some countries there are actually fewer live births than deaths. This is an area we look at more closely in a later table.

The next thing you may want to know about population is just how it's made up. And the first question you'll probably be asking depends on your sex. Where are there more men – or women? Our table here has the answers. And we've got good news for the men. In all but three of the countries listed, there are more females than males. If you want specific information, we've got that too. If you're young and fancy free, you might want to know which country to go to in order to find the highest percentage of women between the ages of 15 and 29.

Our next table gives you the facts on one of the most important of all areas of population – how it's distributed. So whatever your inclinations, whether you're the gregarious type or a hermit, this table should have something for you. First of all, we've got the plain facts about population for different countries. And so as to be right up to date, the figures are the estimates for 1978.

This table also shows the population densities of these different countries. In other words, on average, how many people occupy each square mile of the land. And it's when they're put in this way that the figures are especially revealing. If it's wide open spaces you're looking for, you'll find some of the figures are very encouraging. But there are other parts of the world where you'll find very much less elbow room.

No account of population would be complete without looking at the cities of the world. The next table does this. It tells us just how the largest cities of the world are distributed. First of all, on an overall global scale, from which we can make some interesting comparisons between eastern and western countries, and between the developed countries and those of the Third World. America and China top the list when it comes to the number of super-cities – those with more than one million inhabitants. But when we include the number of lesser-sized cities, we see that many of the less developed areas are surprisingly well represented in the table.

In the final table in this section, we look to the future. What are the prospects for world population over the next 25 years? Estimates of population growth are never easy to compile because conditions can change so quickly – and unexpectedly. But some general conclusions are inescapable. Populations are growing everywhere. And nowhere more than in the countries where overcrowding is already a problem. In comparison with some of the less well developed Third World countries, the population growth rates of some of the developed nations show signs of easing off. But even here, there's no room for complacency. We can see from the table that, within the next century, many of these countries are likely to have doubled their populations.

Shortage of food is a serious problem today in many parts of the world. There is every likelihood that this problem is going to increase – on a global scale – within the next few decades and become the main focus of political attention.

	U.S.A.	CANADA
	1977	1977
1. **Number of live births per 1,000 inhabitants**[+]	15	16
2. **Deaths per 1,000 inhabitants**[+]	9	7
3. **Infant mortality rate**[o] **per 1,000 live births**	16	15
	1975	1975
4. **Life expectancy (years)**[x]		
men	67.9	69.6
women	75.5	75.6
years by which women outlive men	7.6	6.0

If you were married and lived in Dublin, the chances are that you, or your near neighbours, would have a baby each and every year. Ireland has turned birth into a booming industry with 22 live births per thousand inhabitants. When you consider that all of Eire has just over three million inhabitants and that one million are married, it adds up to an incredible one child per year for two out of 15 married couples.

Popular myth has it that Catholicism equals conception. Sometimes it's true, sometimes it isn't. Spain is number two in the league with 18 births per thousand, but Italy's birthrate is relatively modest.

The U.K.'s population is remaining stable: the same number are born as die.

Generally speaking, it would seem Northern Europe is on contraceptives. So much so that wealthy West Germany is actually losing population: there are ten births per thousand inhabitants as compared with 12 deaths.

Clearly, the Australians intend to fill up their wide open spaces. But the availability of land and the birth rate don't necessarily go hand in hand. Crowded Japan has the same birth rate.

The Germans' low birth rate is in line with the low standard of care babies receive. Amazingly, 20 out of every thousand German babies die before they reach the age of one. This is a deplorable record, beaten only by Austria and Italy, both with 21 deaths per thousand births.

Japan not only has a baby boom. It's also a country which is a boon for babies. They really look after their nippers. Only the Swedes provide better infant care, with a rate of eight deaths per thousand.

If you want to reach a ripe old age you're far more likely to do so if you're a woman. There is no equality between the sexes where life expectancy is concerned: in most countries women live at least six or seven years longer than men. In the U.K., women can expect to live six years longer than their husbands.

You'd think that the rigours of Scandinavia's climate would take their toll. Nothing could be further from the truth: they all live longer there. Sweden heads the longevity list for men (an average of

	U.K.	AUSTRALIA	AUSTRIA	BELGIUM	DENMARK	FRANCE	(WEST) GERMANY	IRELAND	ITALY	JAPAN	NETHERLANDS	NORWAY	SPAIN	SWEDEN	SWITZERLAND
	1977	1977	1977	1977	1977	1977	1977	1977	1977	1977	1977	1977	1977	1977	1977
	12	17	12	12	14	14	10	22	15	17	13	14	18	13	12
	12	8	13	12	10	10	12	11	10	6	8	10	8	11	9
	16	16	21	16	12	12	20	17	21	10	11	11	14	8	12
	1975	1975	1975	1975	1975	1975	1975	1975	1975	1975	1975	1975	1975	1975	1975
	69.6	69.7	68.7	70.1	71.6	70.5	68.3	70.1	70.0	71.5	70.8	71.7	70.3	72.2	70.2
	75.6	76.0	75.5	75.8	77.0	76.7	74.3	75.0	75.6	77.3	77.1	77.8	75.4	77.3	75.4
	6.0	6.3	6.8	5.7	5.4	6.2	6.0	4.9	5.6	5.8	6.3	6.1	5.1	5.1	5.2

72.2 years); and Norway for women (a truly remarkable average of 77.8 years).

On average, American men live to be just 67.9 years old and although their wives can look forward to an average of nearly eight more years of sunny retirement in the Sunbelt, they're far worse off in terms of longevity than their counterparts in other countries. The exceptions are Austria, West Germany, Switzerland, Spain and Ireland.

When it comes to the next life, those German-speakers have the highest death rates – notably the Austrians at 13 per thousand population. The U.K., Belgium, and West Germany come next, then Ireland and Sweden.

Then there's Japan. Just six Japanese per thousand die per year. That's half the rate of the U.K., Belgium or West Germany. If current trends continue there are going to be an awful lot of Japanese in Japan.

Sources:

1, 2 & 3 Population Reference Bureau, Washington DC
4 United Nations

+ Estimates for 1977
o The infant mortality rate is the number of infants under the age of one year who died per 1,000 live births.
x Data corresponds to the period 1975-1980, medium variant
Figures rounded up or down

Do men or women make up the larger proportion of the world's population? . . . Every country in our table, except Switzerland, Ireland and Australia has more women than men.

In Austria there are 5.6% more women than men. Male tourists who want plenty of ladies to feast their eyes on should put Vienna at the top of their European itinerary and West Germany with four per cent more women.

In the U.K. the population is almost even – the women just win out at 51.3%. However across the sea in Ireland there are a few less females (49.8%).

In eight of the countries on the table the largest group of women is between the ages of 15 and 29; in another eight the 45 to 64 age group has the most ladies. In Spain, 11.2% of women fall into both categories. In most countries the largest group of men is between 15 and 29 years-old.

The exceptions are West Germany, where 11.4% of men are between 30 and 44; Sweden (11.8%) and women (12%) in the 45 to 64 years-old); the U.K. with 11.1% of men in both the 15 to 29 and 45 to 64 categories; and Switzerland (11.6% in both the 15 to 29 and 30 to 44 age groups). The highest

	U.S.A.	CANADA
	1975	1975
1. Total population (millions)	213.9	22.7
2. Males (millions)	104.4	11.3
3. Males as % of total population:		
0-4 years	4.0	4.5
5-14 years	8.9	9.4
15-29 years	13.6	13.9
30-44 years	8.4	9.1
45-64 years	9.6	9.3
65 and over	4.2	3.7
total	48.8	49.9
4. Females (millions)	109.5	11.4
5. Females as % of total population:		
0-4 years	3.8	4.3
5-14 years	8.6	9.0
15-29 years	13.3	13.6
30-44 years	8.7	9.0
45-64 years	10.6	9.6
65 and over	6.1	4.6
total	51.2	50.1

percentages between 15 and 29 years-old are in the U.S., Canada, Australia, Japan and the Netherlands.

The table therefore shows that in most countries, there are more women than men and most are in an older age bracket

	U.K.	AUSTRALIA	AUSTRIA	BELGIUM	DENMARK	FRANCE	(WEST) GERMANY	IRELAND	ITALY	JAPAN	NETHERLANDS	NORWAY	SPAIN	SWEDEN	SWITZERLAND
	1975	1975	1975	1975	1975	1975	1975	1975	1975	1975	1975	1975	1975	1975	1975
	56.4	13.7	7.4	9.8	4.9	52.9	61.6	3.1	55.0	111.1	13.5	3.9	35.4	8.3	6.5
	27.5	6.9	3.5	4.8	2.4	26.0	29.6	1.6	26.8	54.6	6.7	1.9	17.3	4.1	3.3
	4.0	5.2	3.6	3.7	3.5	4.2	3.0	5.3	3.9	4.7	4.1	4.1	4.6	3.6	3.6
	8.3	9.2	8.5	7.8	7.9	8.2	8.1	10.1	8.3	7.8	9.0	8.2	9.2	7.1	8.1
	11.1	13.2	10.6	11.5	11.8	12.3	10.5	12.4	11.0	12.5	13.0	11.8	11.4	11.1	11.6
	8.8	9.2	9.4	9.2	9.6	9.3	11.4	7.6	10.0	11.5	9.5	8.3	9.0	9.6	11.6
	11.1	10.0	9.4	11.0	10.9	10.0	9.6	9.7	10.7	9.2	9.6	11.4	10.1	11.8	10.3
	5.3	3.7	5.7	5.9	5.8	5.2	5.4	5.1	4.9	3.4	4.6	5.8	4.5	6.6	4.9
	48.7	50.5	47.2	49.1	49.6	49.2	48.0	50.2	48.8	49.1	49.9	49.7	48.8	49.8	50.1
	28.9	6.8	3.9	5.0	2.5	26.9	32.0	1.5	28.2	56.5	6.8	2.0	18.1	4.2	3.2
	3.8	5.0	3.5	3.5	3.3	4.0	2.9	5.1	3.7	4.5	4.0	3.9	4.4	3.4	3.5
	7.9	8.8	8.1	7.5	7.6	7.8	7.7	9.7	8.0	7.5	8.6	7.8	8.8	6.8	7.6
	10.7	12.4	10.3	11.0	11.2	11.7	9.8	11.9	10.6	12.4	12.4	11.2	11.2	10.6	10.9
	8.6	8.6	9.2	9.0	9.3	8.7	10.3	7.4	10.2	11.6	8.9	8.0	9.3	9.0	10.3
	11.9	9.9	12.3	11.6	11.4	10.5	12.4	9.7	11.8	10.5	10.2	11.6	11.2	12.0	10.6
	8.4	4.8	9.4	8.3	7.6	8.1	8.9	5.9	6.9	4.4	6.0	7.8	6.3	8.3	7.0
	51.3	49.5	52.8	50.9	50.4	50.8	52.0	49.8	51.2	50.9	50.1	50.3	51.2	50.2	49.9

– 45 to 64 years-old against 15 to 29.

Sweden has a high percentage of men (11.8%) and women (12%) in the 45 to 64 age bracket. West Germany, Denmark and Sweden have the fewest children under four.

Source:
United Nations

In every country on the table there are more women than men in the over 65 age group.

	U.S.A.	CANADA
	1978	1978
1. Estimated number of inhabitants (millions)	219.8	23.8
2. Inhabitants per square mile	62.3	6.7

How do you feel about people? Are you a gregarious, sociable type, who enjoys the hustle and bustle of living close to your fellow man? Or do you prefer the peace and quiet of wide open spaces? Whatever your answer, this table will show you where you're most likely to feel at home.

The U.K. comes fourth in this table of population with 57.1 million people. Remember our size and you'll see that tiny Britain crams 2½ times as many people into its borders as vast Canada. But one in eight of us actually lives in London. That's why our countryside is still relatively uncluttered.

The U.S. has the highest population estimate for 1978 of the countries in the table. The spacious American landscape is getting more and more crowded! The country is inhabited by 219.8 million people, an increase of 16 million since 1970. That means about two million more Americans are born each year.

Just across the border in neighbouring Canada, there's a completely different picture: with virtually the same area (3.5 million square miles), Canada has a tenth of America's population – a mere 23.8 million. No wonder the crime rate is so much higher in America than in Canada. However, much of Canada's population is concentrated in the south, while vast areas are virtually uninhabited.

America's other neighbour, down in sunny Mexico (not on the table), is packed in more tightly than either the U.S. or Canada. There's a population of over 65 million, but a much smaller land area.

Japan comes second in the table with a population of 115.1 million. When you consider that its land area only equals four per cent of that of the U.S., this is a startling figure. No wonder the Japanese have mastered the art of miniaturising everything – from trees to transistors. The people aren't noted for their great size either. None of the other countries in the table approach the two population giants – the U.S. and Japan.

West Germany comes next, with a population of 65.4 million. West Germany, like Japan, has concentrated on developing industry to provide a high standard of living.

Ireland is at the bottom of the table. The small population there is probably partly due to the massive exodus to America during the 19th and early 20th centuries. In fact, the descendants of those millions of emigrants contribute today to the high U.S. population figures.

However, if you break down the figures into inhabitants per square mile, they take on quite a different meaning.

The U.S. may come top of the population table, but it ranks only 13th when it comes to population density. There's a vast population in America providing a huge work force and so much land that there's room for everybody. It's not surprising that the U.S. has traditionally

22

	U.K.	AUSTRALIA	AUSTRIA	BELGIUM	DENMARK	FRANCE	(WEST) GERMANY	IRELAND	ITALY	JAPAN	NETHERLANDS	NORWAY	SPAIN	SWEDEN	SWITZERLAND
	1978	1978	1978	1978	1978	1978	1978	1978	1978	1978	1978	1978	1978	1978	1978
	57.1	14.6	7.6	10.0	5.1	54.3	61.9	3.2	55.8	115.1	13.9	4.1	36.5	8.4	6.6
	611.4	5.0	237.7	850.7	312.7	257.4	656.8	121.5	491.6	803.7	1065.3	34.3	189.1	53.2	431.1

been the first choice of immigrants fleeing the overcrowded nations of Europe. Holland's the country where elbow room is at a premium: with over 1,000 Dutch per square mile, no wonder most of the country's under sea level. Neighbouring Belgium comes next. You would never be at a loss for company in that part of the world – "getting away from it all" would probably involve leaving the country.

Japan rates high in both tables. Not only is the population large, but the people are also packed in like sardines. However, the Japanese have made the best use of a potentially difficult situation by intensive cultivation of every available square inch of land. They've also mobilised their massive manpower into a highly efficient technological work force.

Similar sized populations, West Germany and the U.K., also have similar population densities. Italy and Switzerland have very different figures. There are almost nine times as many Italians as Swiss, but when it comes to population density, they're almost neck and neck.

By complete contrast, in Australia, there are only five people per square mile. No wonder they're trying to fill the country with sheep. And in Canada, there are fewer than seven. If you need to be alone at any time, sunny Australia or chilly Canada are the places to head for.

Sources:
1 United Nations
2 Heron House estimates; calculated from 1978 population estimates and the land area.

Like Canada, Norway – another northern country – also has a low population density. But Norway hasn't had to rely on immigration to populate its empty spaces. Sweden is another northern country with a low population density. There are about 53 Swedes per square mile. Denmark, further to the south, breaks the pattern. There are over 300 people per square mile, so it's positively overcrowded compared to the other Scandinavian countries.

If you're looking for a happy medium – a country that has neither too few nor too many people – France is the place to go. The French are comfortably distributed; 257.4 per square mile, with a total of 54.3 million in the entire country. That harmonious balance between overcrowding and underpopulation just could contribute to the French *joie de vivre.*

The E.E.C. countries on our table have only slightly more people than the U.S. and Canada, but their average population density is over eight times as much.

The greatest concentrations of big cities are on opposite sides of the world – in North America and China. North America has 30 cities with over a million inhabitants; China has 26. As far as individual countries on our table are concerned, the U.S. has the largest number of great cities. Canada has only three, though this is a significant number for a country with such a small population overall.

Latin America has the next highest total – 21 cities with over a million inhabitants, then Middle South Asia with 16. At first it may seem surprising that this area, with its teeming Indian population of nearly 550 million, is fourth on the list. But the list of large cities of more than 100,000, but fewer than a million people, changes the picture. Here Middle South Asia is surpassed only by North America and Russia. China has so many large cities that only those with over 200,000 people are shown here.

The high figure for Latin America shows how the population there tends to concentrate in large cities. The geography of much of Latin America – with its vast areas of virgin jungle, mountains and pampas – cannot support even comparatively small cities. So people abandon the land for the big cities.

	AFRICA	LATIN AMERICA (inc. the Caribbean)
	1975	1975
1. Number of cities in the world:		
with over 1 million inhabitants	10	21
with over 500,000 inhabitants	27	42
with over 100,000 inhabitants+	132	171

	U.S.A.	CANADA
	1975	1975
2. Number of cities per country:		
with over 1 million inhabitants	27	3
with over 500,000 inhabitants	50	8
with over 100,000 inhabitants	175	21

The U.S.S.R. has the most evenly distributed urban population, with only 12 cities of over a million people. This is as many as there are in the U.K. and West Germany combined. On the other hand their 222 cities with over 100,000 inhabitants are by far the highest number in our tables in this category (with the exception of China).

As in Latin America, there are reasons for this. Travel in the U.S.S.R. is strictly

LARGEST CITIES

NORTH AMERICA	CHINA	JAPAN	OTHER EAST ASIA	SOUTH ASIA	MIDDLE SOUTH ASIA (inc. India)	WESTERN SOUTH ASIA	EASTERN EUROPE	NORTHERN EUROPE	SOUTHERN EUROPE	WESTERN EUROPE	OCEANIA	USSR
1975	1975	1975	1975	1975	1975	1975	1975	1975	1975	1975	1975	1975
30	26	6	5	9	16	6	7	9	9	12	2	12
58	54	7	8	18	44	12	14	23	19	26	6	40
196	105	78	27	68	183	50	63	80	96	138	15	222

U.K.	AUSTRALIA	AUSTRIA	BELGIUM	DENMARK	FRANCE	GERMANY (WEST)	IRELAND	ITALY	JAPAN	MEXICO	NETHERLANDS	NORWAY	SPAIN	SWEDEN	SWITZERLAND
1975	1975	1975	1975	1975	1975	1975	1975	1975	1975	1975	1975	1975	1975	1975	1975
7	2	1	1	1	3	5	0	5	6	3	2	0	2	1	0
17	5	1	2	1	6	11	1	9	7	5	4	1	5	2	1
58	9	5	5	4	50	54	2	43	78	31	17	3	38	10	7

limited, and all moves are controlled. For example, you need a permit to live in Moscow. As a result the authorities can plan population distribution.

In Western Europe, the U.K. tops the list with seven cities of over a million inhabitants – this is just over a quarter of the U.S. number. West Germany and Italy come next with five each while France has just three.

The two largest cities in the world are New York and Tokyo. Both have more than 17 million inhabitants. Tokyo has grown so large that it has swallowed its next door neighbour, the large port of Yokohama.

Source:
United Nations Population Division
+Over 200,000 for China

Overcrowding and soaring birth rates aren't new. Population problems are older than recorded history. During the pre-history of Europe, for instance, there was a continual migration of tribes from overcrowded Asia which eventually peopled the entire European continent. More recently, the colonisation of the New Worlds – like America and Australia – would never have happened without being nudged on by the bulge in the Old.

New nations with plenty of land like a high birth rate: a booming population means a booming work force. A prosperous economy depends on having plenty of people to produce and buy consumer goods.

Until a few decades ago, almost all governments were officially in favour of large families and a high birth rate. However, improved health care and increased life expectancies have changed this picture drastically.

A steady increase of just one per cent a year will double a country's population in less than a century, leaving it to beg, borrow or steal twice as much food and fuel.

The table shows that most countries face a crowded future. There are nearly 220 million Americans – and that number is growing. At its present rate of growth, the population of the U.S. will have doubled by the end of the 21st century.

	U.S.A.	CANADA
1. Projected population (millions)		
1975	213.9	22.8
1980	224.1	24.6
1985	235.7	26.5
1990	246.6	28.4
1995	256.0	30.0
2000	264.4	31.6
2. Years taken to double the population	116	77
3. Year when the population will double	2094	2055

Populations in smaller countries are expanding at an even faster rate. At least in America there's 3.5 million square miles of land. If the population doubles, there will still be only 125 people to every square mile.

Ireland has just over three million people and, by 2041, it'll have twice that number.

Life will be even more crowded in Japan: 115 million Japanese live in just under 144 thousand square miles – over 800 per square mile. Along with the Irish, they're expanding at the fastest rate in the table. By the year 2047, Japan will have to house more than the present population of the U.S. in around a 20th of the area.

None of the countries listed has as big a problem as some Third World nations. For instance the population of Mexico

POPULATION GROWTH

U.K.	AUSTRALIA	AUSTRIA	BELGIUM	DENMARK	FRANCE	(WEST) GERMANY	IRELAND	ITALY	JAPAN	NETHERLANDS	NORWAY	SPAIN	SWEDEN	SWITZERLAND
56.4	13.8	7.5	9.8	5.0	52.9	61.7	3.1	55.0	111.1	13.6	4.0	35.4	8.3	6.5
57.5	15.1	7.6	10.1	5.1	55.1	62.0	3.3	56.3	117.5	14.1	4.1	37.2	8.5	6.7
58.7	16.5	7.7	10.3	5.2	57.1	62.9	3.5	57.5	122.4	14.6	4.2	39.1	8.8	6.9
60.0	17.8	7.9	10.5	5.2	58.8	64.2	3.7	58.7	126.2	15.1	4.3	41.0	9.0	7.1
61.4	19.0	8.0	10.6	5.3	60.5	65.4	3.8	59.8	129.6	15.6	4.4	43.0	9.2	7.2
62.8	20.2	8.1	10.8	5.4	62.1	66.2	4.0	60.9	132.9	16.0	4.5	44.9	9.4	7.4
+	87	+	+	347	231	+	63	173	69	139	231	69	693	231
+	2065	+	+	2325	2209	+	2041	2151	2047	2117	2209	2047	2671	2209

+Countries where the population will not double

(not on the table) has soared from 30 million to over 65 million in the last 25 years. It's expected to double again by the end of this century.

At the other end of the table, the West German population is growing at the slowest rate: by the end of this century, it'll have increased by only seven per cent. And there'll only be eight per cent more people in Belgium – already very densely populated, with more than 850 people per square mile.

Austria and the U.K. are also expanding slowly. Their population will only be around ten per cent larger by the end of the century. Apart from Catholic Spain and Catholic Ireland, European populations are increasing more slowly than those elsewhere in the world. Experts predict that in four of them – the U.K.,

Sources:
1 United Nations
2 & 3 Population Reference Bureau, Inc
1978 estimates

Austria, Belgium and West Germany – the population will never double.

It's true that some countries – notably the U.S., Canada and Australia – still have room to accommodate their growing populations. But as far as population is concerned, no land is an island. Food and energy are exchanged on a worldwide basis, so population is a worldwide problem. Even if we share what we've got, there won't be enough to go round by the end of the century. There could be an international confrontation between the "haves" and the "have-nots".

27

Number of cities with over one million inhabitants

30 —

Americans seem to be the most gregarious people in the world. The U.S. has 27 cities with over a million inhabitants — more than any other nation. Even 800 million Chinese manage to spread themselves more evenly — there are only 26 million-plus cities in China. People tend to think of Asian cities as overcrowded, but in all of Middle South Asia, including India, there are only 16 cities of over a million.

25 —

Some of the oldest and loveliest cities in the world are in Western Europe. Like Russia, Europe has only 12 urban centres with over a million inhabitants. Although you probably don't associate Africa with the word "metropolis", there are ten big-league cities there, almost as many as in Europe.

20 —

They really like elbow-room in Oceania. The whole area, including Australia, has only two million-plus cities.

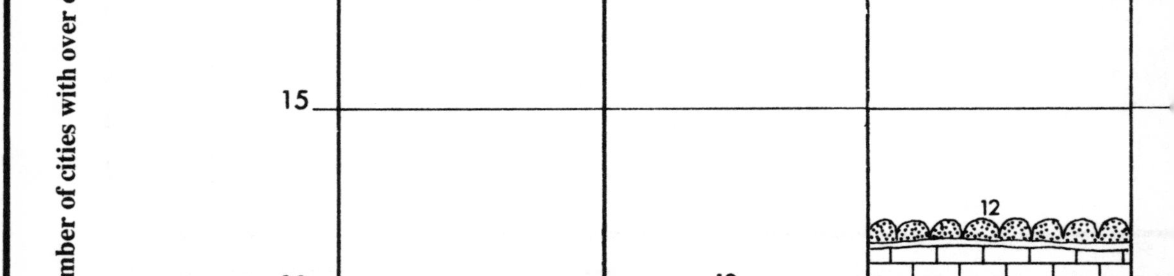

15 —

12

10 —

10

5 —

2

OCEANIA   AFRICA   EUROPE

WORLD'S LARGEST CITIES

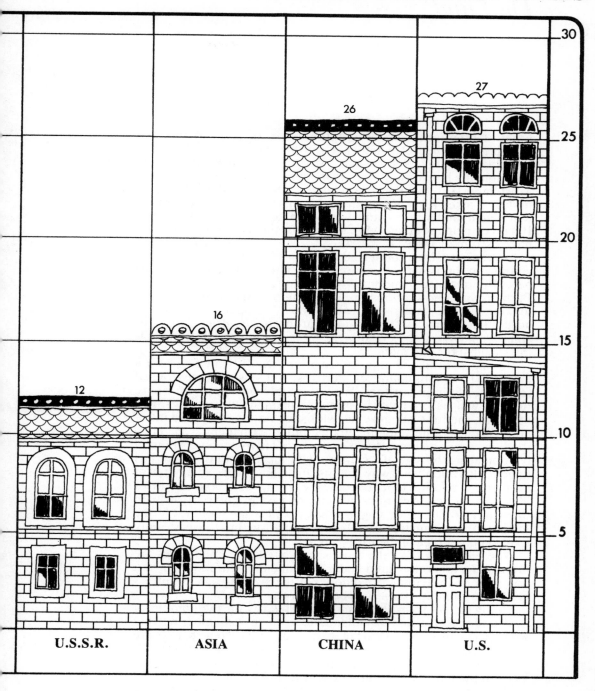

U.S.S.R.   ASIA   CHINA   U.S.

The Land

Land, labour and capital are the basis of any commercial enterprise, no matter whether it's in the Amazon jungle or the heavily populated Netherlands. Stone Age man needed a few tools and the help of some family members in order to sow crops. Most important, however, he needed a patch of land. Land comes first. It is a key form of wealth, and it was for this obvious reason that feudal monarchs assumed the ultimate ownership of all land. It's no coincidence that today's two most powerful countries also have extremely large land areas.

In the first table, you'll find figures for the total area of each country in thousands of square miles. There are some surprises here. For instance, America is not the largest country in the table. Perhaps you know who is on top of the big land league without looking at the table. But what about comparisons? Is the accumulated size of the major European countries larger or smaller than the U.S.? What's the largest country in Europe? Or the smallest for that matter?

Having assimilated this information, you'll find figures on farmland in various countries. Perhaps, you've sometimes longed for a slice of virgin territory. Well, this table will show you where to go. It will also give you information on many other things. For example, the first category has statistics on the total amount of arable land and the amount actually under permanent crop. Then there's a separate breakdown of the two. Have you ever thought of comparing the size of a country with the amount of land it has under cultivation? Norway, for instance, has a very small area of farmland in relation to its overall size. And what about countries like Canada and Australia, which are vast in comparison with those in Europe? They might do very well when it comes to the total number of square miles under cultivation, but these are only a small percentage of their total area.

Where do they have the most sheep or cattle? The figures are there under amount of land under permanent pasture, though you don't get an individual breakdown of the size of each country's herd, you *can* draw your own conclusions.

Following on from this, you'll find figures on the amount of land under forest and woodland. It's worth remembering that wood plays an essential part in many industries. Newspapers are a good example. Every day we take it for granted that they'll arrive on our doorstep. Where does the wood-pulp come from? Look at the figures and you'll be able to work that out.

Moving on from the land, the next table covers water. And

quite rightly so; for without water there would be no farming. First, we have the figures on the number of square miles of water each country has and then go on to give a breakdown of this as a percentage of the total area of the country. To many of us, water is something that is simply there, yet without doubt it's a major resource. For example, Canadians, who have vast quantities, get most of their electricity from water-generated power.

Perhaps the most revealing figures are those for the amount of water as a percentage of the total area of the country. You'll probably be surprised to find who's on top here. The fact that France has the lowest area as a percentage of total land might explain the French aversion to baths (you'll find figures on that in the At Home chapter).

In the same section, you'll also find figures on the amount of land under irrigation. Once again, America is on top. But what about the other countries? You might be under the impression that Europe gets enough rain and therefore irrigation isn't necessary. That's not true. Italy, for instance, has large land areas under irrigation. In fact, if you bear in mind the statistics on land size, you'll find that, proportionally, America doesn't irrigate as much as some European countries do.

Then we have a chapter on the amount of land devoted to national parks. And for you city slickers, there are also figures on the amount of open or green space you'll find in the cities. Here we have something for the holidaymaker. Land might be important when it comes to making a living, but every now and then you want to get out there and simply enjoy it. You'll find it's the New World countries that are way ahead here.

But what about Europe and Japan? These are crowded places. How well do they do in preserving their countryside from urban encroachment?

And finally what about us city dwellers? No country living for us, but every now and then we do like a stroll through the park. Where would be the best bet? You'll find some of the Scandinavian cities head the list here, though certainly not all of them. If you're the type that gets claustrophobic, there are certain cities you'd be best advised to avoid.

	U.S.A.	CANADA
	1975	1975
1. **Total area**[+] (square miles — 000s)	3615	3852
2. **Land area**[0] (square miles — 000s)	3524	3560
as % of total area	97.5	92.4
3. **Arable land and permanent crops (square miles — 000s)**	807.9[xx]	169.0[xx]
arable land[x]	800.7[xx]	168.7[xx]
permanent crops[‡]	7.2[xx]	0.3[‡‡]
4. **Permanent pasture**[**] (square miles 000s)	830.1[xx]	96.9[xx]
5. **Forest and wood**[++] (square miles 000s)	1175[xx]	1244
6. **Other land**[00] (square miles — 000s)	710.6	2050

By far the largest country in the world is Russia – a colossal 8.6 million square miles. That's over double the size of the next largest country, China (3.7 million square miles).

The U.S. and Canada are the giants on our table, with over 3.5 million square miles each. And Australia isn't far behind, with nearly three million. But here the similarities end. The U.S. has more than four times as much arable land (land that produces temporary and permanent crops) than either Canada or Australia – over 800,000 square miles.

However, the latter two countries are better off than the U.S. if you compare populations and arable land. Canada, with nearly 24 million people has just over one-tenth as many mouths to feed as the U.S., and Australia has a mere 14 million (less than the combined populations of New York and London).

Europe is a very different story. The larger countries are about one-tenth the scale of the U.S., Canada and Australia. The smaller ones are about one-hundredth the scale.

Europe's empty countries are Norway, Sweden and Spain. But in terms of land usage, Norway's mountains make the country largely a wilderness. Tiny Belgium, with less than one-tenth the area (and over twice the population), has as much arable land under permanent crops. A massive two-thirds of Sweden is forest and wood. Spain's land is much more amenable to cultivation, but on the whole its quality is poor.

The biggest country in Western Europe is France (over 200,000 square miles), with West Germany and the U.K. both about half that size. The U.K. is worst off for farm land: only 27,000 square miles are arable land.

LAND AREA

	U.K.	AUSTRALIA	AUSTRIA	BELGIUM	DENMARK	FRANCE	(WEST) GERMANY	IRELAND	ITALY	JAPAN	NETHERLANDS	NORWAY	SPAIN	SWEDEN	SWITZERLAND
	1975	1975	1975	1975	1975	1975	1975	1975	1975	1975	1975	1975	1975	1975	1975
	94.5	2968	32.4	11.8‡‡	16.6	211.2	96.0	27.1	116.3	143.8	15.9	125.2	194.9	173.7	15.9
	93.3	2941	31.9	11.7‡‡	16.4	210.8	94.2	26.6	113.5	143.3	13.1	119.0	192.9	158.9	15.4
	98.8	99.1	98.6	99.5	98.4	99.8	98.2	98.0	97.6	99.7	82.1	95.0	99.0	91.4	96.3
	27.0	177.1ˣˣ	6.2	3.2‡‡	10.3	72.6ˣˣ	31.1	4.0ˣˣ	47.5	21.5	3.2	3.1	80.4	11.7	1.5
	26.7	176.4ˣˣ	5.8	3.1‡‡	10.3	66.4ˣˣ	29.1	4.0ˣˣ	36.0	19.1	3.1	3.0	61.1	11.5	1.4
	0.3‡‡	0.7	0.4	0.1‡‡	0.05	6.2ˣˣ	2.0	0.02‡‡	11.5	2.4	0.1	0.05‡‡	19.4	0.2ˣˣ	0.07
	44.9	175‡‡	8.4	2.8‡‡	1.1	51.9ˣˣ	20.2	14.7ˣˣ	20.1	1.7ˣˣ	4.9	0.4	42.8	2.7	6.3
	7.8	531.7‡‡	12.5	2.3‡‡	1.9‡‡	56.4ˣˣ	27.7	0.83	24.3	96.7‡‡	1.2	32.2	57.7	102.0	4.1
	13.6	475.7	4.8	3.4‡‡	3.1	29.8	15.2	7.0	21.6	23.4	3.8	83.3	12.0	42.5	3.5

‡‡Non-official sources   ˣˣFAO estimate

Sources:
Food and Agricultural Organization

+Includes areas under inland water bodies   0Excludes areas under inland water bodies

ˣTemporary meadows for mowing or pasture, land under temporary crops or under market and kitchen gardens or temporarily lying fallow or idle

‡Land cultivated with crops for long periods which need not be replanted after each crop; includes land under shrubs, fruit trees, nut trees and vines, but excluding land under trees grown for wood or timber

**Land used permanently (over 5 years) for both cultivated and wild herbaceous forage crops

++ Land under natural or planted stands of trees

00Includes unused but potentially productive land, built-on areas, wasteland, parks, ornamental gardens, roads, lanes, barren land and all other land not already listed

	U.S.A.	CANADA
	1975	1975
1. Area of water (square miles – 000s)	91.2	291.6
as % of total area of country	2.5	7.6
2. Irrigated land (square miles)	62,942	1831

Water, water, everywhere. Yes, most of the surface of the earth is water. Less than 30% of the earth's surface is land and almost all the rest is ocean.

A tiny percentage of the earth's total surface is made up of rivers, lakes and inland seas. And although these areas might be negligible in comparison with the overall surface area of the world, their significance to the countries in which they are found is enormous.

Imagine America or Canada without the Great Lakes, the Mississipi or the St. Lawrence; Egypt without the Nile, or Amsterdam without its canals.

Although summers in the U.K. must frequently give us the feeling that we're up to our ears in water, nine countries on the table have greater areas of water than we do. In fact we only have 1,200 square miles of inland water. Despite all our famous rivers and the great stretches of water in the Lake District only 1.24% of our total land surface is covered by water.

The country with the greatest area of inland water is Canada, whose lakes and rivers add up to 291,600 square miles.

Right away, you'll probably think of the Great Lakes again. They're well named. Lake Superior, for example, is the largest freshwater surface in the world, with a total area of 32,000 square miles. Large as they are, they by no means account for most of Canada's inland waterways. For one thing, much of them are in U.S. territory. Lake Michigan is entirely Canadian and the border runs straight through the middle of the others.

Most of Canada's largest lakes are in the north and west. There's Lake Winnipeg, and further up in the wilds there's the comparatively unknown ones – Great Slave Lake, Reindeer Lake, Great Bear Lake and hundreds more.

However, Canada's enormous size means that the lakes and other inland water amount to only 7.6% of the total area of the country.

Compare Canada's figure with that of the Netherlands – first in the table. Although the Dutch have a mere 2,800 square miles of water, it amounts to more than a staggering 17% of the country's total area.

Australia is third on the table with a total inland water area of 26,600 square miles. However, Australia has few large freshwater lakes and most of its inland water is the seasonal lakes and salt-marshes of the barren interior.

	U.K.	AUSTRALIA	AUSTRIA	BELGIUM	DENMARK	FRANCE	(WEST) GERMANY	IRELAND	ITALY	JAPAN	NETHERLANDS	NORWAY	SPAIN	SWEDEN	SWITZERLAND
	1975	1975	1975	1975	1975	1975	1975	1975	1975	1975	1975	1975	1975	1975	1975
	1.2	26.6	0.4	0.06	0.3	0.4	1.8	0.5	2.8	0.5	2.8	6.2	2.0	14.9	0.6
	1.2	0.9	1.4	0.5	1.6	0.2	1.8	2.0	2.4	0.3	17.8	5.0	1.0	8.5	3.7
	328[+]	5596	15	4	4	2136[0]	1182	0	13733	10204[x]	282	95	10750	202	114

[+]Excludes Scotland and Northern Ireland   [0]Excludes kitchen and market gardens
[x]Irrigated rice areas only

Source:
Food and Agricultural Organization

For its size, Sweden has more than nine times as much water as Australia. Like Finland, Sweden could well be called the Land of Lakes.

France, Japan and Belgium have the smallest percentage of inland water. France, apart from her famous rivers – most of which have their sources in the mountains to the east of the country – has few areas of inland water.

Although smaller than France, the U.K. has nearly three times as much inland water, a lot of it in Scotland and Northern Ireland.

Since the beginning of civilisation, people have found it necessary to control and direct their water resources. With 62,942 square miles, the U.S. has more land under irrigation than any other country in the table. It covers more than two-thirds of the country.

There's also a lot of irrigated land in Italy – a total of 13,733 square miles – more than its total area of natural water. The same is true of Spain, third in the table. We all know that the rain in Spain falls mainly on the plain, but clearly it doesn't rain often enough.

The economies of both Italy and Spain depend very largely upon their agriculture – and they're the driest countries in Europe. It's hardly surprising that some artificial means of directing their water resources is necessary.

Japan and Australia are other countries where widespread irrigation is vital. In Japan, much is in rice-growing areas.

The scarcity of fresh water in Australia's interior has led to a massive irrigation programme. It may not be so very long before the bare landscape of Western and Central Australia is transformed.

Belgium and Denmark irrigate less land than any other countries – about four square miles each. Ireland doesn't find it necessary to irrigate at all.

The largest countries have the most land set aside for national parks. And the countries with the largest populations (of those on the table) make the most use of their land for recreation.

Although the U.K. is still a "green and pleasant land", we have only 320 square miles of official national park land. West Germany, comparable to us in terms of size and population, has nearly four times as much officially protected land. What we lack in national park land we certainly make up for in local borough public parks and garden space.

		U.S.A.	CANADA
		1975	1975
1.	Number of national parks[+]	252	70
	Total area (square miles – 000s) of national parkland	207.1	88.6
	Population (000s) per square mile of national parkland[o]	1.0	0.26
		1972	1972
2.	Green space in cities:	a	b
	Square yards per inhabitant	23	14[x]

The U.S. leads the field with over 200,000 square miles devoted to national parks and nature reserves (according to the standards of the I.U.C.N.). That's roughly 1/18th of the total U.S.A. land area, or an area almost as large as France.

Canada is next, with just under half as much land devoted to national parks. The other giant, Australia, has one quarter as much or one-sixtieth the total area. But there are few individual parks in Canada. Divide the total national park land figure by the number of parks, and you'll see that each park covers an average of over 1,000 square miles

The surprise in this table is over-crowded Japan. There, 37 national parks cover over 9,000 square miles, the same proportion of park land area as in the U.S.

Between them, Sweden and Norway have 41 national parks and reserves. But neighbouring Denmark is the lowest country on the table, with about ten square miles of national park land. Ireland and Belgium have only one national park apiece, as does Switzerland.

Italy has five national parks, but together they cover only 730 square miles. The U.K. has 320 square miles of park land. West Germany has approximately the same area as the U.K., with a comparable population, but it has nearly four times as much protected land.

We've divided the national park figures by the population figures of each country. In the U.K., over 174,000 people have to share the same square mile of national park land, whereas in Canada the same area caters for only 260 people. In Belgium, there are nearly 700,000 people

	U.K.	AUSTRALIA	AUSTRIA	BELGIUM	DENMARK	FRANCE	(WEST) GERMANY	IRELAND	ITALY	JAPAN	NETHERLANDS	NORWAY	SPAIN	SWEDEN	SWITZERLAND
	1975	1975	1975	1975	1975	1975	1975	1975	1975	1975	1975	1975	1975	1975	1975
	19	239	4	1	1	10	13	1	5	37	21	15	3	26	1
	0.32	50.2	0.43	0.01	0.01	0.96	1.1	0.16	0.73	9.1	0.25	13.2	0.27	2.0	0.07
	174.1	0.28	17.5	698.3	433.3	55.2	54.3	190.9	75.6	12.2	55.0	0.30	133.1	4.3	100.2
	1972	1972	1972	1972	1972	1972	1972	1972	1972	1972	1972	1972	1972	1972	1972
	c	d	e	f	g	h	i	j	k	l	m	n	o	p	q
	14	4^x	12	47	2	12	12	21	11^x	1	36	430	9	88	39

a = New York   b = Montreal   c = London   d = Melbourne   e = Vienna   f = Brussels
g = Copenhagen   h = Paris   i = Frankfurt   j = Dublin   k = Rome   l = Tokyo
m = The Hague   n = Oslo   o = Madrid   p = Stockholm   q = Berne

for every square mile of park. That's about 4.5 square yards of national park for every Belgian.

The Australians are almost as well off as the Canadians, with 280 people for every square mile of national park. Norway has 300 people. The U.S. comes fourth with 1,000 people.

In addition to Belgium, countries with a high ratio of people to national parks include Denmark with over 400,000 citizens for every square mile, Ireland (just under 200,000) and Spain (133,100).

To compare conditions in the cities, we've listed the area of green space available to each inhabitant. Oslo easily tops the table with 430 square yards, five times more than Stockholm. London, a city which prides itself on its parks, has only 14 square yards of park space for each citizen to enjoy. New Yorkers fare

Sources:
1. International Union for Conservation of Nature and Natural Resources
2. Vision, 1973
[0] Heron House estimates
[+] This includes all national parks, national nature reserves and equivalent reserves as defined by the IUCN
[x] National yearbooks

comparatively well; there are 23 square yards of green space for each inhabitant.

Tokyo is easily the worst off for green space – one square yard per inhabitant.

Copenhagen has a mere two square yards of open green space per inhabitant. The Australians must feel a bit cramped in Melbourne – they're each limited to four square yards space.

Square yards per inhabitant

400 —
350 —
300 —
250 —
200 —
150 —
100 —
50 —
23
36
10 —
5 —
1

Tokyo
JAPAN

New York
U.S.

The Hague
NETHERLANDS

If you're looking for wide open spaces, don't head for Tokyo. There's only one square yard of green space per inhabitant there. On the other hand the great record of the northern Scandinavians in Olympic distance races is helped by the fact that even city dwellers have got room to practice. For every 430 Japanese in Shiba Park, Tokyo, there's only one in an Oslo park. If you're planning a picnic in Tokyo, spread out a handkerchief, or be prepared to share your rug with someone else.

Between these extremes there's quite a range in the amount of space available to the inhabitants of the world's larger cities. Stockholm with 88 square yards fares well, while New York is one of the less fortunate. Even with Hyde Park and Regents' Park Londoners are even worse off – they have just 14 square yards.

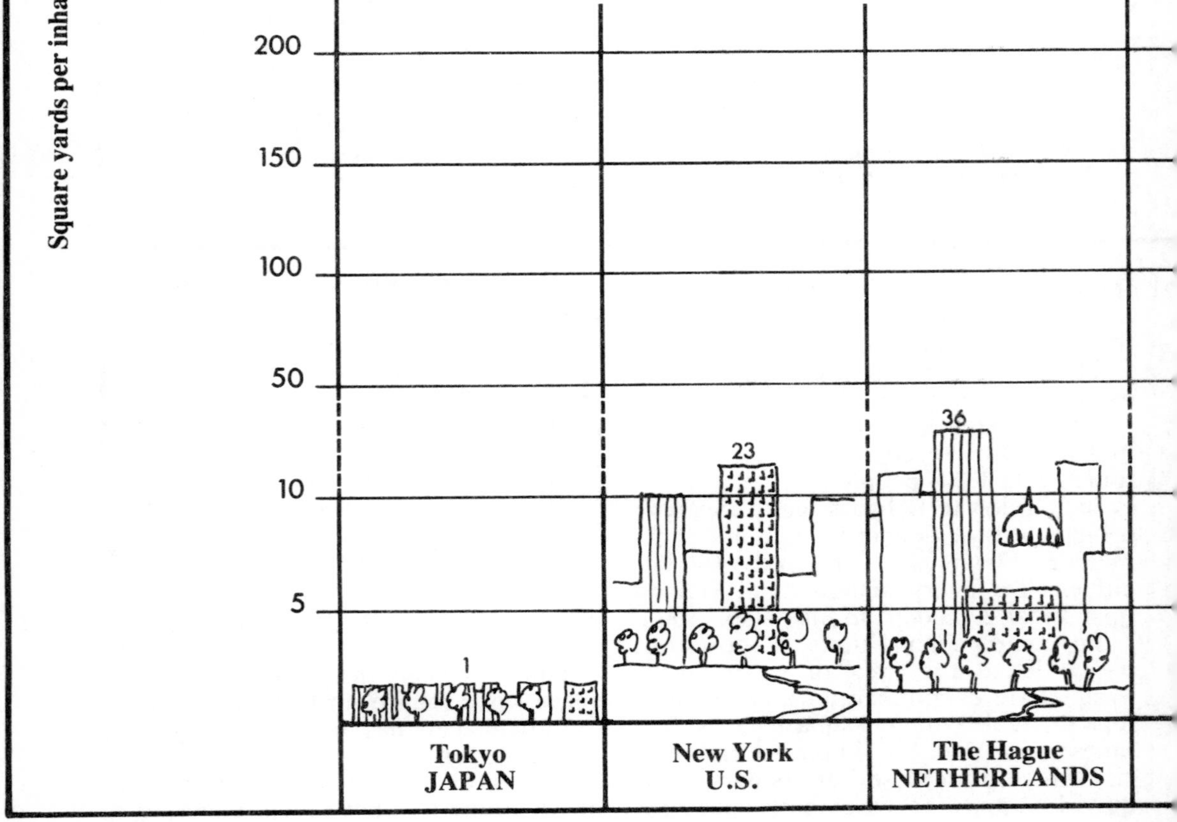

GREEN SPACE IN CITIES

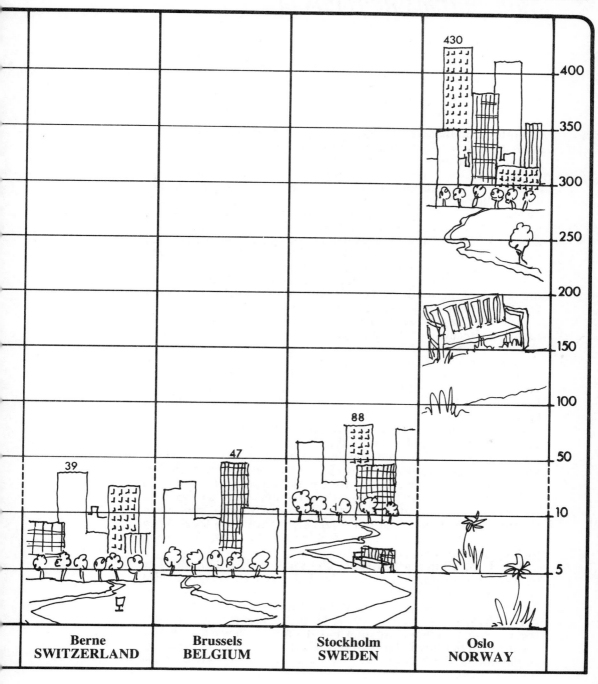

Berne SWITZERLAND 39	Brussels BELGIUM 47	Stockholm SWEDEN 88	Oslo NORWAY 430	

Man and the Environment

It's often said that we're what our environment makes us. If this is so, it's a good idea to take a careful look at the range of different physical environments we find around the world.

Our first table has got all you'll want to know about the amounts of rainfall and sunshine to expect in a number of the world's largest cities. Whether you're planning a long holiday overseas, or just a quick business trip, it could be worth your while to study these. And if you sell umbrellas for a living, the tables will tell you how many wet days you can count on, per year, for each of the cities. Not only that, it'll tell you which months to avoid if you want to miss getting a soaking.

We'd expect Europe's northern cities to have a higher rainfall than those of the south, and that's borne out by the figures. But if you think that by going south all the way to Australia you'll miss out on the grey skies altogether, you could be in for a damp surprise. Because in Sydney you can expect nearly as many showery days as in Oslo.

The next set of figures looks on the bright side. Where are the sunshine cities of the world? And just how many sunny days can you expect in each of these in a year? The table has all the facts here. And it'll tell you which are the sunniest months, and just how sunny they are.

Having dealt with sunshine and rain, in the next tables we turn our attention to temperature. We've taken some of the major countries of the world, and selected a town in each of them where we might expect to find the lowest temperatures. The table has the mean annual temperatures for each of these towns. And just to help in making comparisons between them, there's other data too. Many factors contribute to the climate of an area, and two of the more important are latitude and height above sea level. With the help of these figures you'll be able to make some interesting meteorological speculations.

The U.S. certainly has some chilly cities, like Duluth. The average temperature there isn't much above freezing point. But you can see it gets pretty nippy in some of those northern Scandinavian towns, too. And if you think that it's hot and sunny everywhere in Spain, you could be in for a surprise. Because there are some towns in Spain where it's not much warmer than in the north of Scotland. To enable you to get some idea of the range of climate in these countries, we've got details of some of the hottest towns, too.

Man and the Environment

The next table deals with a different aspect of our environment altogether. Since the arrival of the industrial age, pollution of one kind or another has become a serious problem for most western countries. Nowhere is this more apparent than in the air of our larger cities. We've chosen two measures of air pollution. One measures the accumulation of smoky particles in the atmosphere, and the other provides an estimate of the sulphur-dioxide content of the atmosphere, which is generally associated with heavy industry. The tables are broken down, so you can compare the cities with their suburban neighbours. You'll see that the suburban dwellers don't always get off so lightly. Finally the table enables us to compare the figures over a two year period. And there's good news here. The figures, almost without exception, show signs of improvement.

In the last table we look at the relationship between man and his environment in an entirely different way. We've drawn up a table which offers an original way of seeing how our material environment provides us with the basic needs of life. There's one crude way of doing this, and that's to look at the matter purely in economic terms. How much money does a country's resources generate for each of its members? G.N.P. per capita is the most commonly used measure of this.

But there's another way that we can estimate how well a nation meets the fundamental needs of its people. And that's by making use of what's called the Physical Quality of Life Index. This is explained fully in the text which accompanies the table. Basically this index provides us with a rating which tells us how satisfactorily the members of any country are provided with the necessities of life.

The table shows figures for G.N.P. and P.Q.L.I. for a number of countries. One of the most striking things you'll notice is that the physical quality of life is similar for most western countries, despite some considerable differences in G.N.P. As you'd expect, the low ratings are to be found mostly in the Third World countries. Yet in terms of G.N.P. these are by no means always the poorest countries. In some parts of the world, they seem to be getting very poor value for their money.

Who doesn't keep an eye on the weather? Checking to see if we need a coat on our way to work in the mornings. Or an umbrella in the evenings. Or whether it's going to be a good weekend. Or a good day for planting. For some of us, livelihoods may depend on knowing whether the barometer is going up or down.

If you're a duck, take a holiday on a Dutch canal barge. Amsterdam and nearby Brussels are by far the wettest cities on the table. In each of these cities, it rains, on average, 206 days in the year. That's about four days a week.

If you're spending January in Brussels, you'll be lucky to have more than seven days of dry weather during the entire month. And in December, in Amsterdam, you can expect to have 22 days of rain or snow.

The beautiful city of Copenhagen comes next. If you live with the Danes you'll be wearing your raincoat just about every other day – even more often in December, the wettest month. In Montreal, you can expect 166 wet days in the year.

If you live in one of these cities and are feeling sorry for yourself, take comfort. Things could be worse. In Bahia Felix, Chile (not on our table), it rains, on average, 325 days in the year.

Now where to keep dry? Rome is at the top of the list on the table. It rains only one day in five on their lucky inhabitants (about a third of the rainfall in Brussels). Even in December, the wettest month in Rome, you'll only need your raincoat on eight days in the whole month.

Surprisingly, London is one of the driest cities on our table. It has fewer rainy days than New York, Paris or Sydney, Australia, and roughly the same number of wet periods as Tokyo.

Now let's look on the bright side. Where are the sunny spots? Scientists measure the number of hours of bright sunshine for an area by directing sunlight through a lens onto a rotating drum – the length of the burns corresponds to the

	NEW YORK[x][+]	MONTREAL[+]
1. Days of rain per year[0]	121	166
2. Wettest months	Mar	Jan
days of rain	12.0	18.0[x]
3. Mean hours of sunshine per year	NA	1852
as % of possible hours of sunshine	59	NA
4. Sunniest months	July	July
mean hours of sunshine	66[xx]	253

LONDON	SYDNEY[x]	VIENNA	BRUSSELS	COPENHAGEN	PARIS	MUNICH	DUBLIN	ROME	TOKYO[+]	AMSTERDAM[+]	OSLO[+]	MADRID	STOCKHOLM	ZURICH
113	150	96.0	206	172	164	127.5	142	68.6	115.0[‡]	206	161	84	163	134.6
	Mar/May	July	Jan	Dec	Jan	June		Dec	Sep	Dec	Dec	Mar	Dec	June
11.0**	14.0	9.4	23.0	17.2	17.0	14.0	13.0[++]	7.9	13.0[‡]	22.0	17.0	10.0	17.0	13.0
1514	NA	1891	1551	1603	1840	1862	1386	2491	2019	1651	1632	2843	1973	1694
34	NA	42	35	36	52	39	30	54	NA	34	NA	64	44	38
June	NA	July	June	00	June	July	June	July	Aug	May	June	July	June	July
213	NA	266	223	245	243	261	180	334	203	227	244	366	318	238

[x] Equal to or over a depth of 0.25 millimetres   [xx] % of possible hours of sunshine
[‡] No depth specified
** Jan/April/Dec
[++] Jan/July/Aug/Dec
[00] May/July
NA not available

amount of bright sunshine. The longest burn times are in Madrid, which enjoys an average of 2,843 hours of bright sunshine per year. That's about eight hours of sunshine a day averaged through the year, 64% of the maximum possible. In July, the figure is 366 hours of sunshine for the month – or nearly 12 scorching hours each day.

Rome is second in the sunshine league, with 2,491 hours of happiness a year, or nearly seven hours of sun a day. This is about 54% of the maximum possible, which still means plenty of sun. In July there are nearly 11 hours daily.

If your complexion is delicate and you want to head for the shade, Dublin's the place for you. There, you'll get less than half of the sunshine that they enjoy in Madrid. Dublin sun amounts to less than four hours a day.

London may not have an excess of rain, but it doesn't exactly have a surplus of sun either. The sun shines brightly in London for 1,514 hours a year. That averages out to about four hours a day.

Sources:
Tables of Temperature, Relative Humidity, Precipitation and Sunshine for the World, HMSO
[+] World Survey of Climatology, Elsevier Scientific Publishing Co.
[0] Equal to or over a depth of one millimetre

43

Country	Coldest cities (ft)	Elevation of city (ft)	Length of record (yrs)
U.S.A.[+]	Duluth	610	30
Canada[0]	Saskatoon	1574	30
U.K.	Aberdeen	79	26
Australia	Hobart	177	30
Austria	Salzburg	1427	50
Belgium	Liège	377	30
Denmark	Alborg	10	30
France	Troyes	361	30
Germany (West)	Kiel	46	50
Ireland	Dublin	155	30
Italy	Bolzano	860	9
Japan	Kushiro	108	41
Netherlands	Groningen	16	30
Norway	Vardö	43	40
Spain	Burgos	2808	26
Sweden	Särna	1504	20
Switzerland	Zürich	1539	60

As you'd expect, the coldest cities to be found in the table are in frigid Scandinavian north. Vardö, Norway and Särna, Sweden, both have an annual average temperature at noon of just over 34°F.

North America comes a close second. Over in Saskatoon, out in the bleak Canadian prairies, it's a chilly 35.4°F. And Duluth, Minnesota, isn't much better at 37.9°F. Aberdeen in Scotland is further north than anywhere else on the table, except for those chilly Scandinavian cities. Yet at 47.0°F its average temperature is only four degrees less than Burgos down in sunny Spain, where it's almost 3,000 feet (914 metres) above sea level.

The warmest of the cold spots is Hobart, Australia, with a pleasant average of just over 54°F. Kushiro in Japan is about as far north of the Equator as Hobart is south – but it's quite a bit colder. This is due to the cold Siberian currents from Russia.

But now for the warmest cities. Fort Lauderdale in tropical Florida has an average of over 75.4°F – almost four degrees warmer than Brisbane, Australia, which is located well into the tropics. Malmo, Sweden, has the lowest temperature among the warm cities – an average of 45.9°F. Bergen in Norway doesn't get too much sun either.

COLD AND HOT

Longitude Latitude	Average temp. (°F)	Warmest cities	Elevation of city (ft)	Length of record (yrs)	Longitude Latitude	Average temp. (°F)
46.47N 92.06W	37.9	Fort Lauderdale	7	30	26.07N 80.08W	75.4
52.07N 106.38W	35.4	Vancouver	38	30	49.17N 123.07W	50.8
57.08N 2.06W	47.0	Torbay	27	24	50.28N 3.30W	51.9
42.55S 147.20E	54.2	Brisbane	134	79	27.03S 153.01E	68.8
47.48N 13.02E	47.0	Vienna	663	50	48.12N 16.22E	48.8
50.38N 20.6W	47.0	Ghent	23	30	51.03N 03.43E	50.4
57.03N 09.56E	44.1	Copenhagen	16	30	55.40N 12.35N	46.9
48.18N 04.05E	48.9	Ajaccio	234	46	41.52N 08.32E	60.7
54.30N 10.08E	45.7	Freiburg im Breisgau	912	30	48.00N 7.51E	50.5
53.22N 06.21E	49.0	Cork	56	27	51.54N 08.28W	50.6
46.31N 11.22E	53.0	Palermo	46	25	38.07N 13.22E	64.9
42.58N 144.23E	51.4	Kagoshima	16	30	31.36N 130.33E	62.2
53.13N 06.34E	48.5	Eindhoven	66	30	51.26N 5.30E	51.3
70.22N 31.06E	34.1	Bergen	144	49	60.23N 05.20E	46.0
42.41N 03.42W	51.1	Seville	20	30	37.23N 5.59W	65.8
61.41N 13.07E	34.1	Malmo	10	30	53.46N 13.00E	45.9
47.22N 08.31E	46.8	Lausanne	1831	60	46.32N 6.39E	48.9

Source: National Oceanic and Atmospheric Administration, US Department of Commerce
[+]Excludes the North West territories [+]Excludes Hawaii & Alaska

	ST. LOUIS	VANCOUVER
The sulphur dioxide count[+] in:		
suburban residential areas		
1973	NA	NA
1974	NA	NA
suburban industrial areas		
1973	NA	NA
1974	NA	NA
The suspended[x] particulate count[+] in:		
suburban residential areas		
1973	112.0	63.8
1974	122.5	64.5
suburban industrial areas		
1973	127.2	78.5
1974	143.5	75.0

As late as in the mid-1950s, London was regularly paralysed several times each winter by two and three-day pea soupers. The sky would turn dim and dark yellow at noon, and fogs would reduce visibility to 20 yards or less.

In the 1940s and 1950s, the city of Pittsburgh, Pennsylvania could be seen from many miles away as a dirty black cloud on the horizon. Nowadays, Pittsburgh is almost sparkling clean. And Londoners spend entire winters without a serious fog.

London achieved its results by establishing "smoke-free" zones in which no pollution was allowed and Pittsburgh by imposing stringent controls on smoking chimneys. Similar steps have been taken in other major industrial cities – also with success – but air pollution still remains a menace. Each year it claims its victims by causing serious illness and even death – especially among children, the elderly, and people with weak hearts and lungs. This table shows the levels of air pollution in selected cities. And it's worth noting that in almost every case they're going down. The sulphur dioxide figures give an indication of gas pollution, and the suspended particulate figures show how much solid pollution (dust, sand, coal-dust and so on) there is. In most of the cities listed, industrial areas have more sulphur dioxide pollution than residential ones. This is only to be expected. However, there are exceptions. In Prague, the air in the residential suburbs is almost twice as polluted with sulphur dioxide as in the industrial districts. Frankfurt is worse with nearly four times as much sulphur dioxide pollution in the suburban residential areas. The residential suburbs of Frankfurt are easily the worst for sulphur dioxide pollution in our table.

Madrid has the highest level of sulphur

AIR POLLUTION

LONDON	BRUSSELS	PRAGUE	FRANKFURT	ROME	TOKYO	AMSTERDAM	MADRID	NYKOPING
110.8	102.7	125.8	116.5	22.6	50.9^0	41.5^0	73.3	NA
80.0	84.7	106.8	119.3	16.2	37.4^0	NA	62.6	NA
197.4	123.2	75.4	17.6	57.4	83.3^0	57.0^0	192.7	30.6^0
128.1	102.9	56.0	33.0	107.5	88.4^0	33.7^0	177.7	NA
36.9	31.2	139.0	NA	30.5	72.3^0	NA	72.3	NA
25.7	20.0	134.6	NA	27.5	64.0^0	NA	66.3	NA
46.8	28.7	NA	NA	53.1	77.1^0	NA	303.5	NA
29.7	23.6	120.8	NA	60.3	63.2^0	NA	306.8	NA

NA not available

dioxide pollution in industrial areas. London comes second (despite strict regulations) and then Rome.

It is said that you can barely breathe on a bad day in Tokyo because of car fumes. They do have street corners where you can stop for a blast from an oxygen machine. However, on our table, the figures for Tokyo are only average.

Madrid has the dubious honour of being most polluted by suspended particulates, which clinches its position as the most polluted city on the table. St. Louis comes second. It's the only American city that can come close to European competition.

Which cities have the lowest levels of solid pollution? London and Brussels, both high in sulphur dioxide, are the lowest in suspended particulates.

Source:
World Health Organization
+Micrograms per cubic metre in 24 hours (annual mean)
0Micrograms per cubic metre in one hour
xDirt: sand, dust, coal dust, etc

	U.S.A.	CANADA
1. GNP per person[+] (£)	4418	4206
2. Physical quality of life index[0]	94	95

Money isn't everything. And at last even the statisticians have come to realise this. The richest country in the world isn't necessarily a pleasant place in which to live. We know how to measure the wealth of a country, but how do we measure its "quality of life"?

Recently, the Overseas Development Council (a world-renowned Washington organisation devoted to the improvement of conditions in under-developed countries) worked out a "Physical Quality of Life Index". This combines life expectancy, infant mortality and literacy rates in a single figure – and thus provides us with an insight into how people's basic human needs are being met in any given country. The ratings run from one to 100.

The G.N.P. only shows the annual wealth per capita generated in a country. For example, although the G.N.P. per person in Kuwait is nearly £9,000 – almost four times ours – the P.Q.L.I. figure is only 74, compared with 94 for us. On the simplest level, this shows that in spite of the massive wealth generated by the citizens of Kuwait, the physical quality of life there is considerably lower than ours.

The tables give the P.Q.L.I. ratings and also the G.N.P. per person. Thus we can see at a glance how much real value people are getting for their money.

Sweden has the highest P.Q.L.I. rating – 97. And it comes second in G.N.P. per person. But what about the Swiss? Their G.N.P. per person is highest but their P.Q.L.I. rating is only 95 – about average. However, only two points separate Sweden and Switzerland – which

shows how high the P.Q.L.I. ratings are amongst the countries on the main table. Between them are three, the Netherlands, Japan and Denmark, all equal at 96. It's interesting to see how the G.N.P. figures differ for these countries: in Japan, for instance, it's just under £3,000 per person whereas in Denmark it's just over £4,000 – a substantial difference.

Spain and Norway are lowest. Next lowest in the P.Q.L.I. are the Italians, and then come five countries which all have a rating of 93 – including affluent West Germany, and Ireland, which has the lowest G.N.P. on the table. There seems to be little correlation between the P.Q.L.I. and the G.N.P. – take the figures for the U.K. and the U.S.: the per person G.N.P. for the U.S. is nearly twice that of the U.K. – but they both have a P.Q.L.I. rating of 94.

The accompanying list is full of surprises. The P.Q.L.I. figure for the U.S.S.R. is equal to that of Spain – the lowest in the main table. South Africa – with a G.N.P. per head approximately half that of the U.S.S.R. – has a P.Q.L.I. rate of 48, a mere seven points higher than India. And India has a G.N.P. per person only one-tenth that of South Africa. In Gabon, West Africa, they have a G.N.P. per head of £1,450 – but their P.Q.L.I. is only 21, just over half India's. So all that money must be going into the pockets of a select – and very wealthy – few.

PHYSICAL QUALITY OF LIFE

	U.K.	AUSTRALIA	AUSTRIA	BELGIUM	DENMARK	FRANCE	(WEST) GERMANY	IRELAND	ITALY	JAPAN	NETHERLANDS	NORWAY	SPAIN	SWEDEN	SWITZERLAND
	2251	3416	2985	3797	4172	3668	4133	1434	1708	2750	3472	4155	1635	4855	4973
	94	93	93	93	96	94	93	93	92	96	96	91	91	97	95

	U.S.S.R.	ARGENTINE	INDIA	SRI LANKA	SOUTH AFRICA	GABON	KUWAIT
1. GNP per person$^+$ (£)	1546	868	78	112	750	1450	8669
2. Physical quality of life index^0	91	85	41	82	48	21	74

Source:
Overseas Development Council, Washington DC
$^+$Preliminary 1976 World Bank data
^{0}Composite of life expectancy, infant mortality and literacy figures, each rated on an index of 1 to 100

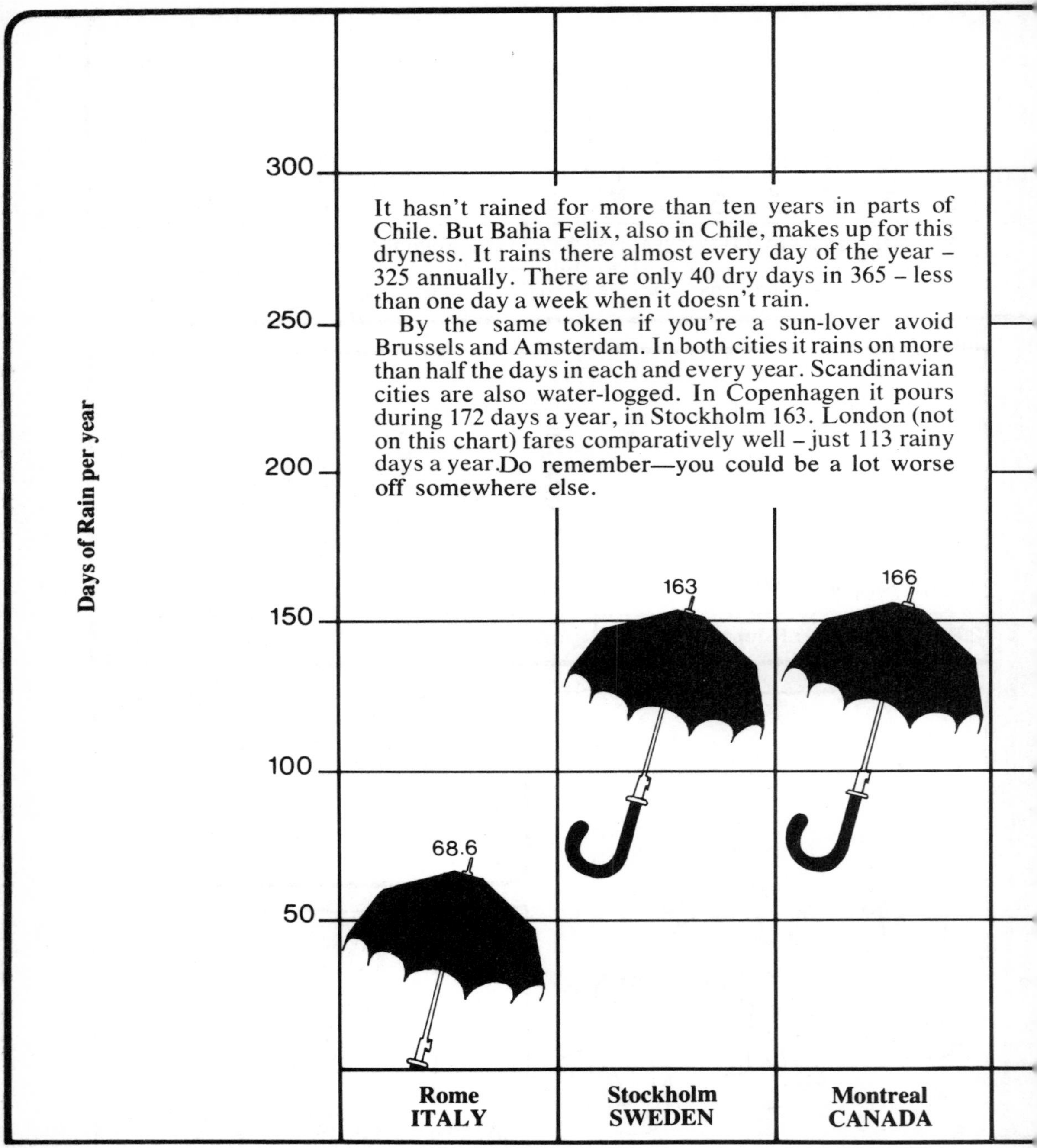

Days of Rain per year

It hasn't rained for more than ten years in parts of Chile. But Bahia Felix, also in Chile, makes up for this dryness. It rains there almost every day of the year – 325 annually. There are only 40 dry days in 365 – less than one day a week when it doesn't rain.

By the same token if you're a sun-lover avoid Brussels and Amsterdam. In both cities it rains on more than half the days in each and every year. Scandinavian cities are also water-logged. In Copenhagen it pours during 172 days a year, in Stockholm 163. London (not on this chart) fares comparatively well – just 113 rainy days a year. Do remember—you could be a lot worse off somewhere else.

300

250

200

150

163

166

100

68.6

50

Rome
ITALY

Stockholm
SWEDEN

Montreal
CANADA

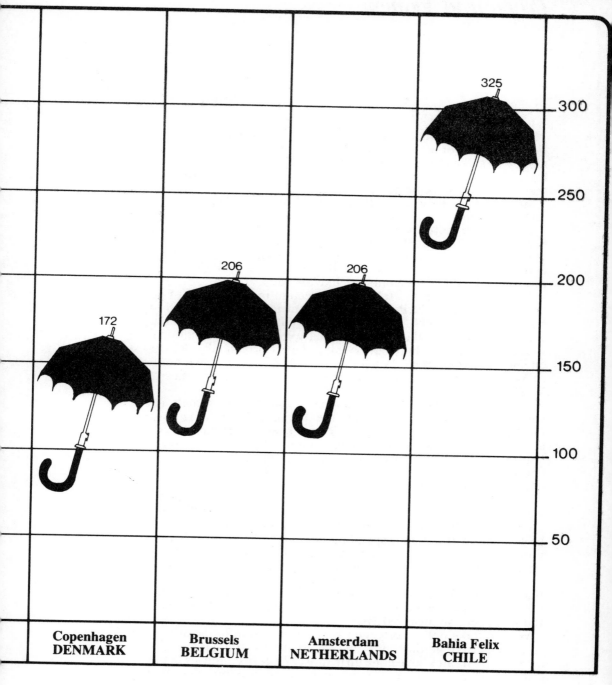

300

250

206

206

200

172

150

325

100

50

| Copenhagen DENMARK | Brussels BELGIUM | Amsterdam NETHERLANDS | Bahia Felix CHILE |

People and Money

There's one thing that no one can do without and that's money. For better or worse, we're all involved in the old game of making money to make the world go round. So read on and find out who the richest people are, who has the most diamonds, or for that matter, who spends the most on steak.

In the first table you will find statistics on per capita income, and interesting they are too. Americans, for instance, don't, on average, earn more than the people of most other countries. In fact, the statistics show that to be wealthy, economic might is not everything. The figures support the idea that small is beautiful. Of course it helps if you produce oil, like the United Arab Emirates, or provide facilities for gambling, like Monaco.

Having looked at the table on per capita income, you'll find statistics on the G.N.P. of various countries. And the word to bear in mind here is "gross" if you'll pardon the expression. Before, we were looking at individual wealth, now it's national wealth, and that includes everything that is produced from car engines to hamburgers. America is in a league of its own here, followed a long way behind by the post-war boom economies of West Germany and Japan.

Is Britain's growth rate in as shameful a state as we sometimes suspect? The table shows you the major industrial nations' G.N.P. figures for several years. You can see the general trends and speculate as to why a particular country's growth rate might rise or fall dramatically. Are the economically successful countries maintaining their positions? Why should Norway and Sweden have such different rates of growth?

Of course, in these matters there comes a time of reckoning. For you and me, it's the bank statement, but for each individual country, it's the balance of payments, and this is what the next table deals with. Who's spending beyond their means? Who's putting money aside for a rainy day? Look carefully at the statistics and you'll be able to draw some interesting conclusions.

But bear in mind one very pertinent factor – inflation. The figures in this table are evidence that no country is immune to the effects of rising prices. In fact, this, above all else, is the thorn in the side of economic progress.

Want to take out a mortgage? You'll find that inflation affects your chances of getting one – and the interest you'll have to pay on it. Interest rates have gone up in most Western countries in the last seven years. There are, however, certain anomalies. For example, why has West Germany's interest rate dropped? A look at the tables might give you an answer. And you don't have to be an economic expert to figure this one out.

Perhaps you've noticed that all the tables so far, except the first one, have dealt with what is termed, in economics, the macro-picture, i.e. the country as a whole. Now we take a look at how people are doing on an individual basis: how does the financial situation vary from the inhabitants of one country to another? Where would you be best off?

And just to remind us who's who in the per capita income league, we've introduced some more up-to-date figures. These only cover the major economic countries of the West so if you want to remind yourself about how people in Liechtenstein, say, are doing, you'll have to refer to the first table again.

But, for those of us who like to imagine we're hobnobbing with the rich, we have a couple of tables on gold, diamonds, furs and expensive cars. Which country do you think buys the most Rolls Royces in a year? And, for that matter, who's likely to be wrapped in sable?

Then we come back to the world of everyday living. Have you ever wondered how your cost of living compares with that of people in other countries? This table will show you just that. The next table gives basic food prices. Where can you buy the cheapest pound of coffee or cheese? Or for that matter, the most expensive egg? Perhaps you feel you already know. Read on and see if your suspicions are confirmed.

The next table deals with household expenditures. The table gives a breakdown of the various items on which people spend their money. Who spends the most on food? Why is it that Americans spend such a high proportion of their incomes on transport? Is it just the size of their country or is there some other factor involved? Do we manage to spend as much as other countries on recreational activities? Are our household expenditures greater than other countries or is it not so bad after all?

The final table deals with socio-economic groups. It tells you what percentage of the working population is in the professional and executive category; what percentages are white-collar, skilled, or unskilled workers.

	U.S.A.	CANADA
	1974	1974
National income per person (£)	2555	2439
Countries in order of wealth	8	9

Which are the richest countries in the world? If you think America comes anywhere near the top of the list, you're wrong. As might be expected, the tiny oil-producing Arab nations around the Persion Gulf head our wealth parade. The United Arab Emirates, with a population of around 800,000 and a land area one-ninth the size of Texas, has a per capita income of £6,880 – almost three times that of the U.S. Qatar and Kuwait aren't far behind. But before you start envying the Arabs, ask yourself if you'd really like to live in Abu Dhabi (if it drove you to drink you'd be out of luck – alcohol is prohibited).

The table shows that industrialisation and technical expertise are no guarantees of wealth. Three great industrial powers, the U.K., France and Italy are all conspicuously missing from the list. And Belgium, another industrial nation, barely qualifies. The top half of our table is dominated by fairly small countries.

The richest country in Europe is the dot-sized principality of Liechtenstein (all 20 square miles of it) with an income of £3,440 for every inhabitant. Liechtenstein derives much of its income from printing postage stamps which are seldom used for mailing anything. Instead they're sold to stamp collectors around the world. The principality also serves as nominal headquarters for many foreign corporations which register there to avoid being taxed in their home countries.

Switzerland, with £2,726 per capita, is the second richest European country. The Swiss export watches, dairy products and chocolate. They also have a thriving tourist industry, but their principal import is money. Untold billions in foreign funds lie on deposit in Swiss banks, making the country one of the great financial powers in the world.

After the Swiss come the Swedes, who enjoy £2,645 per capita. Sweden is a highly industrialised nation with extensive natural resources. Not only do the Swedes earn a high income, they're also covered by one of the most comprehensive social welfare systems in the world. Citizens of a welfare state *can* be hard-working.

The famous gambling tables of Monte Carlo – and perhaps the presence of glamorous Princess Grace – attract enough gamblers and tourists to give the people of Monaco the fourth largest per capita income in Europe – £2,580.

Next, we come to the U.S. – number eight on the world-wide list, with £2,555 per capita. It's curious that this figure should be so low, since their Gross National Product is more than £3,010 for every man, woman and child. Perhaps government spending and the reinvestment of profits by large corporations account for much of the disparity. The Canadians are close behind in ninth place.

A real oddity on our list is country number 13, New Caledonia. It's a small

	AUSTRALIA	ANDORRA	BELGIUM	DENMARK	(WEST) GERMANY	KUWAIT	LIECHTENSTEIN	LUXEMBOURG	MONACO	NEW CALEDONIA	NORWAY	QATAR	SWEDEN	SWITZERLAND	UNITED ARAB EMIRATES
	1974	1974	1974	1974	1974	1974	1974	1974	1974	1974	1974	1974	1974	1974	1974
	2348	2150	2169	2323	2349	4257	3440	2171	2580	2322	2112	5375	2645	2726	6880
	11	16	15	12	10	3	4	14	7	13	17	2	6	5	1

Source:
United Nations

group of South Pacific islands, discovered by Captain Cook and now administered by France. When its 100,000 inhabitants aren't busy basking in the tropical sun, they mine their mineral resources, including gold and silver, and cultivate something equally precious nowadays – that brown gold substance called coffee. The happy New Caledonians have become so prosperous that they're now good customers of the Rolls-Royce Company.

The tiny state of Andorra (190 square miles), tucked away in the Pyrenees between France and Spain, is 16th in prosperity. Tourism is one of Andorra's principal sources of revenue: 500,000 tourists pass through the country every year.

Finally, high income *per se* doesn't always mean affluence. To find out which country has the most buying power, you'll have to consult the table on taxation. It may be that those happy-go-lucky New Caledonians are the richest of all.

What does Gross National Product mean? It's a cash figure for the value of everything that a country produces or earns each year – from nuts and bolts to the tourist trade. So the figure for the Gross National Product (or G.N.P. as it's usually called) is the measure of the real economic strength of a country.

The U.S. is generally considered the largest economic power in the West, but its sheer magnitude is staggering. The U.S.'s G.N.P. is the equivalent of its four closest free-world competitors combined (Japan, West Germany, France and the U.K.) – more than £670 thousand million. That's the equivalent of over £4,000 a year generated for every man, woman and child in the U.S., or nearly £115 per head for the entire population of the world. And the U.S.'s G.N.P. is growing at a rate of nearly five per cent – more rapidly than all the other countries shown except for Japan and Ireland.

Next come the two boom economies of two nations defeated in the Second World War: Japan, with a G.N.P. of nearly £222 thousand million, and West Germany with just over £184 thousand million. Both these countries benefited from a fresh start after the war. Compare their figures to those of the U.K. Formerly second only to the U.S., the U.K. is now producing about half as much as either West Germany or Japan.

But, in some ways, the U.K. is a special case. Since the war, we have shed the burdens and the benefits of an empire. Membership of the E.E.C. has meant that we have had to break into tough European markets, instead of relying on Commonwealth members to buy our goods. These factors, plus a reputation for industrial inefficiency, strikes and outmoded business attitudes have helped cause the country's industrial decline.

France comes after Japan and West Germany, but ahead of the U.K. For years, France suffered from appalling labour relations, unstable governments and the loss of an empire. But with de Gaulle the country began to enjoy increasing stability and prosperity.

Italy has not yet found its economic solution. In fact, its political and industrial unrest is on the upswing. There may be an economic boom in the northern cities – Milan and Turin – but the south of Italy is still one of the most backward and poverty-stricken areas in Europe.

Next comes Canada. Canadians rival the U.S. in terms of production effi-

	U.S.A.	CANADA
	1975	1975
1. Average annual GNP (£ – thousand millions):	679.0	68.4
2. Average rate of economic growth (%):		
1964 – 74	3.3	5.2
1975	−1.8	0.6
1976	6.0	4.9
1977	4.9	2.3

U.K.	AUSTRALIA	AUSTRIA	BELGIUM	DENMARK	FRANCE	(WEST) GERMANY	IRELAND	ITALY	JAPAN	NETHERLANDS	NORWAY	SPAIN	SWEDEN	SWITZERLAND
1975	1975	1975	1975	1975	1975	1975	1975	1975	1975	1975	1975	1975	1975	1975
97.0	34.2	16.0	26.7	15.7	137.0	184.0	3.4	73.8	222.0	34.3	11.7	43.0	29.0	23.1
2.4	4.7	5.1	4.8	3.8	5.2	4.1	3.9	4.7	8.8	4.9	4.3	6.5	3.4	3.3
−1.8	0.5	−2.0	−1.9	−0.7	−1.3	−3.2	−0.5	−3.7	2.1	−1.1	3.3	0.8	0.6	−7.0
1.2	3.5	5.2	2.3	5.0	5.2	5.6	3.0	5.6	6.3	4.6	6.0	2.1	1.5	−2.1
0.6	2.0	3.5	2.8	1.0	2.3	2.4	5.3	1.8	5.7	2.5	4.5	2.4	−2.3	3.5

ciency: nearly £3,500 per head of population. Compare this with the U.K.'s £2,285 per head. And Canada, with all those natural assets in the far north, is certainly a country to watch in the 1980s.

Spain may be the land of the siesta, but it's actually producing nearly half as much as the U.K.

Ireland has easily the lowest G.N.P. figure in the table – just £3,400 million. That's only a third of the figure for Norway. But Ireland's rate of economic growth is now second only to Japan's.

The 1973 oil crisis made a nasty dent in everyone's figures. In most places the growth rate went into the red. Only Japan and Norway sustained a positive rate of over one per cent during the worst of the recession. Japan's high oil expenditure was counter-balanced by massive exports of low priced goods. Norway was saved by the beginning of the North Sea oil bonanza. Switzerland was hit worst of all, with a negative rate of seven per cent in 1975.

What are the prospects for the future?

Sources:
1. *World Bank Atlas*
2. Organization for Economic Co-operation and Development

Along with Japan and Ireland, the U.S. has a healthy rate of expansion at nearly five per cent. And Norway's rate, like that of the U.S., will probably continue to grow. With a G.N.P. of nearly £11,800 million – less than half that of Switzerland or Belgium – Norway is still a long way from being a world economic power. But if North Sea oil revenue is put to good use, the country could lead the field in its growth rate.

Things don't look too rosy for its Scandinavian neighbours. Sweden's growth rate, never high at the best of times, is now minus 2.3%. And Denmark's G.N.P. is barely growing at all. But at least Denmark is doing better than we are. With a growth rate of just over one half of a per cent, the U.K.'s economy still isn't out of the woods.

We hear a lot about the balance of payments when people diagnose the state of our economy. If a country exports more than it imports, the result is a profit for the country – and the balance of payments is a positive figure. If a country, like a shop, stocks (imports) more than it sells (exports) it's in deficit and the balance is negative. So the balance of payments is a financial account of a country's overseas trade.

Balance of payments (£ – millions):	U.S.A.	CANADA
1972	−2329	−155
1973	+2826	+44
1974	+502	−662
1975	+7982	−2123
1976	+1942	−2342

Imports mean consumer items and raw materials brought in from outside the country. Exports mean goods sold abroad. But something known as "invisible exports" is another factor to consider. These are services – such as banking, insurance and tourism. Though no actual goods are exchanged, the money paid for these services comes into the country.

As you can see, many of the countries on our table had a surplus (or positive) balance of payments in 1972 – the exceptions were Norway, Denmark, Canada, Ireland, Austria and the U.S., which had the biggest deficit – £2,329 million, caused by the enormous expense of the Vietnam War. When the U.S. withdrew from Vietnam in 1973, the balance shot up to a surplus of £2,826 million.

The oil crisis caused by the 1973 Yom Kippur War lies behind many of the changes in the figures on our table. The Arab countries embargoed oil as a result, and the effect was catastrophic.

Our figures reflect this. Price rises began to bite into the spending power of people in the West in 1973 and 1974. The U.S., Switzerland, the Netherlands, West Germany and Belgium were the only countries who managed to avoid going into deficit.

The U.K. was easily the worst hit country of all. After two years of the oil crisis, a £154 million trade surplus had crashed to a record £3,366 million deficit.

In 1972, Japan – which relies totally on imported oil had a trade surplus of £2,650 million. By 1973, the booming Japanese export economy was running at a deficit of over £56 million. And in 1974, this had plummeted to a deficit of £2,018 million. Italy, another country without internal oil resources, went from a healthy surplus of £906 million to a deficit of £2,597 million in those two years.

It would be interesting to compare these figures with the balance of payments in oil-rich, exporting countries. If Iran were shown, we would see that it was one of the main beneficiaries from the oil price rise. In 1972, it had a trade deficit of £218 million (almost the same as the Australian surplus for that year). In 1973, this deficit was transformed into a

U.K.	AUSTRALIA	AUSTRIA	BELGIUM	DENMARK	FRANCE	(WEST) GERMANY	IRELAND	ITALY	JAPAN	NETHERLANDS	NORWAY	SPAIN	SWEDEN	SWITZERLAND
+154	+224	−75	+457	−25	+119	+299	−54	+906	+2650	+516	−24	+261	+107	+88
−739	+194	−135	+473	−190	−283	+1792	−96	−1029	−56	+960	−150	+233	+498	+114
−3366	−1124	−196	+392	−422	−2555	+2030	−295	−2597	−2018	+886	−480	−1389	−401	+73
−1673	−262	−151	+317	−221	−1	+1753	−31	−238	−307	+746	−1105	−1317	−738	+1164
−1227	−770	−823	−166	−1055	−3340	+1880	−138	−1581	+2037	+1320	−2064	−2207	−1334	+1936

Source:
International Monetary Fund

£86 million surplus. In 1974 the surplus suddenly rose to a gigantic £6,896 million.

What are the general trends in our table? The U.K. has gradually recovered from its record deficit. Though the negative figure of £1,227 million for 1976 is still far from healthy, we did break even in 1977.

Other countries haven't managed to stage a similar recovery – although no one else had quite so far to go. The figures for all the Scandinavian countries show an increasing decline in their balance of payments. There are many reasons for this. But above all, neither Norway or Sweden is a member of the Common Market, and as a result, both countries have been cut out of their traditional markets and they're beginning to feel the pinch now that the E.E.C. tariff barriers are going up. Denmark, on the other hand, *is* a member of the E.E.C. and still faces a large deficit. Norway does have the hope of selling North Sea oil to improve their exports: the Norwegians are expected to find themselves the wealthiest people in Europe by 1982.

France, Canada, Spain and Austria still have to halt the slide in their balance of payments. France and Canada are worst off but France improved in 1977. And the U.S. economy has also taken a steep downturn. Some people think that weak countries should devalue their currencies to improve their foreign trade balance. Devaluation would mean that imported raw materials cost more. But it would also mean that a country's goods are cheaper and, thus, more competitive overseas. Devaluation can therefore be beneficial for some economies, although it is a drastic step.

Now for the success stories. Japan, West Germany and Switzerland have the best records. Switzerland, as always, has a large invisible exports surplus. But West Germany and Japan have won through by the sheer strength of their economies. Japan weathered the oil crisis despite a couple of bad years. And West Germany seems to be able to maintain a pretty stable trade advantage. In 1977, Japan's balance of payments surplus was nearly £10,000 million.

Inflation goes up, up, up. And the value of the currency goes down, down, down. When prices rise, wages rise too. Then prices go up again. Soon we're caught in a vicious circle.

In the U.K. – and many other countries – inflation has been the most troublesome economic problem for the past half-dozen years. The rising rate of inflation has wiped out most increases in pay. It continues to undermine the standards of profit for consumers and labour as well as for businesses.

Inflation also goes hand-in-glove with higher rates of unemployment. Economies running at full tilt tend to be inflationary, while those that keep inflation to a minimum often do so at the expense of full employment. Thus any severe anti-inflationary measures often come at the expense of those who can't get work or lose their jobs.

Of the major Western industrial countries, the U.K. was worst hit by inflation between 1970 and 1975. Our rate leapt a record 11% after the oil crisis and reached an astounding 24% during 1975 a year remembered for its rapid price rises. Ireland followed close behind at 21%.

The rate of world-wide inflation was given a sharp boost by the quadrupling of oil prices by the O.P.E.C. in 1974. This affected nearly every major Western industrial economy. In the Netherlands, all Middle Eastern oil imports were cut off for a time. Italy, too was hard hit and was forced to introduce rationing. Soon the whole world was feeling the pinch. As

The rate+ of inflation (%):	U.S.A.	CANADA
1964-75	4.7	4.8
1970-75	6.3	7.3
1975	9.1	10.8
1976	5.8	7.5
1977	6.5	8.0

the table shows, inflation started to spiral.

The exceptions? West Germany's post-war economic miracle rode out the storm and its inflation rate actually decreased in 1977 to a lower level than in 1964-75. Switzerland's dropped to an almost non-existent 1.3%.

By the standards of some European countries, the U.S. has kept its inflation rate within acceptable single digit limits. But for most Americans, it's still too high. Maintaining the rate below seven per cent has been one of the Carter administration's chief priorities.

Italy was also badly hit by the oil price rise. It started to recover in 1976, but inflation began to spiral again in 1977. With high unemployment and lack of confidence in the economy – as well as the government's inability to deal with increasing political unrest – the economic prospect in Italy looks bleak indeed.

In 1975, the Western industrial nations began a concerted effort to bring down inflation. Everywhere except Spain and Sweden things began to improve, though with varying degrees of success. Spain's economy was adversely affected by the political unrest which accompanied the

INFLATION

U.K.	AUSTRALIA	AUSTRIA	BELGIUM	DENMARK	FRANCE	(WEST) GERMANY	IRELAND	ITALY	JAPAN	NETHERLANDS	NORWAY	SPAIN	SWEDEN	SWITZERLAND
6.9	5.6	4.9	5.1	7.5	5.6	4.1	7.7	5.8	7.9	6.1	5.9	8.2	5.7	5.2
13.0	10.2	7.3	8.4	9.3	8.8	6.1	13.3	11.3	11.5	8.6	8.2	12.1	8.0	7.7
24.2	15.1	8.4	12.8	9.6	11.7	6.0	20.9	17.0	11.8	10.2	11.7	16.9	9.8	6.7
16.5	13.5	7.3	9.2	9.0	9.6	4.5	18.0	16.8	9.3	8.8	9.1	17.6	10.3	1.7
15.9	12.3	5.5	7.1	11.1	9.5	3.9	13.6	19.3	8.0	6.7	9.1	24.5	11.4	1.3

Organization for Economic Co-operation and Development
+ Annual average rate of increase

transformation from dictatorship to parliamentary government after Franco's death in 1975.

If the U.K. was the country worst hit in the first half of the 1970s, we have also made the greatest strides in the worldwide battle against inflation. Heavy cuts in public spending, wage restraint and the revenue from North Sea oil brought our inflation rate down to below ten per cent in early 1978.

Australia's inflation rate is still dangerously high. Its traditional markets in the U.K. and other E.E.C. countries are being cut off by rising tariffs. Until it finds new markets for its exports it will have trouble reducing its rate of inflation.

Sweden's figures also point to an uncertain future. By keeping its traditional neutrality, it has not joined the E.E.C. As a result, it is losing ground in many traditional European markets.

Norway also voted not to join the E.E.C. But the development of off-shore oil has contributed to the health of its economy.

Which countries are doing well? The world needs a Switzerland – a stable, invisible safety deposit box for its savings, that somehow escapes the pressures of the world's economic ups and downs. As long as the Swiss continue banking everyone else's money, they won't have much to worry about. West Germany too has managed to ride out the world economic recession. When you compare the size and complexity of the German economy with the comparative simplicity of the Swiss, you can see that in hard economic terms it is the Germans who have done best of all. Austria has also survived well. Even though it's not in the E.E.C. and can never become a full member for treaty reasons, its economy is tied to that of West Germany.

The Japanese economy is buoyant but its inflation figures are far from excellent – especially compared to those of West Germany. Japan's export figures soared when inflation hit the Western economies. But now Western economies are taking protective measures against the unlimited flow of underpriced Japanese products.

	U.S.A.	CANADA
Rate of interest(%) at end of:		
1965	4.5	4.7
1970	5.5	6.0
1975	6.0	9.0
1976	5.2	8.5
1977	6.0	7.5

When we think of banks we often think of getting a loan – but banks are also heavy borrowers. And when banks borrow, they turn to the central government bank: the Bank of England in the U.K., the Federal Reserve system in the U.S., and similar monetary sources in other countries.

The interest rate that the government charges the banks is a crucial figure for the whole economy: it's the base upon which the whole pyramid of other interest rates is built. This figure is generally referred to as the minimum lending rate or bank rate.

During the economic uncertainties of the recent past, the minimum lending rate in the U.K., as elsewhere, has tended to fluctuate several times in the course of a single year. Whether the rate moves up or down depends a great deal upon how much the government is trying to attract foreign investment capital. Other factors being equal, investors are most eager to lend where they can get the highest return for their money. The disadvantage for the borrower, of course, is that he must pay more for the privilege of using the money.

In times of economic strength, a country tends to lower its minimum lending rate. But a nation that is having problems – a heavy deficit in its balance of payments – will probably offer a high rate of interest as part of its attempt to attract foreign capital. Thus the minimum lending rate is one reliable guide to the general health of a country's economy.

Minimum lending rates often don't vary up or down by more than a fraction of a percentage point. But such a seemingly small shift can still make a costly difference. Investments affected by the change in the rate often run to tens or hundreds of millions of pounds. An adjustment upwards of a half per cent means an extra £50 thousand in the interest payments on a £10 million loan. When interest rates make a soaring leap of nine per cent as they did in Italy between 1975 and 1976 – it means £900 thousand extra interest added on to the original £10 million borrowed.

Italy with a rate of 11.5% wins the dubious distinction of having the highest minimum lending rate for 1977. But even that figure is a significant drop from the 1976 rate of 15% – the highest figure in the whole table. These punishing interest rates give some indication of the massive economic problems the Italians have had to contend with in recent years. Only 1976 figures for Ireland (14.7%) and the U.K. (14.2%) approach Italy's rates.

The second highest lending rate figure in 1977 is for Australia. Since Australia doesn't have an official government

U.K.	AUSTRALIA[+]	AUSTRIA	BELGIUM	DENMARK	FRANCE	(WEST) GERMANY	IRELAND	ITALY	JAPAN	NETHERLANDS	NORWAY	SPAIN	SWEDEN	SWITZERLAND
6.0	4.8	4.5	4.7	6.5	3.5	4.0	5.9	3.5	5.5	4.5	3.5	4.6	5.5	2.5
7.0	6.3	5.0	6.5	9.0	7.0	6.0	7.3	5.5	6.0	6.0	4.5	6.5	7.0	3.7
11.2	8.5	6.0	6.0	7.5	8.0	3.5	10.0	6.0	6.5	4.5	5.0	7.0	6.0	3.0
14.2	8.7	4.0	9.0	10.0	10.5	3.5	14.7	15.0	6.5	6.0	6.0	7.0	8.0	2.0
7.0	9.7	5.5	9.0	9.0	9.5	3.0	6.7	11.5	4.2	4.5	6.0	8.0	8.0	1.5

lending rate, the numbers here are based on rates for short-term government bonds, which give a roughly equivalent figure. The rate of 9.7% reflects Australia's recent trade problems: higher tariff barriers set up by the European Economic Community have cut her off from many of her traditional markets.

France (9.5%), Belgium (nine per cent) and Denmark (nine per cent) follow Australia. All three countries had to wrestle with both high unemployment and balance of payments deficits. Though they were not struck by the recession of 1975-6 as hard as the U.K., Ireland and Italy, their recovery has been slower.

Three countries stand out for their markedly low interest rates – particularly Switzerland with an eye-catching 1.5%. After Switzerland, West Germany (three per cent) and Japan (4.2%) have the lowest lending rates.

The huge balance of payments surpluses that both countries are enjoying are becoming something of a political embarrassment. Western countries have been applying pressure on Japan and West Germany to revalue the yen and the

Sources:
Bank of England
International Monetary Fund
[+]No official interest rate. The figures shown are the rates on short term government bonds for the average period involved.

mark to slow their economic expansion. This measure does have a built in backlash however; when investors suspect that a currency is to be revalued, they may pour in funds in order to profit from an improved exchange rate.

On the whole, interest rates all over the world have been rising and don't seem likely to return to 1965 levels in the foreseeable future. In most countries, rates peaked in 1976 when the world-wide economic recession was at its worst. In 1977, only Switzerland, West Germany and Japan had lower lending rates than in 1965 – another indication of their persistent economic strength, based in the first case on international banking services, and in the latter two, modern industrial methods.

	U.S.A.	CANADA
	1976	1976
National income		
per household (£ – 000s)	13.3	12.4
per person (£)	4302	3722

National income figures are designed to show how much a country produces in goods and services in a year. It's a systematic way of looking at all the manufacturing, servicing, buying and selling that's done. It's usually expressed in money terms as the simplest and most direct way of covering a wide variety of economic activities. The per capita income is the national income divided by the population and the household income is the same figure divided by the number of households. Just because a country has a high per capita income, however, don't assume that everyone has the same amount of money to spend. It's only an average figure and doesn't show how that money is spread among the population. The wealth may be in the hands of a very few – as it is in Spain, for instance – and the gap between rich and poor may remain considerable.

After years of hearing about our gloomy trade figures and of being told that we have to tighten our economic belts, it probably comes as no surprise to see how badly the U.K. fares. Our per capita income is £2,168 and our income per household is £6,465. In both cases, this places us 14th in the league. Our figures are well under half of those for the most affluent countries and that means only Italy, Spain and Ireland are less well-off than we are. We're right down there with the countries that are traditionally regarded as the underdeveloped, poor cousins of Europe.

In contrast Switzerland, which has never been a major world power in other areas, is certainly the leader when it comes to money. While the rest of the world has only associated them with cuckoo clocks, skiing, chocolate and cheese, the Swiss have been building up a reputation as serious and respected bankers. It's this, more than anything else, that puts them right at the top of our table. From all over the world, money has found a home in the secret and inviolate vaults of the Swiss banks. The fabled "gnomes of Zurich" have obviously been taking care of their own interests, too. Today, the Swiss enjoy a per capita income of £5,235 and an income per household of £14,867.

Sweden comes a close second. The Swedes have some of the richest ore deposits in the world which they have no trouble turning into cash. They also have abundant timber and one of the highest agricultural yields per acre in Europe, so they don't have to spend their money abroad. All this adds up to a handy £5,040 per person.

With Norway and Denmark coming fourth and fifth in the table, it's clear that the Scandinavians know how to make money. Denmark has a healthy £4,171 per capita income and it's fairly certain that Norway's figure of £4,197 is more likely to go up than down now that the

INCOME

	U.K.	AUSTRALIA	AUSTRIA	BELGIUM	DENMARK	FRANCE	(WEST) GERMANY	IRELAND	ITALY	JAPAN	NETHERLANDS	NORWAY	SPAIN	SWEDEN	SWITZERLAND
	1976	1976	1976	1976	1976	1976	1976	1976	1976	1976	1976	1976	1976	1976	1976
	6.4	13.1	8.9	12.0	11.4	11.1	10.6	5.8	5.1	8.6	11.7	12.4	6.4	13.3	14.8
	2168	3781	3045	3879	4171	3647	4108	1413	1644	2477	3548	4197	1609	5040	5235

Sources:
Government sources

North Sea oil is flowing steadily. Generally speaking, the Scandinavians feel that things have improved in the last five years and whatever problems the future holds, it doesn't look as if money is going to be one of them.

Right at the bottom end of the scale come Italy, Spain and Ireland. These are primarily agricultural countries, with no great banking houses to invite foreign capital and very little industrial development in relation to the population. The small slice of the economic cake that these countries have been able to get is reflected in their per capita income figures. Italy has £1,644 per person, Spain £1,609 and Ireland comes right at the bottom with £1,413. Ireland's figure is only just over a quarter of the Swiss figure. The vicious circle that such poverty sets up has been seen in all three of these countries. Having very little at home to swell the coffers, they've been forced to export their most valuable asset – manpower. Many of their workers have emigrated and helped to enrich the rest of the world.

The most interesting disparities between per capita and household incomes occur at the bottom of the table. The household income for Italy, for instance, is £5,146 which is lower than either Spain or Ireland (£6,436 and £5,779 respectively.) In the per capita table, Australia comes eighth and Canada ninth but in the household table they have changed their positions to fourth and sixth. West Germany, which is in sixth place in the per capita income table has, with a household income of £10,681, dropped to 11th.

Before you rush off to live in Switzerland or Sweden, it might be worth checking with the Cost of Living table. After you've paid for the *fondue* or *smorgasbord* will there be enough money left for the rent, not to mention the *schnapps*? It looks as though the best deal might still be in the U.S. Though no longer the richest country in the world, the cost of living there is still comparatively modest.

If "diamonds are a girl's best friend", it's no surprise to find that the U.S. – where the song was written – has the most women with these expensive baubles: 44% of American women over the age of 15 own diamond jewellery – and that figure doesn't include diamond engagement rings.

At 19%, the U.K. figure is average, but even so, nearly one in five British ladies owns a diamond.

The Canadians and West Germans come second – 24% of the women in their countries have some diamond jewellery. Again, the Canadian figure doesn't include engagement rings. The Spanish and Belgians follow with 20% each.

Pity the poor girls of Italy: there, only 13% of women own diamonds.

Japan is the top country for average retail price at £457. Although comparatively few Japanese women manage to get their hands on a diamond, when they *do*, it's a whopper.

At the other end of the scale, the U.K. has the lowest average figure: £108.

Belgium is second highest in the list. The average price there is £308. Their high prices aren't surprising when you consider that Antwerp is the diamond market of the world. But the Belgian ladies were reserved about the prices of their jewels: 53% were unable – or unwilling – to put a value on their

	U.S.A.	CANADA
	1976	1976
1. Diamonds:		
women over the age of 15 who own diamond jewelry (%)[+]	44	25
average retail price (£) of a piece of jewelry[+]	134	196
total number of pieces of diamond jewelry sold (000s)[+]	6,227	400
2. Gold:		
made into jewelry (lbs - 000s)	147.5	26.9
made into medals, medallions and coins[σ] (lbs)	1,543	221
made into official[x] coins (lbs)	3,086	24,471

diamonds. This vagueness could also be true of women in other countries.

The most reliable facts about diamonds are probably in the figures for the total number of pieces of diamond jewellery sold. Again the U.S. is way ahead: they bought over six million pieces in 1976.

Some fascinating facts emerge when you compare the number of pieces of diamond jewellery sold each year with the country by country female population over the age of 15. If diamonds were distributed evenly – which is most unlikely – about one in 12 American women would buy or be given a piece of

DIAMONDS AND GOLD

	U.K.	AUSTRALIA	AUSTRIA	BELGIUM	DENMARK	FRANCE	(WEST) GERMANY	IRELAND	ITALY	JAPAN	NETHERLANDS	NORWAY	SPAIN	SWEDEN	SWITZERLAND
	1976	1976	1976	1976	1976	1976	1976	1976	1976	1976	1976	1976	1976	1976	1976
	19	NA	NA	20	NA	16	24	NA	13	18	18	NA	20	18	NA
	108	NA	NA	308**	NA	252	189	NA	252	457	NA	NA	NA	400	NA
	420	NA	NA	64	NA	359	1,388	NA	395	862	73	NA	315	495	NA
	43.2++	9.7	4.9	14.6	3.7	53.8	79.4	NA	390.2	104.5	5.3	1.8	100.3	4.0	24.0
	882++	NA	221	661	NA	221	6,614	NA	12,566	2,204	661	NA	6,614	1,323	1,984
	6,614++	221	62,611	6,614	NA	882	NA	NA	1,323	NA	441	NA	NA	NA	7,496

++ Includes Ireland   NA not available

**This figure is only approximate, as 53% of people questioned were unable or unwilling to give prices

jewellery every year. The U.K. figure is much lower: only one in 37 British women would enjoy the glitter of a diamond. Other comparable figures are Canada, 16.5 women, and France, 47.4.

The other – perhaps more traditional – indicator of wealth is gold. Even the U.S. comes behind crisis-ridden Italy, where nearly 400,000 pounds of gold are used.

The U.S. figure is still substantial: over 147,000 pounds of gold are melted or beaten down. And in Spain and Japan over 100,000 pounds of it were devoted to these end products. The U.K. figure is meagre, at least in comparison with Italy or the U.S.: just over 43,000 pounds.

The lowest figures may say something about Scandinavian taste: Norway is lowest. It has only 1,800 pounds of gold made into jewellery. Denmark and Sweden come next.

Sources:
1 Market Research Surveys
2 Consolidated Gold Fields Ltd, London
+Excludes diamond engagement rings
0Not official government issue
xOfficial government issue

If you're thinking of changing your car and have thirty thousand pounds to spare, you could treat yourself to a Rolls-Royce. Or why not keep out the winter chills with a sable coat – a mere snip at from four to eight thousand pounds. Most people find such luxury isn't quite their style – if only because they can't afford it. The figures in this table show the countries where those with a few thousand to fritter away are only too eager to give house-room to such status symbols. The figures for sable (winter warmth for people who like to be pampered) are for the number of raw sables *imported*, rather than sold. Not all that fur necessarily finds its way on to the backs of the ladies in the countries concerned – some coats will be exported.

We've all heard that we have been suffering from economic difficulties. Even so, 1,450 people in the U.K. still manage to ride high in the comfort of a Rolls. And those 5,700 imported raw sables must help a few of us to keep out the economic draughts.

It seems that a lot of the wealthy in the U.S. also like to sink their money in a Rolls. A thousand of them opted for luxury English-style, rather than something home-grown.

They seem to have a passion for sable too. They imported 76,700 raw pelts to be made into coats. These must grace the backs of quite a few lucky people.

The Swiss also seem to favour sable – they imported 9,400. At least a few Alpine ladies must be defying the weather in a truly chic fashion. But when it comes to Rolls-Royces, it looks as if those thrifty Swiss just aren't interested. They bought no Rolls-Royces at all during 1977.

The Arabs have no problem as far as running a Rolls is concerned – most of them shouldn't have to go too far to get petrol.

The people of New Caledonia don't go for sables either. They're in the tropics, and are sitting on a gold mine of minerals. They have substantial revenue from coffee, too. In 1977, 93 New Caledonians chose to travel to work in nothing less comfortable than a Rolls-Royce – compensating them for the discomfort of the tropical heat.

On the other hand, the millionaires of

	U.S.A.	CANADA
	1977	1977
1. Number of cars bought:		
Panther de Ville	3	0
Rolls-Royce	1000	64
Aston Martin (V8 and V8 Vantage)	90^x	0
2. Number of raw sables bought$^+$	76,700	NA
as % of all sables bought	76.7	NA

WEALTH INDICATORS

U.K.	ENGLAND	NEW CALEDONIA	AUSTRIA	BELGIUM	DENMARK	FRANCE	(WEST) GERMANY	IRELAND	ITALY	JAPAN	HONG KONG	MIDDLE EAST	NETHERLANDS	SWITZERLAND	OTHERS
1977	1977	1977	1977	1977	1977	1977	1977	1977	1977	1977	1977	1977	1977	1977	1977
4	NA	0	1	0	0	1	2	0	0	0	2	5	NA	2	NA
1450	NA	93	0	12	1	70	70	82	55	0	0	173	NA	0	NA
90^x	NA	0	0	0	0	0	0	0	0	30^x	0	0	NA	0	NA
NA	5700	NA	500	NA	NA	400	1200	NA	1000	2400	600	NA	1000	9400	1100
NA	5.7	NA	0.5	NA	NA	0.4	1.2	NA	1.0	2.4	0.6	NA	1.0	9.4	1.1

^{x}Estimates   NA not available

Sources:
1 Aston Martin Lagonda Limited, Rolls-Royce Motors Limited, Panther Cars Limited
2 Sojuzpushnina, Moscow
$^+$A total of 100,000 raw sables were sold by auction in Russia in 1977. Calculations were based on invoicing and shipping instructions: it was assumed that home consumption of sables in Europe is higher as some dressed skins and ready-made coats are sold back from the U.S.

Japan just aren't interested in our Rolls-Royces. And even when it comes to sables they only bought a moderate 2,400 pelts. Perhaps that famous oriental asceticism dies hard – even for millionaires.

Canada bought 64 Rolls-Royces, according to our table, but no sable at all. (But then, Canadians have plenty of sable of their own, plus minx, lynx and foxes.)

Austria, Denmark, and Switzerland come lowest in our luxury table. They only have one Rolls between them. Perhaps they prefer something a little less showy.

You might think a Rolls is costly, but the Panther de Ville costs about £20,000 more than a Rolls. Only 20 were produced in 1977. Five of these found their way to the Middle East, four to the U.K. and three to the U.S.

Ninety Americans plumped for an Aston Martin along with 90 British. The Japanese bought 30: so what *have* they got against the Rolls? Perhaps "Rolls-Royce" is harder for them to pronounce.

69

	NEW YORK	MONTREAL
	1978	1978
Cost of living+	98	78

Ten of the World's Least Expensive Cities

London, U.K.		100
1.	Moscow, U.S.S.R.	49
2.	Colombo, Sri Lanka	49
3.	Belize, Belize	59
4.	Alexandria, Egypt	64
5.	Valletta, Malta	64
6.	Cook Islands	69
7.	Bratislava, Czechoslovakia	69
8.	Tarawa, Gilbert Islands	69
9.	Kingston, Jamaica	69
10.	Mexico City, Mexico	69

If you work in London, and your boss tells you you're being transferred to Germany with a whopping 30% increase in salary, don't celebrate – consider quitting instead. Your standard of living will actually drop by 26%. That's what these fascinating figures show. This data is compiled by the ultimate international organisation – the United Nations. Their personnel people use it to equitably determine salary rates amongst professional and higher categories of staff all over the world. It's a professional lifestyle cost of living index which is radically different (and we believe much better) than your normal cost of living rating.

There are a number of reasons why this is so. A normal cost of living index is generally more accurate for low income families and is based on a limited number of basic cost factors (meat, bread, etc.). Not so the U.N. data. Instead it's based on a comparable 300 items for each of their 164 countries. And the expenditure on those items (the kinds of things an upper middle class family of three buys) is actually researched amongst their own staff. Secondly, all living standards are expressed in one currency – U.S. dollars – then adjusted for local currency fluctuations, so the index deals with comparable data. Finally the figures are constantly reviewed and are adjusted no less than three times a year. Those shown are for the end of June 1978. For convenience we've indexed all of them to London.

So much for the background. What do the figures say? The first big surprise is that today, London is actually more expensive than New York and sig-nificantly more expensive than Washington, D.C. when you include decent rented accommodation, public transport, petrol, car costs, drink and other items. Our governments have successively told us that while we have lower salaries and higher inflation here in the U.K., this is compensated for in many ways by lower living costs. The figures certainly support this. In the E.E.C.'s headquarter town, Brussels, you'd need 58p added to every pound you earn today just to stand still. In Geneva you'd very nearly have to double your salary to level peg, while in Tokyo you'd need £20,600 to equal a £10,000 London salary's purchasing power. Of all 164 U.N. countries' posts fully 40% are more expensive to live in than London.

If you're thinking of fleeing to an island, don't necessarily flee to Bermuda or the Bahamas to save your sterling –

COST OF LIVING

LONDON	SYDNEY	VIENNA	BRUSSELS	COPENHAGEN	PARIS	BONN	DUBLIN	ROME	TOKYO	THE HAGUE	MADRID	STOCKHOLM	GENEVA
1978	1978	1978	1978	1978	1978	1978	1978	1978	1978	1978	1978	1978	1978
100	103	143	158	150	139	156	95	88	206	153	98	132	178

Ten of the World's Most Expensive Cities

London, U.K.		100
1.	Tokyo, Japan	206
2.	Geneva, Switzerland	178
3.	Kinshasa, Zaire	176
4.	Brussels, Belgium	158
5.	Manama, Bahrain	156
6.	Bonn, Germany	156
7.	The Hague, Netherlands	153
8.	Copenhagen, Denmark	150
9.	Muscat, Oman	147
10.	Djibouti, Djibouti	147

Source:
United Nations
+Based on index rating of 100 for London

each is 3% more expensive than London.

The two real bargain cities of the world are shown on the other table – Moscow and Colombo (in Sri Lanka – as if you didn't know). Obviously accommodation is a large part of the equation. In Moscow, while rents may be low, you don't choose your apartment – it's chosen for you. So here our diplomatic cost of living presents a relatively false picture. In either city we think you would find that doubling what your pounds can buy is cold comfort when there's not much choice of what to spend it on.

Food prices vary greatly from country to country – as many tourists have found to their cost. And they don't always differ in the way you'd expect.

First let's look at steak. It's one of the things most of us would like to be able to buy cheaply. The U.K. and Ireland are both good beef countries, yet steak prices are about average – £1.33 a pound in the U.K. and £1.17 in Ireland. The best place to go for steak is Australia. Here a sirloin costs only 52p a pound. It's so cheap they even barbecue it on the beach at picnics and have it in sandwiches.

Portugal is the next best place to buy steak – but you'd have to pay almost double the Australian price. The U.S. and Canada also do well: steak there costs under £1.15 a pound.

Japan is the place *not* to go for a cheap steak. A pound of sirloin will set you back more than £4.30. No wonder a lot of Japanese mainly eat vegetarian meals – paying that price for steak is enough to turn the most dedicated carnivore into a vegetarian.

If you can't eat steak, how about fish? Unfortunately – and unexpectedly – it's still the same sorry story in Japan. Fresh fish costs £2.75 a pound – almost double what it costs in Portugal and Italy, the next most expensive countries in our table. These high prices are surprising – all these countries have long coastlines, and are renowned for their excellent fish dishes. In fact, in Japan, fish is one of the major sources of protein.

For good, inexpensive fish, go to Holland. The Dutch have easily the cheapest in our table – 40p a pound. Denmark is next best, at 45p.

You'd expect fish to be less expensive in countries with long coastlines than in countries with no coastline at all. This just isn't borne out by the facts. Though fish is inexpensive in the U.K. (just under

The cost (£) of:	U.S.A. 1976	CANADA 1976
boned sirloin[+]	1.10[‡]	1.13[‡]
fresh fish[+]	.87	.79
white wheat bread[o]	.45	.36
potatoes[+]	.07	.04
1 egg	.04	.04
cheese[+]	1.01	.92
pasteurized milk[x]	.15	.19
butter[+]	.78	.64
sliced smoked bacon[+]	1.04	1.04
onions[+]	.12	.12
oranges[+]	.25	.15
sugar[+]	.13	.10
coffee[+]	1.23	1.37

FOOD BASKET

U.K.	AUSTRALIA	AUSTRIA	BELGIUM	DENMARK	FRANCE	(WEST) GERMANY	IRELAND	ITALY	JAPAN	NETHERLANDS	NORWAY	PORTUGAL	SWEDEN	SWITZERLAND
1976	1976	1976	1976	1976	1976	1976	1976	1976	1976	1976	1976	1976	1976	1976
1.33	.52	2.12	2.40	2.04	1.86	2.94	1.17	1.67	4.32	NA	1.98	.97	3.08	1.88
.74	.61	.63	1.12	.45	1.09	.80	.57	1.37	2.75	.40	.67	1.47	.98	1.22
.22	.29	.64	.48	.61	.45	.52	.22	.32	.47	.32	.47	.08	.75	.39
.11	.13	.03	.09	.14	.12	.11	.08	.12	.11	.11	.11	.06	.12	.09
.03	.05	.07	.06	.06	.05	.04	.04	.04	.03	.04	.09	.03	.07	.08
.51	.56	.76	1.00	1.14	NA	1.30	.65	2.13	.98	.90	.74	.61	1.07	1.42
.10	.11	.12	.12	.14	.12	.13	.08	.12	.28	.12	.09	.06	.11	.17
.48	.49	.81	.96	.92	.96	.90	.52	1.12	1.20	.88	.64	.46	.63	1.35
.81	1.41	.38	.94	NA	.83	.94	.67	.16	1.73	NA	2.09	.14	1.70	1.04
.14	.16	.09	.13	.26	.17	.16	.15	.12	.14	.14	.28	.04	.33	.18
.15	.11	.16	.20	.22	.20	.21	NA	NA	.21	.18	.26	NA	.23	.20
.25	.08	.13	.15	.03	.14	.16	.12	.16	.21	.16	.17	.15	.20	.18
1.45	NA	1.57	1.65	1.98	NA	2.27	NA	1.50	NA	1.46	1.56	NA	1.57	1.39

‡With bone   NA not available

75p a pound) it's even cheaper in landbound Austria (63p).

If you look at the different prices for bread and potatoes, you'll get a general idea of basic food costs. With a few regional exceptions, they are the staple foodstuffs in the countries in our table.

Portugal is easily the cheapest country for bread, at 8p a large loaf. Ireland, home of the famous soda bread, is also low on this table. On the other hand, Sweden is worst off. There, bread costs over 75p a loaf, more than three times the price of Irish bread.

Source:
Bulletin of Labor Statistics
+1lb weight
0Large loaf
xPint

The Irish were once the world's greatest potato growers and eaters. But the potato blight caused widespread famine and mass emigration to the U.K. and the U.S. Today, Austrian, Canadian, and Portuguese potatoes are cheaper than Irish.

Where does all our hard-earned money go? Mostly on food and rent. This is true of 16 out of the 17 nations studied. Six spend the most on food, and nine on rent. Norway spends equally on both. With less than four per cent of its land under cultivation, it has to import most of its food. Groceries consume £3.00 out of every ten spent by the average Norwegian housewife. But that's including tobacco and alcohol.

The U.K. and Ireland also have to import a lot of food, which pushes our prices up and adds to the high grocery bill. In the U.K., almost £1 in every £3.00 goes for food, drink and tobacco. Ireland, which has an enormous 43.6% expenditure on food, tobacco and drink, spends 11.9% of that figure on alcohol. That's over £1 out of every £10.00 – more than the Irish spend on clothing and shoes. We spend a comparatively small amount on rent, fuel, furnishings, household goods and health compared to other countries on the table. Only Ireland spends less. Our comprehensive health service might account for part of this low figure but Sweden, who also has a national health system, pays considerably more in this section.

You might expect the French, with their renowned cuisine and wines, to be next on the table. But the Italians beat them to it. A higher percentage of lira goes for *fettucine* and *linguine* than francs go for truffles and *foie gras*. The Italians spend £37.20 out of every hundred pounds they earn, on food. Almost six per cent of that goes on alcohol.

Ireland and Italy are the big spenders in the food and drink category, with most other countries spending between 24% and 33%. There's a big drop in the figures for North America. Canadians spend only 18.5% in this category. And the U.S. spends even less – only 15 pence, out of every pound earned in income, is spent on food, drink and tobacco. This doesn't mean they don't eat as well as the French or Italians. They just produce more. Both the U.S. and Canada are more than able to fill their own food requirements.

Where do Americans spend the bulk of their incomes? Not surprisingly – on transport. And a major portion of their transport bill is for cars. The U.S. is the most mobile nation in the world (see table on cars). A car is regarded as more of a necessity than a luxury. A whole subculture of services and activities has

Household expenditure (%) on:	U.S.A. 1976	CANADA 1976
food, drink and tobacco	15.3	18.5
rent, fuel, furnishings, household goods and health	28.7	36.6
education, recreation, personal effects and financial other services;	18.8	22.9
car and other transport	30.3	15.0
clothing and footwear	6.9	7.0

HOUSEHOLD EXPENDITURE

	U.K.	AUSTRALIA	AUSTRIA	BELGIUM	DENMARK	FRANCE	(WEST) GERMANY	IRELAND	ITALY	JAPAN	NETHERLANDS	NORWAY	SPAIN	SWEDEN	SWITZERLAND
	1976	1976	1976	1976	1976	1976	1976	1976	1976	1976	1976	1976	1976	1976	1976
	31.0	24.9	27.5	28.3	31.1	25.9	28.1	43.6	37.2	31.0	26.4	30.9	33.5	28.5	26.0
	26.6	28.5	32.2	37.8	30.0	34.3	32.1	20.9	27.7	29.2	35.1	30.9	27.1	31.8	31.7
	20.1	16.0	16.0	12.2	20.8	21.0	17.0	14.7	15.1	15.8	18.9	17.2	20.8	18.4	23.5
	13.5	21.0	12.2	14.2	12.1	10.7	11.9	11.1	10.9	13.5	9.9	11.6	9.4	14.0	12.6
	8.8	9.6	12.1	7.5	5.8	8.1	10.9	9.7	9.1	10.5	9.7	9.4	9.2	7.3	6.2

grown up around the car – drive-in movies, drive-in restaurants, drive-in banks, drive-in churches, even drive-in funeral parlours. To support this culture they spend almost one out of every three pounds they earn on cars – and other forms of transport.

Apart from the U.S., only the Australians spend a significant chunk of their incomes on transport. Twenty-one per cent of their incomes goes towards getting around their vast, sparsely populated country.

Most of the other European countries on the chart spend an average of about 14% on transport – a much lower rate than the Americans or Australians.

An overall look at the table shows a definite pattern for most countries – the cost of eating and maintaining a roof over the head accounts for about 60% of the total expenditure. Belgium spends the most with the combined costs of these two categories totalling over 66%. The

Source:
Euromonitor

cost of housing in Belgium is unusually high – not surprising in a country with an average 1,000 people per square mile.

Most of the householders surveyed spend between 15% to 20% of their total outlay on personal needs, recreation and educational expenses. Again the exception is Belgium, this time with the lowest expenditure (12.2%). Presumably, the overcrowded Belgians don't have enough elbow room to indulge in recreation – or enough money left over after buying food and paying the rent.

Clothing expenses come at the bottom of everybody's list. The French, surprisingly, maintain their famous chic while spending less money on clothing than most countries on the table. The Austrians hand over considerably more for their clothing expenses.

When Scott Fitzgerald told Ernest Hemingway: "The rich are different from you and me", he was making a class distinction in terms of money. There are other ways of defining class. In most countries a poverty-stricken poet or artist would be granted higher class status than a materially secure car worker. The table looks at how people earn their living across the world and places

	U.S.A.[+]	CANADA[+]
	1976	1977
Working population [0] divided into social classes:		
A professional and executive[x] (%)	24	20
B white-collar workers[‡] (%)	23	26
C skilled workers** (%)	33	29
D workers[++] (%)	16	17

them in four categories: professional and executive; white collar workers; skilled workers; and unskilled workers and manual labourers. The figures only refer to heads of households.

In the U.K. things are fairly well balanced. While 15% rise into the professional and executive class, and 27% stick to unskilled or semi-skilled work, 55% opt for skilled manual work, office jobs or small business enterprises.

There isn't any doubt as to which country wins for the highest number of professional and executive workers – the U.S., home of "the man in the grey flannel suit", is way out front with 24%.

Almost an identical number (23%) are white collar workers. That's the third lowest figure on the table. On the other hand, the figure for skilled workers (33%) is the second highest of the countries shown. In other words, well over half the population falls into either the professional and executive or the skilled worker category.

The Canadian figures closely parallel the situation in the U.S. There's a slightly lower percentage (20%) in top-level jobs, and four per cent fewer skilled workers.

The Scandinavians boast the second largest number of executive and professional workers, with nearly one in six heads of households in jobs with status. But the figures show a surprisingly broad base of unskilled labour when you consider that Scandinavia is a relatively affluent area. In Sweden, 39% of family heads (35% in Norway and 33% in Denmark) are unskilled labourers.

Italy has more unskilled workers and labourers than any other nation in Europe.

In Ireland, life also seems to be a matter of brute strength – at least for the 45% of household heads who make their money by their muscle. Many of them move temporarily to the U.K. to earn higher wages in the British construction industry. Only four per cent of Italians and seven per cent of Irish make it into the top category of teachers, doctors, management executives and so on. Earning one's bread by the sweat of one's brow has always been a feature of Italian and Irish life. Only 18% of Italian

	AUSTRALIA[+]	AUSTRIA	BELGIUM	DENMARK	FRANCE	(WEST) GERMANY	IRELAND	ITALY	JAPAN[+]	NETHERLANDS	NORWAY	SPAIN	SWEDEN	SWITZERLAND
U.K.														
1970	1971	1970	1970	1970	1970	1970	1970	1970	1976	1970	1970	1970	1970	1970
15	16	9	5	16	9	5	7	4	11	8	16	14	16	13
26	23	41	35	28	32	44	29	18	29	44	21	36	27	47
29	37	18	21	23	24	29	19	23	37	26	26	32	17	27
27	15	27	39	33	30	24	45	52	20	20	35	15	39	15

Source:

"Survey of Europe Today", *Readers Digest*, 1970

[+]Heron House estimates based on ILO figures

[0]Heads of households

[x]People in professions and specialist occupations; employers of ten or more people; employees at top management or senior executive level

[‡]Employers of less than ten people; self-employed; non-manual workers and non-manual employees of supervizory or lower grade

[**]Skilled manual workers; self-employed semi-skilled workers

[++]Semi-skilled and unskilled manual employees; self-employed laborers

workers fall in the scope of the second category on the table: white collar workers. (This category includes small employers, self-employed non-manual workers and non-manual employees of a supervisory or lower grade.) Another 23% find work as skilled manual workers and self-employed, semi-skilled workers.

Spain is one of the poorer, less industrialised countries in Europe, yet 14% of Spanish heads of households find their way into some professional or administrative position. Though more than 50% of Spain's income comes from the agricultural products which fill the nation's dinner plates and pull in foreign currency, a mere 15% of Spanish heads of family are manual employees.

Judging by the table the Germans have a pretty hard time making it to the top. In 1970 only five per cent of them reached the upper echelons of management and the professions. They appear to have a comfortable middle class with 44% in the ranks of small employers, self-employed or white collar workers. Another 29% earn their living as skilled manual, or self-employed semi-skilled, workers.

Number of cars bought

1,400			
1,300			
1,200			
1,100			
1,000			
900			
800			
700			
600			
500			
400			
300			
200			
100			

We hear a lot about our economic troubles in the U.K. but we like to suffer in comfort. We bought 1,450 Rolls Royces in one year, more than anywhere else in the world.

Americans were the next biggest customers for these instant status symbols. One thousand were sold in the U.S. That's still only one for every 200,000 people, so it can hardly be called a fad.

No prizes for guessing who's next. The Middle East oil sheikhs bought 173 Rolls Royces. At least the petrol is cheaper there.

Ninety-three went to New Caledonia. Just in case you're wondering, that's a Pacific island were the inhabitants must be doing *something* profitable.

Ireland isn't a very prosperous nation, but 82 Rolls Royces were bought there. Somebody must be finding the pots of gold that the leprechauns hide.

Only one lonely Rolls ended up in Denmark. Either the Danes just aren't interested, or they prefer to spend their kroner on local luxuries like *aquavit*.

DENMARK	FRANCE	IRELAND
1	70	82

ROLLS ROYCES

NEW CALEDONIA	MIDDLE EAST	U.S.	U.K.
93	173	1,000	1,450

Where the Money Goes

"Government spending reaches new high" scream the headlines. But what does that really mean? Government budgets are written in figures so huge they seem almost meaningless. When you've only thousands to deal with, it's relatively easy to figure out if you'll have enough money to buy a new car next year. But how *does* one decide if it's worth spending £24 thousand million on a new missile programme?

However, if you want to know what your government is up to, you should at least try to comprehend these colossal sums. And it's not really that difficult to think big – as you'll see if you look at the tables in this chapter.

The first table deals with government expenditure. And instead of using those mind-boggling thousands we've split the figures up into percentages. This way you can see at a glance what your government's priorities are – as well as how other governments choose to spend their money.

The first list of percentages is for government defence spending. This is something a lot of us question. Why waste so much money on defence? Does the world really need so many intercontinental ballistic missiles, nuclear defence systems, and such a sophisticated arsenal of weaponry? Well, perhaps if the world was a better place it wouldn't. But at the moment world peace seems to be maintained in part by fear – fear of what the other side can do to you. Every time the U.S. comes up with something new like the neutron bomb, Russia feels obliged to devise something equally awesome. And vice versa.

But the two super powers are the only big spenders in this field. Surprisingly little is spent on defence by most other governments. At least, so it appears. The average for the countries on our list is around ten per cent. The U.S., however, shells out a quarter of its budget on military purposes. Still, even ten per cent of a multi-million pound budget is an awful lot of money – especially when it could be spent on something positive and useful.

Some may look at defence spending in another way – that it's the price we pay for freedom and independence with the government of our choice. So, you could say that most countries use that ten per cent of their national budget on giving teeth to democracy – since all the countries in our list are democracies of one kind or another.

Now let's see what else the government squanders our money on. Besides defence, we've also got percentages for government spending in the following categories: health, social welfare and security; public works and housing; law and order, police and justice; public debt; finance; and "other spending".

Surprisingly, almost every country spends more for health, social welfare and security than anything else. The U.S. government is the biggest welfare spendthrift of all with almost half the national budget in this category – 13% more than socialised Sweden, for instance.

Other big figures are for education and public works – and also that anonymous category "other spending", which accounts for over half the government spending in Spain and Belgium. Some countries seem to spend nothing at all on certain activities. Actually this means they don't want to specify the amounts they disburse and put them under "other spending", or turn them into budgets for something totally different. Spain, for instance, is loath to reveal how much she spends on her police, and America's C.I.A. budget has never been disclosed.

Now that you know (almost) how governments of different countries spend their money, it's interesting to find out how much they pass on to the rest of the world. The next table deals with foreign aid, and it shows how generous certain countries are to each other.

Obviously, countries with bigger budgets and a bigger G.N.P. can afford more money. For this reason, we've split up our figures to show spending in terms of percentage of the G.N.P., which gives a picture of the generosity of governments in terms of how much they can actually afford. The table shows how much different countries really care about the rest of the world, especially those less fortunate than themselves.

There's a catch here, however, foreign aid isn't always altruistic. There are often strings attached – such as friendship treaties, trade agreements and so on. Foreign aid doesn't necessarily mean a country wants to be helpful but that it desires to have its influence felt in the world. Today, we no longer have empires, and dominance is a subtle matter of "spheres of influence" and "economic interdependence". So you'll have to decide whether these figures really represent government generosity, or just disguised self-interest!

GOVERNMENT EXPENDITURE

One indication of what type of government you have is the way it spends all those hard-earned pounds, dollars, francs or marks that its citizens pay in tax.

However, figures can be misleading. According to the table, the U.S. is the biggest welfare state; a massive 43.7% of their outlay goes on health, social welfare and security. Yet many other countries have more comprehensive health and welfare systems than they do, even though these countries spend less of their budgets on them. The U.K. lays out 32.5% of its budget for health and welfare and Sweden 30.4%, yet we both offer far more extensive welfare coverage to our citizens than America does.

In the U.K. we spend a lot more on community development and housing than any other country in the table. At 10.7%, our figure is a lot higher than the next country, the Netherlands (6.6%).

The Australians have the highest figure for expenditure on education, over 30%. Both the West Germans (5.2%) and the Danes (8.1%) appear to have low figures, but seeing as they have large, well-equipped school systems, the figures must be incomplete. The U.S. – at 5.1% – is low because the figure represents only federal money. We come eighth in the list, spending 12.5% of our total expenditure on education.

The U.S. is easily highest when it comes to expenditure on defence, with a figure of 24.6% followed by West Germany and Switzerland. The large military budgets of these two countries are understandable. West Germany is right on the front lines should a Third World War ever start in Europe. And Switzerland has always been anxious to defend its traditional neutrality.

The U.S. are at the top of the table when it comes to subsidising agricultural programmes. Their figure is a mere 0.7%. The Swiss are the big spenders in this area – 9.1% of their budget goes on agriculture followed by the Irish (7.1%) and the Spanish (5.3%). With 2% of our budget going on agriculture we are in tenth place.

	U.S.A.	CANADA
	1976	1976
Government expenditure (%) on:		
defence	24.6	9.0
health, social welfare and security	43.7	26.9
public works, transport	4.8	3.6
education	5.1	NA
agriculture	0.7	NA
community development, housing	1.5	NA
law and order, police, justice	0.9	NA
public debt	NA	NA
finance	NA	20.8
other	18.7	39.6

GOVERNMENT EXPENDITURE

	U.K.	AUSTRALIA[+]	AUSTRIA[+]	BELGIUM	DENMARK	FRANCE	(WEST) GERMANY	IRELAND	ITALY	JAPAN	NETHERLANDS	NORWAY	SPAIN	SWEDEN	SWITZERLAND
	1976	1976	1976	1976	1976	1976	1976	1976	1976	1976	1976	1976	1976	1976	1976
	10.6	14.0	3.73	7.2	6.4	18.0	20.6	4.4	6.4	6.5	8.2	9.9**	14.9	10.1	20.4
	32.5	23.7	25.2	19.8⁰	32.0	17.5	36.6	29.9	8.0	19.3	19.2	26.5**	NA	30.4	18.0
	4.7	19.2	26.5	NA	1.5	5.8	7.1	3.6ˣ	2.2	15.9	7.3	12.5**‡	10.3	4.8	18.2
	12.5	30.5	9.2	NA	8.1	25.5	5.2	13.5	11.6	12.9	21.6	11.4**	15.3	14.0	9.7
	2.0	3.4	2.4	NA	1.1	3.7	1.2	7.1	2.2	NA	4.5	NA	5.3	4.3	9.1
	10.7	1.3	0.5	NA	NA	5.3	1.3	NA	NA	NA	6.6	NA	NA	4.9	NA
	1.9	NA	3.5	NA	2.3	NA	NA	3.7	0.8	NA	11.3	NA	NA	4.0	NA
	9.3	NA	19.5	7.2	NA	NA	NA	20.2	NA	NA	6.7	NA	NA	5.0	NA
	1.3	NA	NA	NA	0.1	NA	NA	NA	10.6	16.0	NA	NA	NA	7.0	NA
	14.5	7.9	9.4	65.8	48.5	24.2	28	17.6	41.8	29.3	14.6	49.7**	54.2	15.5	24.6

NA not available   ⁰Includes education   ˣTransport only   ‡Communications
**1978 budget estimates

When it comes to law and order only Italy, with a stingy 0.8%, spends less than the U.S. (a meagre 0.9%). The U.K. figure for this programme is double that of America's. We spend 1.9%. At the top of the table the Dutch spend 11.3%, and judging by their low rate of road accidents (see dangerous drivers table) their rigid laws on drink and driving are worth every guilder.

As you look at this table, bear in mind that every country has a slightly different way of reporting its expenditures. For instance, funds the U.S. reports under welfare might be put under the category of community development and housing by another country. In some cases, governments will say they have no

Sources:
Europa Publications Ltd
[+]National statistical offices

expenditures in certain categories. This doesn't mean they spend nothing on the activity. Instead, the expenditure is listed elsewhere. Some nations don't like to be too specific: about half the budgets of Belgium, Spain, Denmark, Norway and Italy are listed under the heading "other expenditure."

In general, most governments spend the highest percentage of their budgets on health, social welfare and security. Australia, the Netherlands and France are exceptions.

International aid is global charity on a grand scale with rich countries spending from small to colossal sums of money to help poorer ones.

But countries are as human as the people in them. Strings are often attached to the aid – trade agreements, mutual co-operation pacts, or treaties of alliance. In other words, international aid is often tied to performance clauses.

Aid given through United Nations' organisations is an exception. The United Nations administers their funds independently and its agencies set up and develop schemes of their own.

This is why we have two tables. One gives total governmental development contributions to international aid, the other lists contributions to United Nations' agencies. Sheer size of aid is not the real measure of generosity. The real indicator is the country's contribution as a percentage of its G.N.P.

In 1976, the U.S. gave more than £2.4 thousand million. But they're by far the largest and richest country in the table with a G.N.P. of almost £15 billion. So the amount they give in aid is only 0.25 per cent of their G.N.P.

The U.K. is only tenth in terms of percentage contribution from its G.N.P.

		U.S.A.	CANADA
		1976	1976
1.	Governmental development aid (£ – millions)	2427	496
	as % of GNP	0.25	0.46
2.	Countries in order of:		
	% of GNP allocated to development aid	12	7
	per capita GNP	4	3
3.	Extra amounts needed (£ – millions) to equal Sweden/Netherlands	5409	389
4.	Government contributions to UN agencies (£ – millions):	183.1	66.7
	as % of development aid	7.5	13.0
5.	Government food aid contribution (£ – millions):	721	106.2
	as % of development aid	30	21

Yet our percentage of contribution is higher than that of the U.S. – which has traditionally been considered generous.

Sweden and the Netherlands are the most generous countries in terms of percentage of G.N.P.: both give more than 0.8 per cent. That's £403 million from the Netherlands and £340 million from Sweden. Both countries give around a fifth of their contributions through United Nations' organisations.

Norway is next, giving more than 0.7

	U.K.	AUSTRALIA	AUSTRIA	BELGIUM	DENMARK	FRANCE	(WEST) GERMANY	NEW ZEALAND	ITALY	JAPAN	NETHERLANDS	NORWAY	FINLAND	SWEDEN	SWITZERLAND
	1976	1976	1976	1976	1976	1976	1976	1976	1976	1976	1976	1976	1976	1976	1976
	468	216	22	190	120	1202	775	30	127	619	403	122	29	340	63
	0.38	0.42	0.10	0.51	0.56	0.62	0.31	0.43	0.13	0.20	0.82	0.71	0.18	0.82	0.19
	10	9	17	6	5	4	11	8	16	13	1=	3	15	1=	14
	16	9	13	8	6	10	7	15	17	14	11	5	12	2	1
	539	207	163	116	55	388	1275	27	657	1926	—	19	101	—	209
	28.9	9.2	3.9	13.7	38.7	11.6	42.7	3.7	5.2	42.1	64.1	40.3	7.7	69.7	14.2
	6.0	4.0	18.0	7.0	32.0	0.96	5.5	13.0	4.0	7.0	16.0	33.0	27.0	20.5	23.0
	18.6	20.0	0.6	9.9	12.6	28.1	50.7	1.5	11.2	4.5	25.3	7.8	5.6	14.0	7.7
	4.0	9.0	3.0	5.0	10.5	2.0	6.5	5.0	9.0	0.7	6.0	6.0	20.0	4.0	12.0

percent of its G.N.P., about a third of which goes to United Nations' organisations (the highest proportion shown on our table).

France comes fourth in terms of real generosity. It gives over £1 thousand million in aid – second only to the amount given by the United States.

The U.S. is generous only in the actual amount given. Canada gives almost twice the percentage they do. In terms of "percentage" generosity, only Japan,

Source:
Organisation for Economic Co-operation and Development

Switzerland, Finland, Italy and Austria give less. Along with these countries they surely qualify as the Scrooges of international aid. In order to equal the generosity of the Swedes and Dutch, the U.S. would have to give over three times what it presently does.

85

Governmental development aid (as % of G.N.P.)

.8

.7

.6

.5

.4

.3

.2

.1

It isn't the biggest, wealthiest nations who dig deepest into their pockets for aid to less-developed countries. Tiny Holland and Sweden tie for first place as the most generous in terms of aid as a percentage of G.N.P. They both contribute 0.82% of their wealth to benefit other countries. Neighbouring Norway is next, with 0.71% earmarked for this purpose. That's almost double our 0.38%.

Affluent France coughs up 0.62% of its G.N.P. No doubt a lot of that aid goes to its former colonies. Another Nordic country is high in the international generosity league — Denmark gives 0.56% of its G.N.P. to the less developed countries. It's followed by Belgium, with 0.51%.

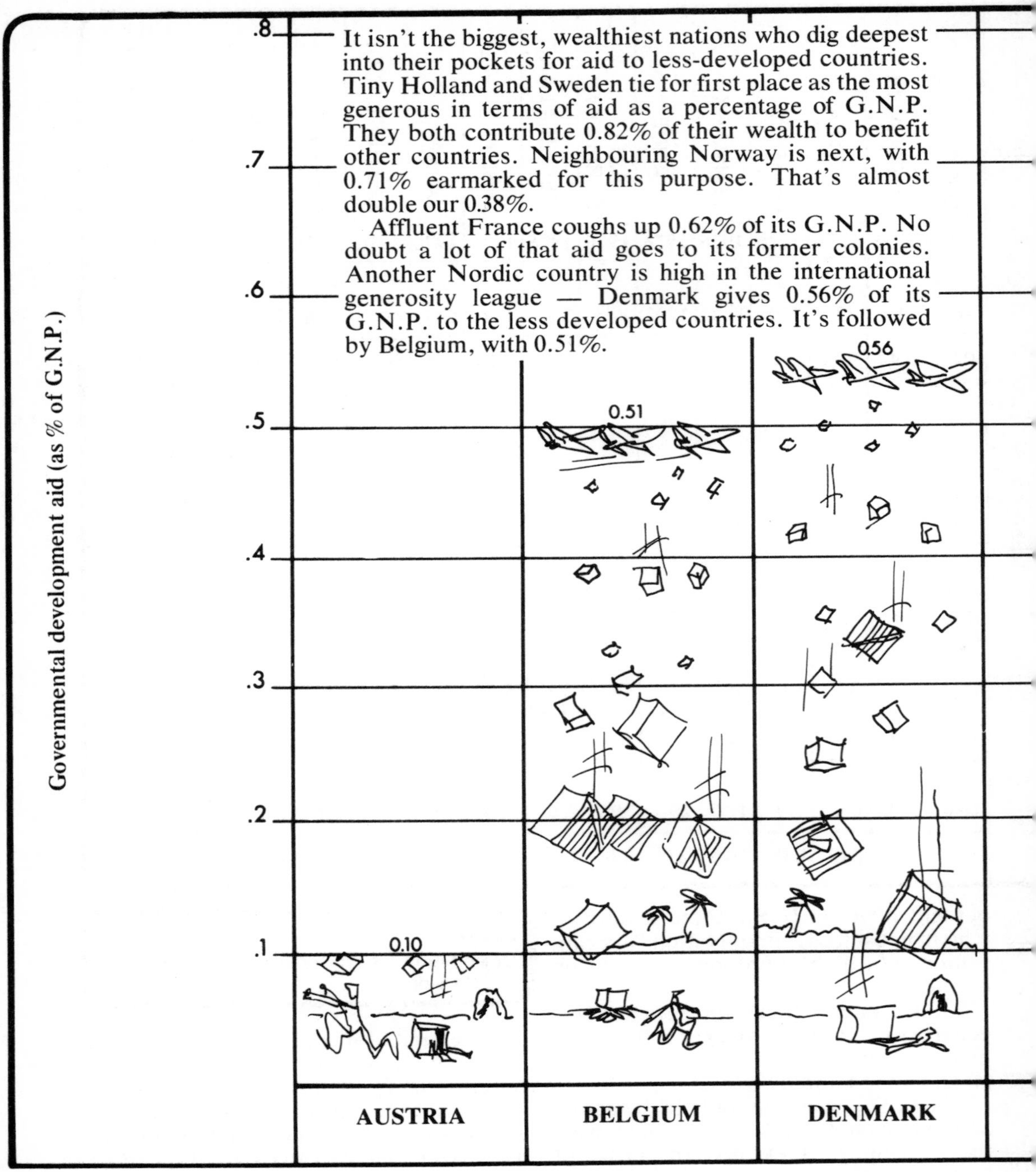

0.56

0.51

0.10

| AUSTRIA | BELGIUM | DENMARK |

FRANCE	NORWAY	SWEDEN	NETHERLANDS
0.62	0.71	0.82	0.82

The Taxman Cometh

Have you ever wondered if there's anything you can be absolutely sure of in this fickle world of ours? Well, one thing is tax. Unless you are very poor or rich enough to avoid him in style, the tax man looms large in all our lives. Even if you avoid income tax, there's always V. A. T. Then, of course, there are death duties, gift taxes, excise taxes. Read on and find out what your friendly taxman has in store for you. You might as well, you'll end up paying him anyway.

The first table shows tax revenue per capita in £ sterling. In other words, the total sum of money raised in taxation is divided by the number of people in the population. And remember this includes both direct and indirect tax, i.e. both income tax and sales taxes. So if you've sometimes wondered if anyone could possibly shell out more than you do, here's your chance to find out if it's true.

Moving on, you'll find the next figures in this table show what percentage of the total revenue is raised by income tax. Perhaps you are under the impression that having paid your dues to society, the money you put in the bank is yours and no one else's. Well, if you look at the figures, you might be rather shocked. Only Denmark raises more than 50% of its revenue by direct means. Or put another way, having paid your income tax, there's still a lot more to pay.

What about tax raised on personal income as a percentage of G.N.P.? These figures are also on this table. And remember G.N.P. is a way of showing the monetary value of everything that has been produced in a year from pins to tinned tuna, and cars to multi-storey blocks. These statistics, then, will show how much G.N.P. the government is taking in the way of income tax, so you can see what your contribution is towards the country's economic well-being.

The next figures follow directly on from this and show the total tax revenue as a percentage of the G.N.P. This includes both direct and indirect taxation as well as social security. It will probably come as quite a surprise when you find out how much the friendly social security people are salting away each year!

Having digested all this, would you like to know the actual rates of income tax in various countries? The first figures in this table show what percentage of a single person's income goes in taxes. In the same table, you will also find figures on what percentage social security takes, and then a total of these two as a percentage of the average income. For those of you who're married and – in this instance – it's married with two children, there are other figures. You can draw your own conclusions about whether it's better to be single and pay more tax, or married and pay less. Presumably, that depends on how much it costs to support a family. You can also compare rates and will probably find, no matter how much your tax bite hurts, that you're not paying the highest. Still, the table might cause you to consider emigrating – to the country with the lowest tax rate, of course. Which is? Have a look and find out.

And talking of finding out, have you given much thought to the problem of inheritance tax? As they say, people come and people go, but the taxman he always cometh – bless his little heart. In fact, many more people end up paying inheritance today, than, say, 25 years ago. Not just because the government has possibly raised the rates, but simply because people own more. For example, far fewer people owned a house before World War II, and if they did, it certainly didn't cost £17,500 or £25,000.

If you look at the table, you'll find it's divided into three categories. The first one shows what percentage you would pay the taxman on a £5,000 inheritance and what you'd have left after that, the second on £50,000 and the last one on £500,000. You'll probably be quite surprised at the number of countries that don't even have an inheritance tax on the first £10,000, or for that matter, those that do! For millionaire readers, this interesting information is worth the price of the book.

Sales taxes are the hardest to avoid: impossible really, unless you want to give up smoking, drinking and just about everything else. If not, you can't help but feed coffers, seeing as just about everything you're likely to buy has sales tax or V. A.T. slapped on top of the basic purchase price.

The next table gives a breakdown of the tax revenues raised from the sale of tobacco and liquor. And if you sometimes wonder what percentage of the total revenue raised by the government comes from taxes on these two items, you'll find figures on that, too. The information's all there, but perhaps you'd rather leave it for another night. Anyone for a drink?

Nobody loves them. But almost everybody has to pay them.

The difference between taxes and other forms of revenue is that taxes are "unrequited". A lot of money goes out – but few tangible goods come back, except in the case of direct welfare or other government payments. What do come back, theoretically, are a lot of social services. Roads, schools, and social service programmes of many types. Americans also pay billions to finance the government bureaucracy, including thousands of civil servants in the Bureau of Internal Revenue.

Although taxes are collected for the benefit of all tax-payers, the amount paid by each individual is by no means equal to the benefit received. A millionaire industrialist pays much more towards the building of a highway than a small shopkeeper – but they're both entitled to use the road (or hospital, or public swimming pool). On the other hand, the poor and the rich get docked the same amount for taxes on consumer goods and services.

There are two main classes of taxes – direct and indirect.

Direct taxes are essentially the taxes you pay to state or federal governments. They include individual income tax, levied on personal net income, usually when it rises above a certain basic minimum.

There are four main categories of indirect tax: on consumer goods, including customs duties on many imports; on consumer durables (cars, appliances and other consumer goods); on intermediate goods and production factors (including raw materials), and on certain legal and financial proceedings, such as stock transactions and capital gains. Few stones are left unturned by our alert friends at the Inland Revenue.

Who pays the most taxes? The long-suffering Scandinavians cough up the most money – Norwegians, Danes and Swedes each pay an average of more than £1,100 per year, with Sweden at the top of the list with £1,298. The West Germans, the Dutch, the Belgians and the Canadians come next. They are assessed at well over £900.

Americans are comparatively well off, paying a per capita average of under £900. If you're pathological about the tax collector, the best place to live is Spain. The Spanish pay an average of

	U.S.A.	CANADA
	1974	1974
1. Tax revenue per capita (£)	823.7	946.6
2. Tax on personal income:		
as % of total tax revenue	33.96	34.95
3. as % of GNP	9.82	12.16
4. Total tax revenue as % of GNP:		
including social security	28.93	34.79
excluding social security	22.77	31.61

	U.K.	AUSTRALIA	AUSTRIA	BELGIUM	DENMARK	FRANCE	(WEST) GERMANY	IRELAND	ITALY	JAPAN	NETHERLANDS	NORWAY	SPAIN	SWEDEN	SWITZERLAND
	1974	1974	1974	1974	1974	1974	1974	1974	1974	1974	1974	1974	1974	1974	1974
	519.7	630.1	713.5	902.1	1193.8	813.6	994.9	306.3	367.9	404.2	992.8	1114.3	197.2	1298.7	854.7
	35.17	40.07	24.20	29.36	53.37	10.83	30.49	22.92	15.24	24.17	27.61	27.49	12.93	45.05	34.72
	12.51	10.89	9.23	11.19	24.91	4.06	11.47	7.43	4.86	5.36	12.48	12.44	2.43	19.90	9.10
	34.56	27.18	38.14	38.13	46.68	37.50	37.64	32.43	31.86	22.18	35.18	45.27	18.83	44.21	26.11
	29.47	27.18	28.41	26.11	44.00	21.77	24.39	28.65	18.55	17.60	27.76	32.09	10.42	35.68	18.81

£197.2 per person. But then they don't earn much money.

The figures for taxes on personal income as a percentage of total taxation show how much of the per capita tax revenue is levied in the form of direct income tax. The results largely echo those of the previous table. The people who provide the highest proportion of tax revenue tend to pay a higher percentage of it as income tax. The Danes are at the top of the list – over 53% of the taxes they pay are as income tax. The Swedes are second, paying 45%. The Spanish pay less than 13%. And the French lowest, less than 11% – although they rank tenth in amount of tax revenue per capita.

The next figures on this table show what percentage of the taxes paid by the citizens of each country contribute to the Gross National Product. In Sweden, income and other personal taxes account for almost 20% of the entire GNP. Spain's figure is under three per cent.

Source:
Organisation for Economic Co-operation and Development

91

This table shows how much of what you earn goes to the taxman – and how much people in other countries lose out of their wage packets.

Contrary to popular myth, we in Britain aren't the most hard done by on this score. The U.K. tax rates, though slightly above average, are far from being the highest on this table.

The Danes have the worst deal. In Denmark single people turn over an average of 43% of their paycheque. If you are a Dane earning £100 a week, you'd take home only £57. Even if you were married and had two children, you'd still give 38% of your salary to the state. That's the same rate a single person pays in Sweden – the second highest rates on the table. If you were a married Swede with two children, you'd get a bit of a better deal, paying 33%.

Where does it pay to have marriage and a family? Austria gives a massive average tax credit of 12% to married couples with children (compared to the average tax rate for single persons). Germany is next with ten per cent, followed by the U.S. and Ireland with nine per cent and the U.K. with eight per cent. Surprisingly, two of the most family-oriented countries in the world – Italy and Spain – offer no tax credit for families.

At 33% average single tax rate, West Germany and Norway run a close fourth place on the table. The U.K. is only sixth on the table. A single person here pays only five per cent more than his American counterpart.

For an accurate comparison, tax rates should be compared to wage rates and the cost of living index for each country. Higher wages and lower tax rates in the U.S. mean more money in the pocket than in most European countries. When it comes to buying groceries and paying rent, it's take-home pay that counts. A primary school teacher in the U.S. earns about twice as much as one here or in

	U.S.A.	CANADA
	1974	1974
1. Tax paid by a single person:		
average rate of taxation on personal income (%)	20	19
social security rate (%)	6	2
total (%)	26	21
2. Tax paid by a married couple with two children:		
average rate of taxation on personal income (%)	11	14
social security rate (%)	6	2
total (%)	17	16

TAX RATES

	U.K.	AUSTRALIA	AUSTRIA	BELGIUM	DENMARK	FRANCE	(WEST) GERMANY	IRELAND	ITALY	JAPAN	NETHERLANDS	NORWAY	SPAIN	SWEDEN	SWITZERLAND
	1974	1974	1974	1974	1974	1974	1974	1974	1974	1974	1974	1974	1974	1974	1974
	25	21	13	12	39	8^x	19	20	5	10	15	25	3	36	12
	6	—	13	11	4	8	14	4	7	4	20	8	4	2	10
	31	21	26	23	43	16	33	24	12	14	35	33	7	38	22
	17	16	1	10	34	NA	9	11	5	5	11	20	3	31	9
	6	NA	13	11	4	8	14	4	7	4	20	8	4	2	10
	23	16	14‡	21	38	8	23	15	12	9	31	28	7	33	19

^{x}Does not take account of communal taxes (ie Taxe d'Habitation)
‡Tax credits wholly or partially cancel out the tax liability

Source:
Organization for Economic Co-operation and Development

Germany, and pays five to seven per cent less in taxes.

Which country is best in terms of keeping your income in your pocket? In Spain a single person pays only seven per cent of his total earnings. But he doesn't get a tax reduction if he's married with two children.

The Italians have the second-lowest tax rates. Single people pay only 12% of their wages, but get no relief if they're married with two children. Low-paid workers with heavy family expenses pay the stiffest penalty – a fact which may contribute to social unrest. Italy also suffers from a marked disparity in wages. The top of the wage scale isn't particularly high compared to other countries – but the lower end of the scale is very low. A primary school teacher in Italy earns a mere 50% of someone doing the same job here in the U.K.

How much tax do you pay? Everybody complains about paying taxes. But this table shows you just how your situation compares to fellow-sufferers in other countries. First, let's take a look at what happens to a fairly average income of £4,331 after the taxman gets his hands on it.

If you live in the Netherlands, you're certainly entitled to moan the loudest: Dutch people with an annual income of £4,331 have to pay a staggering 26.1% in tax. Maybe it costs a lot to maintain all those dykes and windmills.

The British wallet suffers almost as badly. We'd lose 25.1% of that £4,331: £1,189 gets snatched away by the Government. Of course, some of that money comes back to you in services like the National Health, which provides virtually free medical care.

In this income bracket, it's our lucky American cousins who have least cause to complain about taxation. A mere 8.2% of their £4,331 would vanish, leaving almost £4,000. One reason for the low rate of taxation in the States is the large population – over 200 million citizens provide quite a broad tax base.

Japan doesn't put too much strain on the taxpayer's wallet either: someone earning £4,331 would pay only 8.4% to the Government.

Now we'll look at what happens to

Income tax and social security contributions (as % of total earnings) paid on: [+]	U.S.[0]	U.K.
	1977	1977
£4331	8.2	25.1
amount left	3975	3242
£8662	17.5	40.4
amount left	7146	5164
£12,993	23.0	56.0
amount left	10005	5709

someone earning an income of twice our original example. Here in the U.K., there's a real increase in the amount that's earmarked for taxes – it jumps 15.3% to 40.4% of the income being assessed. It's by far the highest taxation level of any nation in our chart.

If you lived in France and earned the same amount of money, you'd lose only 11.6% of it and again, the Japanese rank second in the low taxation league, with a rate of 14.1%.

The outlook gets even bleaker when we turn to the figures for the executive income bracket. On an average salary of £12,993, a beleaguered Briton would shell out a massive 56% in taxes. That means he'd be left with a paltry £5,709, not much more than his middleclass counterpart who was left with £5,164. There may be something in those arguments about heavy taxation destroying the incentive

COMPARATIVE TAX RATES

AUSTRALIA	BELGIUM[0]	DENMARK[0]	FRANCE	(WEST) GERMANY	IRELAND	ITALY	JAPAN[0]	NETHERLANDS
1977	1977	1977	1977	1977	1977	1977	1977	1977
9.8	18.3	18.7	9.2	24.7	21.8	17.0	8.4	26.1
3906	3538	3521	3932	3261	3387	3595	3967	3200
23.3	28.2	34.8	11.6	25.3	29.6	24.3	14.1	30.1
6644	6219	5648	7657	6470	6098	6557	7441	6055
30.7	33.6	43.9	14.4	28.4	39.7	28.6	16.7	35.5
9004	8627	7289	11122	9302	7835	9277	10823	8380

to work harder and get ahead.

In the U.S., someone in the same bracket would be left with £10,005. In fact, the taxation rate there rises less than five per cent from the lowest salary we looked at to the highest, whereas in the U.K., it rises a whopping 30.9%.

And look at the figures for France. The French big earner pays the lowest tax percentage in our tables: he's left with £11,122.

Once again, the Japanese get off comparatively easy. They pay only 16.7% on an income of nearly £13,000.

It's an interesting – or should we say depressing – fact that the U.K. definitely tops the charts when it comes to heavy taxation. British taxmen impose heavier tax penalties than any other country. As if the weather wasn't enough to cope with!

Source:
Inland Revenue, London
+Amount paid by a married man with two children
0Figures include deductions for local income taxes

Inheritance taxes aren't easy to calculate or compare. Most countries have systems which allow for innumerable deductions and exceptions before tax is payable. This can reduce the amount considerably. And many people (perhaps the wisest) leave such small sums that no tax has to be paid on them.

Inheritance tax[x] (as % of total estate) paid on:	U.S.A.[+]	CANADA[o]
£5,000	0	11.0
amount left (000s)	5.0	4.4
£50,000	15.6	11.0
amount left (000s)	42.2	44.5
£500,000	33.4	19.0
amount left (000s)	332.9	405

Different rates of tax may apply depending on who inherits the money. Generally, the tax system favours close relatives. More duty may be payable if an estate is left to distant relatives or friends. Our table shows minimum inheritance taxes. In many cases the amounts are much higher.

Which countries demand the highest death taxes? Look first at the tax payable on the lowest of the figures in our table – an estate worth £5,000.

In the U.K. you don't pay anything on an estate of £5,000 – something to remember when you're on your last shopping spree.

Japan and Canada have the highest levels of inheritance tax (11%) with Spain (ten per cent) close behind. If a Japanese or Canadian leaves £5,000 to his children, they'll end up with £4,400 after paying taxes to the government.

In Australia, Denmark, Ireland, the U.S. and the U.K. no tax is paid on estates worth £5,000.

It's a different story if you leave a £50,000 estate. This figure isn't as large as it seems, since it's tax based on the value of the whole estate, not just cash in the bank. A small business could be worth that. Or a home with furnishings.

In the £50,000 range, the Swedes pay the highest rate of inheritance tax, an astonishing 53.5%. If a successful Stockholm businessman dies and leaves £50,000 to his wife, she'll end up with only £23,200. With the cost of living what it is in Sweden, she'll need every kroner of it.

The U.K. comes second with a death duty of 35%. This may be less than Sweden, but it still means that only £32,500 is left from a £50,000 inheritance. In other words it costs £17,500 for family assets or business to be passed on to the next generation. American beneficiaries pay 15.6% and end up with £42,200 Canadians are even better off: they pay 11% and are left with £44,500.

Ireland is at the top or bottom of the league, depending on whether you're the heir or the tax collector. Over there you keep every penny of the £50,000 left to you. Other countries with low death duties are Switzerland (4.5%) and West Germany (six per cent). In these countries there's little difference between tax rates on small and medium estates.

INHERITANCE TAX

U.K.	AUSTRALIA	AUSTRIA	BELGIUM	DENMARK	FRANCE	(WEST) GERMANY	IRELAND	ITALY	JAPAN	NETHERLANDS	NORWAY	SPAIN	SWEDEN	SWITZERLAND
0	0	2.5	3.0	0	5.0	3.0	0	5.0	11.0	9.0	8.0	10.0	7.5	3.0
5.0	5.0	4.9	4.8	5.0	4.7	4.8	5.0	4.7	4.4	4.5	4.6	4.5	4.6	4.8
35.0	7.0	7.0	10.0	15.7	2.0	6.0	0	16.0	24.2	15.0	30.0	13.0	53.5	4.5
32.5	46.4	46.5	45.0	42.1	40.0	47.0	50.0	42.0	37.9	42.5	35.0	43.5	23.2	47.7
65.0	26.3	12.0	17.0	29.9	20.0	12.0	29.1	45.0	52.5	17.0	35.0	19.0	53.5	6
175	368	440	415	350.3	400	440	354	275	237.5	415	32.5	405	232.1	470

The most striking contrast between progress rates of inheritance tax is found in Switzerland and the U.K. In Switzerland, inheritance tax is paid at a low rate regardless of the size or value of the estate. In the U.K. tax is minimal on small estates – then soars.

On estates of £500,000 the inheritance tax rate in the U.S. more than doubles to 33.4%. But it means you'd only pay the sum of £167,000 if you were lucky enough to die in such a wealthy condition. The Canadian wealthy do even better, paying a mere £95,000 for the privilege of passing on.

A U.K. estate worth £500,000 is taxed at a rate of 65%. If an Englishman dies and leaves his £500,000 castle to his heirs, only £175,000 will remain to be shared out after tax – or about the value of your average turret.

Switzerland is at the low end of the table; inheritance tax on £500,000 is just six per cent, up only three per cent from the tax rate on £5,000 This means that a Swiss left a £500,000 estate pays only £30,000 in tax – about £355,000 less than his U.K. counterpart. Inheritance taxes

Sources:
Official government documents
×1978 figures
Figures are illustrative calculations, by Heron House. They show how much inheritance tax would be paid if the entire estate were left to the spouse and/or children — the lowest-rated category.
+State death tax, unified rate schedule
0In Ontario

are a way of dividing a country's wealth among more of its people, and Switzerland's considerable riches are shared comparatively equally in the first place. In the U.K. a much higher proportion of money and property is in the hands of relatively few individuals.

Whatever the reason for inheritance taxes, the contribution they make to a country's total tax revenue is small. In the U.S., where just over 33% of a £500,000 estate goes in tax, inheritance taxes account for about 1.4% of the total federal tax. Even in the U.K. it only amounts to about 1.3% of the total tax income – a very low figure.

This table is all about booze and cigarettes – how much they cost and why.

Take cigarettes, for instance. On average, the cost of manufacturing a packet of cigarettes runs between 18 and 35 pence. In Spain, France and Italy, costs are even lower (Spain produces 20 cigarettes for about only 4p). Norwegian cigarettes are the costliest to manufacture at 38p a packet. If you buy cigarettes in any of these countries, the prices you'll pay will vary widely.

Taxes on tobacco are an important source of revenue to all nations' governments.

Where are cigarettes cheapest? In Spain you can buy a packet for just 12p. Japanese cigarettes are cheap as well, about 29p – 3p less a packet than in Italy.

Prices in the U.S. vary from state to state and city to city because of differences in local taxes, but they average 30p a packet, a low price compared to other countries.

A packet of cigarettes costs about 55p in the U.K., which is average for our list – although this doesn't stop us complaining.

Perhaps we who moan should try visiting Denmark. There the average packet of cigarettes costs a colossal £1.33. In Norway, the second most expensive country in our list, cigarettes cost £1.08. Sweden charges 82p. Then comes West Germany at 71p and Belgium at 67p.

In Ireland, nearly eight per cent of national revenue is supplied by smokers, almost double the rate for the next country in the list – the U.K.

Who pays the most for another pleas-

	U.S.A.	CANADA
	1974	–
1. Tax revenue[+] (UK £ – millions) from the sale of:		
tobacco	2674	NA
liquor	3605	NA
2. Tax revenue[+] (as % of total revenue) from the sale of:		
tobacco	1.53	NA
liquor	2.07	NA
	1978	1978
3. Average retail price (UK £) of a pack of 20 cigarettes[0]	0.30	0.37
excluding taxes	0.10	0.13
4. Estimated tax as % of the retail price	55-75	55-75
5. Average retail price (UK £) of bottle of Scotch whisky[x]	4.10[xx]	4.29[‡]
excluding taxes	2.54[xx]	1.45[‡]
6. Estimated tax as % of retail price	38[++]	00 66

LIQUOR AND TOBACCO TAXES

	U.K.	AUSTRALIA	AUSTRIA	BELGIUM	DENMARK	FRANCE	(WEST) GERMANY	IRELAND	ITALY	JAPAN	NETHERLANDS	NORWAY	SPAIN	SWEDEN	SWITZERLAND
	1974	–	–	1974	–	1974	1974	1974	–	1974	–	1974	–	1974	1974
	1282	NA	NA	135	NA	593	1748	75	NA	1022	NA	79	NA	234	69
	1107	NA	NA	98	NA	625	1165	113.4	NA	1483	NA	146	NA	418	72
	4.40	NA	NA	1.54	NA	1.39	2.83	7.96	NA	2.30	NA	1.78	NA	2.21	1.78
	3.80	NA	NA	1.11	NA	1.46	1.89	12.06	NA	2.78	NA	3.30	NA	3.94	1.30
	1978	1978	–	1978	1978	1978	1978	1978	1978	1978	1978	1978	1978	1978	1978
	0.55	0.34	NA	0.67	1.33	0.23 ++	0.71	0.43	0.32	0.29 ++	0.64 00	1.08	0.12 ++	0.82	0.45
	0.16	0.12	NA	0.19	0.17	0.06	0.20	0.16	0.08	0.14	0.20	0.38	0.04	0.28	0.16
	70	55-75	NA	72	87	73	71	62	74	50	69	55-75	55-75	55-75	55-75
	4.44	5.39	5.43	3.69	8.92	4.19	4.30	4.90	2.74	5.60	3.74	9.58 ++	5.86	8.54 ++	9.45
	0.97	2.69	NA	1.58	NA	1.59	2.40	NA	1.34	3.30	NA	NA	2.93	NA	NA
	78	50	NA	57	NA	62	44	NA	51	41	NA	NA	50	NA	NA

++ NA not available xx For New York ‡ For Ontario
++ State monopoly 00 A pack of 25 cigarettes

ure in life – Scotch? Again, the poor Norwegians seem heaviest hit by the taxman. They pay £9.58 for a bottle compared to £9.45 in Switzerland and £8.92 in Denmark. Americans get off cheaply at £4.10, while Canadians pay £4.29. In the U.K., where Scotch is actually produced, the price averages £4.44 a bottle. That's over four times as much as it need cost.

Sources:
1 & 2 Organization for Economic Co-operation and Development
3, 4, 5 & 6 Confidential industry sources
+ All taxes levied on production, sale, leasing, delivery of goods, etc.
0 Popularly-priced brand
x Standard brand; 26 ounces

Estimated tax as % of retail price

This chart is a sort of "drinking man's guide to the world". If you like to tipple, head for the U.S. Uncle Sam is the most lenient taxman on the tables when it comes to whisky. Under 40 pence of every pound spent on a bottle of Scotch goes to the government. That may sound a lot, but you'll pay considerably more anywhere else.

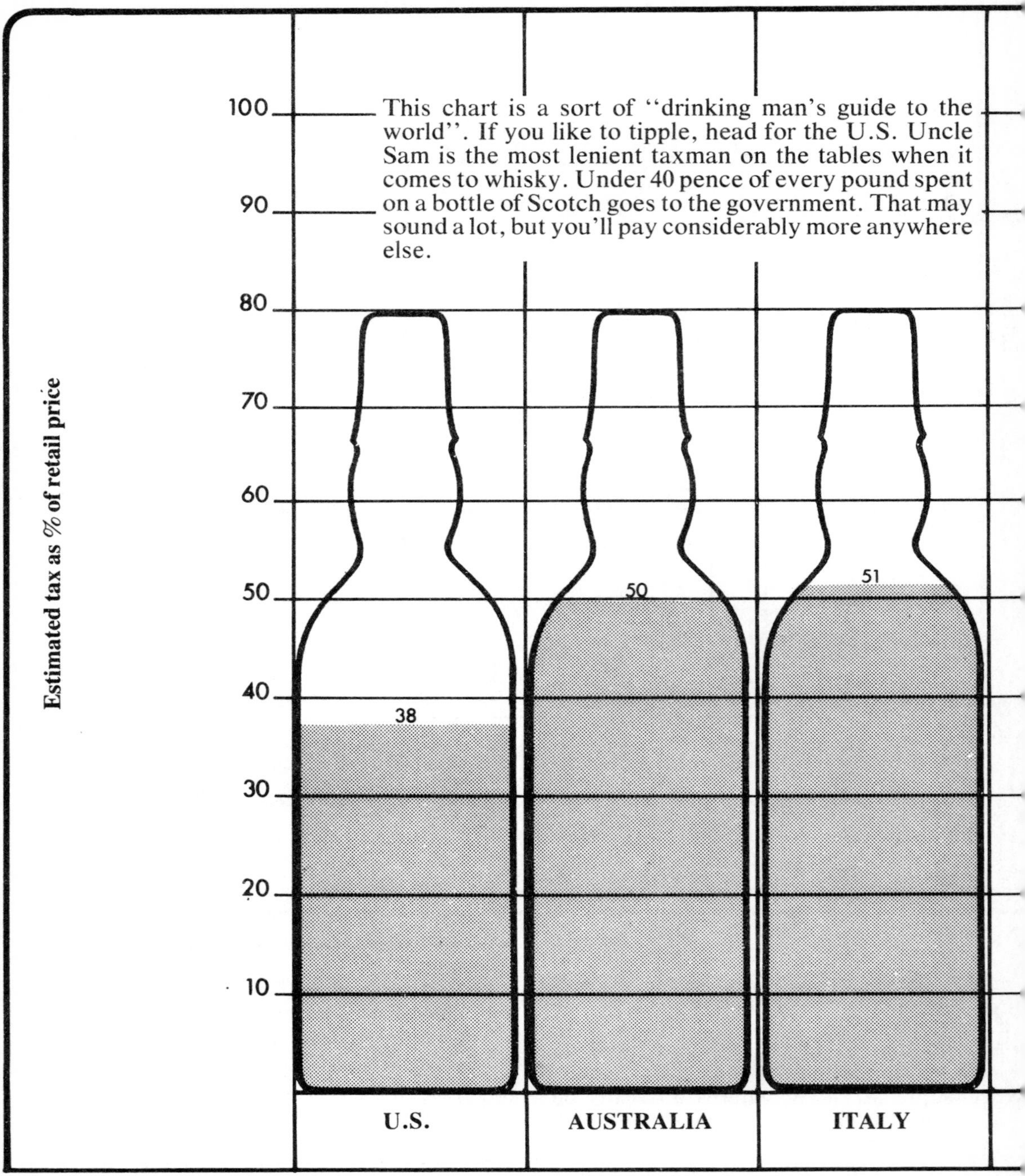

100
90
80
70
60
50
51
50
40
38
30
20
10

U.S.   AUSTRALIA   ITALY

TAX ON WHISKY

No wonder those hard-drinking Aussies are so fond of lager — the whisky tax down under is 50%. The French are better off sticking to wine: a 62% tax makes a bottle of Scotch *très* expensive. In Canada, home of rye whisky, the taxman takes a massive 66% cut.

We long-suffering British are hardest hit by alcohol taxes. We pay 78% tax — over twice as high as the U.S. figure. It's enough to drive a man to drink.

BELGIUM	FRANCE	CANADA	U.K.
57	62	66	78

At Work

If there's one thing that's virtually as certain as taxes and death, it's work. Most of us spend more of our lives working than involved in any other single activity. So it's a subject worth finding out about. This chapter shows you, around the world, who works, and at what and for how much. It also gives all sorts of comparative data, so, if you're foot-loose and fancy free with top secretarial skills, you may well read the working chapter with special interest. Even if your primary aim in life is to avoid work, there's still a few useful facts – you can find out where unemployment is highest, for instance.

Our first table shows what people do in the way of work in the 17 countries we cover. For example, how many people work on the land, compared with those who work in manufacturing, in mines or in construction.

The table also gives you an insight into the working priorities each country has and its stage of development. For instance, it's interesting to see that the U.K. is now a service oriented society. More than 56% of the work force is employed in some form of service. With our ever-growing bureaucracy you might expect the U.K. to have one of the largest service sectors. But do we?

The next table gives you facts on the total workforce and divides it into three revealing categories. First, it shows how many people are wage earners and salaried staff – in other words how many work for someone else. Next, it shows how many people are employers or are self-employed – that is, how many bosses there are, and how many prefer to remain their own boss. And then it shows how many people are unpaid family workers – mostly family members and dependants who help out around a business or farm but receive no specific remuneration.

Now that you know how many people work and what they actually do in the way of work, it's interesting to know how many people there are in each country who can't get work. That's what the third table in the chapter is about. It has quite a few shocks. For example, over in the U.S., the world's largest and strongest economy, they have the highest rate of unemployment amongst all countries covered.

Then there's the question of productivity – how much each individual worker produces. In other words, in raw cash terms, how much does each worker actually contribute, on average, to the total earnings of his country? That's what the next table is all about. This isn't an easy topic to discuss. All kinds of factors play a part – worker efficiency, industrial plan efficiency, management planning, incentives and (in some cases) restrictions on the number of hours which can be worked.

Causes aside, the results speak for themselves – they show which countries are the most efficient in the world today. How does the British worker rate?

Mention steel production at a cocktail party and you're apt to encounter complete silence. But steel is the basis of the West's industrial might, so the figures in the following table on steel production and consumption are an important indicator of the relative strength of various countries' economies. As in almost all of our tables, there are one or two surprises here. And the accompanying text provides some interesting insights.

One of the reasons we work is to go on holiday. So we've included some interesting comparisons on paid holidays. But one of the most basic reasons most of us work is to earn money. *That's* what the next tables are all about. Where can you get the highest average earnings per hour? Where are people paid the least? And the accompanying text also shows you how to get an idea of how much this cash is worth in terms of actual spending power.

The next table breaks down these overall figures by comparing the earnings in different countries for six different occupations. Here, for instance, you'll find out how much a female textile worker in London earns in comparison with a construction worker, and also how much she earns in comparison with her counterparts in Madrid or Zurich.

No chapter on employment would be complete without an account of industrial strife – strikes. Which people are most strike-prone? Where do most workers go on strike? And which countries lose the largest number of working days through strikes? The section which follows has all the details.

Finally, we've tables which profile how safe it is to go to work and also how much you'll have squirrelled away when you retire in any of the countries on our list.

We've divided types of work into several sections to compare how many people do what type of work in which countries. The categories are divided broadly by agriculture, mining, manufacturing, utilities, construction and services – with a catch-all category called "other".

Most of the countries on the table are industrial nations. But in Ireland one person in every four lives off the land or sea – the highest figure in that category on the table. Almost as many Spaniards are country-dwellers. Compare these percentages with the U.K., where about one person out of every 37 in the workforce farms or fishes for a living. Or the U.S., where a scant four per cent of the workforce supplies enough food, fish and timber to support the entire country. That means that one U.S. farmer or fisherman feeds about 62 people.

After Ireland and Spain, Italy, Japan, Austria and France have the highest percentage of the workforce engaged in agriculture, fishing and forestry. They've got more open land than most of the other countries on the table. In France and Italy the agricultural tradition has maintained in several areas, while industrialisation has developed elsewhere.

Mining and quarrying occupy few workers in all the countries on the table. (The list would change dramatically if South Africa and Zaire were included.)

	U.S.A.	CANADA
	1975	1975
1. Total working labor force+ (000s)	84,783	9171
2. Percentage of working labor force+ employed in:		
agriculture, hunting, forestry, fishing	4.0	6.1
mining, quarrying		1.4
manufacturing	29.0	20.9
electricity, gas, water		1.1
construction		6.5
services[0], others	67.0	64.0

But over a quarter of the workforce of the countries listed is involved in manufacture – producing the volumes of hard and soft goods that have changed the nature of our values and our ways of life.

Switzerland has almost half of the workforce involved in manufacture, power industries and construction. (The rest are either herding their Swiss Brown over the edelweiss or minding everyone else's money in the banks.) West Germany has 36% involved in manufacturing – roughly one in every three workers. We all know what they're up to – making sure the post-war economic miracle survives into the 21st century.

In the U.K., Austria, Belgium, France, Italy and Sweden, about three in ten workers are employed in manufacturing. The lowest figure on the table

U.K.	AUSTRALIA	AUSTRIA	BELGIUM	DENMARK	FRANCE	(WEST) GERMANY	IRELAND	ITALY	JAPAN	NETHERLANDS	NORWAY	SPAIN	SWEDEN	SWITZERLAND
1975	1975	1975	1975	1975	1975	1975	1975	1975	1975	1975	1975	1975	1975	1975
24632	5726	2943	3748	2332	20764	24828	1030	18818	52230	4535	1694	12692	4062	2784
2.7	6.7	12.5	3.6	9.8	11.3	7.3	24.5	15.8	12.7	5.0	10.2	22.1	6.4	7.9
1.4	1.4	1.2	1.0	0.1	0.9	1.4	1.0	1.8	0.3	0.2	0.7	0.8	0.5	
30.9	23.6	30.1	30.1	22.7	27.9	35.8	20.4	32.6	25.8	24.0	23.8	26.7	28.0	45.0
1.4		1.2	0.9	0.6	0.8	1.0	1.4	NA	0.6	1.0	1.0	0.7	0.8	
7.1	8.8	11.9	7.9	8.1	9.1	7.7	6.9	9.8	9.2	9.6	8.7	10.0	7.1	
56.5	59.5	43.1	56.5	58.7	50.0	46.8	45.9	40.0	51.5	60.1	55.6	39.7	57.1	47.1

NA not available

for manufacture is Ireland – consistent with the high Irish rate of fishing and agriculture. Surprisingly, in industrial Japan, only one in four in the workforce is involved in manufacturing – ten per cent less than in West Germany.

Neither Canada nor Denmark employs a lot of people in manufacturing. But many Canadians and Danes work in the service industries – covering everything from wholesale and retail sales to restaurants, banking, social services and tourism. In Canada two out of every three workers are employed in a service industry.

But the percentage in the U.S. is even higher. Sixty-seven per cent of the U.S. workforce provides a service of some kind. This may be an indication of the growing "softness" of the U.S. and other economies (as compared to the "hardness" of manufacturing actual goods). But it also could be evidence of a world economic trend away from essentials to "luxuries" of one sort or another. First you need food. Then you need cars, detergents and appliances. Then you need services that make life a little easier and more indulgent – such as having fish and chips delivered to your door.

Source:
Organization for Economic Co-operation and Development

+Civilians only; excludes members of the armed forces
0Wholesale and retail trade, catering, business services, etc

Most people work, and most people work for someone else. Are you a boss, or a mere cog in the machinery? Self-employed, or a wage-slave?

The first category in our table deals with the total number of people at work (excluding members of the armed forces). Not surprisingly, the U.S. tops the table. There are a lot of potential workers in a population of over 200 million. Of course, capital has to be added to this potential – in the form of equipment and know-how, to transform it into economic wealth.

What better proof of this than the second-ranking country on the table – Japan? It wasn't until the post-war years and the introduction of modern technology that the vast labour force there was equipped to create material prosperity.

Of course, modern methods of production involve highly organised workforces directed towards a common goal. And that's the antithesis of self-employment.

The highest proportion of people who work for someone else is in the U.K., where 92.2% of our work force are employees. One of the criticisms of British society today is that fewer and fewer people work for themselves, and more and more are employed by the government doing jobs that produce no tangible goods that can be sold to increase our prosperity.

It's harder to explain why the U.S.

		U.S.A.	CANADA
		1975	1975
1.	Total working civilian[+] labor force (millions)	84.7	9.1
2.	Total working civilian[+] labor force divided into:		
	wage earners; salaried staff (%)	90.3	88.9
	employers; self-employed (%)	8.7	9.5
	unpaid family workers[0] (%)	1.0	1.6

should be second on the list (90.3% employees) – notably since they're reputed to be a nation of rugged individualists. They certainly don't have the same huge proportion of government workers as we do. The explanation for the low number of bosses and self-employed is probably that most U.S. businesses today are corporations. Even the smallest business or partnership is apt to incorporate if it becomes at all successful.

Sweden comes third in the employee sweepstakes. Not much doubt why – it's their socialist welfare state. All that government involvement in the life of the nation means mountains of paperwork. And that means thousands of workers, in the form of government clerks, to keep the bureaucratic wheels moving.

The Italian, more than anyone else on our table, likes to be his own boss. More than one in five Italians are self-employed in a business or profession. This figure reflects the way of life in much of Italy,

LABOUR FORCE: STATUS

	U.K.	AUSTRALIA	AUSTRIA	BELGIUM	DENMARK	FRANCE	(WEST) GERMANY	IRELAND	ITALY	JAPAN	NETHERLANDS	NORWAY	SPAIN	SWEDEN	SWITZERLAND
	1975	1975	1975	1975	1975	1975	1975	1975	1975	1975	1975	1975	1975	1975	1975
	24.6	5.7	2.9‡	3.7	2.3	20.7	24.8	1.0	18.8	52.2	4.5	1.6	12.6	4.0	2.7
	92.2[x]	85.7	80.5	83.2	81.4	80.9	84.2	70.8	72.4	69.8	84.5	84.4	69.8	89.3	NA
	7.8[x]	14.0	19.5	13.1	13.8	19.1	10.1	29.2	21.4	18.0	15.5	11.8	30.2	10.4	NA
	NA	0.3		3.7	4.7		5.7		6.2	12.0		3.8		0.3	NA

[x] This figure is slightly inflated, as the number of unpaid family workers is not available
‡ Includes armed forces. NA not available

where there are still plenty of small shops and independent craftsmen. Japan (18%) is second on the list of self-employed workers, in spite of the fact it is now the home of gigantic industrial corporations. Japan, like Italy, still has many small tradesmen – peddlers, operators of small food stands and shops, etc. Australia, Denmark and Belgium also have high proportions of the self-employed.

It's an obvious conclusion that the British way of life is actually eliminating the small businessman and other self-employed workers.

The third category of workers is people who work without pay. These are mostly family members and dependants who help out around a business or farm and don't receive any specific remuneration. Spain tops the table here with a combined figure of 30.2% for self-employed plus unpaid family help. Think of all those small bars and cafes in Spain, open at all hours, with everyone from Grandma to

Source:
Organization for Economic Co-operation and Development
+Excludes members of the armed forces
OPeople who assist in the operation of a business or farm, and who have worked at least one-third of the period covered

the kids lending a hand. Japan is second with a combined figure of 30% and Ireland close behind with 29.2%. All three countries have strongly conservative family traditions. Italy is next – about one in three of the self-employed has a family member working without pay. The combined Italian total is 27.6%. This is followed by Austria (19.5%) and France (19.1%).

Australia and Sweden have the lowest number of unpaid workers – about three per thousand in the working population. The U.S. figure is ten per thousand, and in Canada, 16 per thousand are unpaid.

The world trade recession has caused the most feared of all economic diseases – high unemployment. Although we British believe that unemployment is our major problem, the U.K.'s unemployment rate of just under 3.4% is below average for the countries listed.

It is the U.S. which has suffered worst of all. Over eight per cent of their total labour force was out of work in 1975. That's nearly one person in 12 without a job. In a country the size of the U.S. that's nearly eight million people out of work.

The proportion of people unemployed in Ireland is almost as high as it is in the U.S. But there the situation is different, because Ireland has always relied on emigration to solve the worst of its unemployment problems.

Third in our table comes Canada, where unemployment is running at seven per cent of the total labour force. Canada, the U.S. and Ireland are way ahead of the other countries on our table. Canada, for instance, has double the British rate.

A surprise on this table is Italy's low figure. At 3.3% it's even fractionally lower than the U.K.'s. But there's a hidden factor here. The Italians traditionally have a large migratory labour force. Many Italian workers head for the factories of Germany, Switzerland or other parts of Europe, and work there for several years before returning to Italy. A similar situation exists in Spain. There the rate of nearly five per cent is high enough – without taking into account the migrant workers in the factories of Paris or the hotels of London who aren't included in the figures.

The country with the lowest unemployment rate is Switzerland. However, its figure of less than half of one per cent only takes account of the registered wholly unemployed. But when migrant

	U.S.A.	CANADA
	1975	1975
1. Unemployment:		
Total unemployed (000s)	7,830	694
Unemployed as % of labor force	8.3	7.0
Unemployed men as % of male labor force	7.6	7.3
Unemployed women as % of female labor force	9.3	6.4
2. Youth unemployment (15-24 year olds):		
1970	9.9ˣ	10.3
1973	9.8ˣ	9.7
1975	15.2ˣ	12.2
1976	14.0ˣ	12.5

UNEMPLOYMENT

	U.K.	AUSTRALIA	AUSTRIA	BELGIUM	DENMARK	FRANCE	(WEST) GERMANY	IRELAND	ITALY	JAPAN	NETHERLANDS	NORWAY	SPAIN	SWEDEN	SWITZERLAND
	1975	1975	1975	1975	1975	1975	1975	1975	1975	1975	1975	1975	1975	1975	1975
	866[+]	254	52	168	121	889	1,074	90	654	1,000	206	40	625	67	10[0]
	3.4[+]	4.2	1.7	4.2	4.9	4.0	4.1	8.0	3.3	1.9	4.3	2.3	4.7	1.6	0.4[0]
	4.4	3.4	1.4	3.1	4.7	2.7	3.7	8.6	2.8	2.0	NA	1.8	4.8	1.3	NA
	1.6	5.7	2.2	6.3	5.1	6.1	4.6	6.0	4.6	1.7	NA	3.0	4.2	2.0	NA
	2.7[‡]	2.5	1.4[*,*]	NA	NA	1.5	0.3[++]	NA	10.2[00]	1.9	NA	NA	2.3	2.8[x]	NA
	2.9[xx][‡]	3.8	1.4[**]	NA	NA	2.9	1.0[++]	NA	12.6[00]	2.2	NA	NA	6.7	5.3[x]	NA
	7.4[xx][‡]	8.9[xx]	1.4[**]	NA	NA	7.6	5.8[++]	NA	12.8[00]	3.0	NA	NA	10.5	3.6[x]	NA
	13.1[xx][‡]	9.0	1.4[**]	NA	NA	9.9[xx]	5.2[++][xx]	NA	14.4[00][xx]	3.1	NA	NA	12.5	3.6[x]	NA

[+]Registered wholly unemployed excluding school leavers [0]Registered wholly unemployed [x]16-24 year olds [‡]Under 25 year olds — labor force 16-24 year olds [**]Under 30 years old — labor force 15-29 year olds [++]Under 25 years old — labor force 15-24 year olds [00]14-24 year olds [xx]OECD estimates NA not available

labourers are out of work the government doesn't support them. So they're not registered as unemployed.

Sweden has the second lowest unemployment rate — just over 1.5%. Austria comes next with a rate of only 1.7%. These countries use immigrant labour which isn't included in national figures. On the other hand immigrant labour *is* included in the U.K. figures.

As far as unemployment among the young is concerned, Italy has the highest rate. The figures for the U.K. show that

Source:
Organization for Economic
Co-operation and Development

there has been a dramatic increase in youth unemployment. We now have the third worst rate in this category, and this has been one of the government's main priorities. The special youth employment schemes have undoubtedly kept the number of young unemployed lower than it otherwise would be.

	U.S.A.	CANADA
	1975	1975
1. National economy[+] (£ 000s) per person employed	8.72	8.10
2. Comparison of national economy[0]	131	122
3. Comparison of 1975 national economy with 1970[x]	105[‡]	114[**]

Labour productivity figures are one way to measure workforce efficiency. If you look at the overall production of a country, you can see how much wealth its people actually create. Our figures show how much the members of a country's working population earn for their country each year.

With characteristic Swiss efficiency, each worker there produces an average of £9,070 worth of goods a year. And this in a country with little heavy industry, and few raw material resources. Most of this revenue comes from what's called "invisible earnings" – money generated without producing tangible goods. Banking and tourism, for instance, are both cost-efficient and capable of earning a lot without employing lots of labour – especially if planned with Swiss know-how.

In Switzerland, tourism brings year-round profits because it caters not only to the winter tourist who enjoys skiing and winter sports, but also the summer visitor captivated by the scenery.

And the wealthy, wherever they obtained their money, often deposit it in Zürich banks, secure in the knowledge that Switzerland has a high level of political and social stability.

In most statistics, America tops the league in terms of overall production, but their workers produce only £8,720 worth of goods a year – about four per cent less than the Swiss.

This may be because some elements of their workforce – for instance, migrant farm workers or the urban poor – don't have high productivity levels.

Next in terms of labour productivity is the Netherlands (£8,240) followed by Canada (£8,100) and West Germany, close behind at £8,070.

Holland and West Germany are known for their efficient industries, arising from the modernisation which followed World War II. Canada has enjoyed a steady and undisturbed growth rate, helped by rich mineral and fuel resources available for export. Furthermore, all three have prosperous agricultural industries.

Sweden is also high in this table. It's sixth, with £7,790. Modern industrial methods, good labour relations, social stability and an efficient farming industry all play their parts in this high figure. Belgian workers are only £20 behind the Swedes. Norway and Denmark also do well. Their workers produce over £7,000 worth of goods every year.

France comes next with £7,180, surprisingly low, considering France's pre-

LABOUR PRODUCTIVITY

	U.K.	AUSTRALIA	AUSTRIA	BELGIUM	DENMARK	FRANCE	(WEST) GERMANY	IRELAND	ITALY	JAPAN	NETHERLANDS	NORWAY	SPAIN	SWEDEN	SWITZERLAND
	1975	1975	1975	1975	1975	1975	1975	1975	1975	1975	1975	1975	1975	1975	1975
	4.27	6.52	5.91	7.77	7.35	7.18	8.07	3.61	4.27	4.65	8.24	7.58	3.69	7.79	9.07
	64	98	89	117	111	108	122	54	64	70	124	114	56	117	137
	104	NA	118	108++	111	118	116	121	111	127	118\00	120	NA	106	109

NA not available   ‡Total private economy   **Excluding public administration
++1972 figure   001974 figure

sent economic prominence. However, France is recovering from previous decades of economic difficulty and its figures are improving year by year. Only Japan, Ireland and Norway are moving upward at a greater rate.

Ireland, with £3,610 annual production per worker, is at the bottom of the list, while Spain at £3,690 is little better. In both countries nearly a quarter of the working population is employed in agriculture – more than any other European country. Not only do relatively few people work in industry, but farms are comparatively small. Both countries are unsuitable for large-scale mechanised farming, although the picture may soon change. Spain proposes to join the Common Market, and massive foreign investment is pouring into Ireland.

Italy with Britain are near the bottom of the table, tied at £4,270 annual production per person.

A high percentage of Italy's workforce is in manufacturing, but its worker-management relations are appalling. It's

Sources:
1 International Labour Organisation
2 & 3 Heron House calculations based on ILO figures
+Ratio of output (GNP) to input (civilian employment).
0Index of 100 as the average of the specified countries.
xIndex of 100 – GNP or GDP per employed person

a land of contrast. Modern Italy is concentrated in the north, while in the south life has hardly changed since the last century.

	U.S.A.	CANADA
	1976	1976
1. **Total crude steel production** (tons — millions)	128.0	14.4
2. **National consumption** (tons — millions)	141.5	13.5
3. **Steel consumption per capita (lbs)**	1316	1173

There would be no modern industry without steel. It's the foundation of our present industrial system and the chief basis for our high standard of living. The U.S., as you'd expect, is the leading steel producer (128 million tons). Perhaps the U.S.S.R. makes almost as much but we don't know – for security reasons the Russians refuse to release any reliable statistics.

Japan is the second largest producer (118.3 million tons) on our table, a difference of just about ten million tons. This might seem like a lot, but in the overall figure it's not.

Japan probably has the most efficient steel industry in the world and, therefore, is able to capture a large part of the world market. The Japanese produce almost twice as much as they consume (118.3 million tons compared with 65.7 million tons consumed). This isn't true of the U.K., which has to import nearly one million tons a year.

Germany comes next in terms of crude steel production (46.7 million tons), followed by France (25.6 million tons). The U.K. is in sixth place with 24.5 million tons. The German steel industry is thoroughly modernised and almost as efficient as the Japanese. It's the old story. Both countries were completely devastated at the end of World War II and had to rebuild, making use of the most modern methods and equipment available. The U.S., France and the U.K. are barely managing to hold their own in the steel market. The U.S. industry is ham-pered by rising labour costs and outdated, deteriorating manufacturing plants. They can't produce steel as cheaply as the Japanese, even though Japan imports much of its iron from them.

The efficiency of a country's steel production can be estimated by adjusting the population figures. For instance, Canada has just over one inhabitant to every nine in the U.S. Multiply Canada's steel output by nine and you get 129.6 million tons – showing that the Canadian steel industry has a per capita production almost identical to the U.S.'s (128.0 million tons). If the U.K. were the same size as the U.S., theoretically we would produce about 98 million tons a year. By the same calculation, the Germans would produce 163 and the Japanese an amazing 236 million tons a year. Japan is by far the largest steel exporter – 52.6 million tons annually. Of the other countries on the table, only West Germany exports any significant quantity of steel: 6.3 million tons.

High steel consumption indicates that a country is heavily industrialised. The table shows that the U.S. is most

	U.K.	AUSTRALIA	AUSTRIA	BELGIUM+	DENMARK	FRANCE	(WEST) GERMANY	IRELAND	ITALY	JAPAN	NETHERLANDS	NORWAY	SPAIN	SWEDEN	SWITZERLAND
	1976	1976	1976	1976	1976	1976	1976	1976	1976	1976	1976	1976	1976	1976	1976
	24.5	8.6	4.9	13.3	0.8	25.6	46.7	.06	25.8	118.3	5.7	1.0	12.0	5.6	0.6
	25.2	7.8	3.0	5.2	2.5	25.8	40.4	NA	23.8	65.7	5.0	1.9	11.4	6.6	2.1
	899	1146	805	1023	983	979	1314	NA	851	1166	736	968	637	1614	650

NA not available

industrialised, followed by Japan, West Germany, France, the U.K. and Italy. Countries with the lowest steel consumption tend to have the least industry: Spain, Switzerland and the Netherlands, for example, where you'll find fewer landscapes marred by ugly factory chimneys and cooling towers.

The heaviest consumers of steel are the Swedes, who get through 1,614 pounds per person every year. The U.S. is next at 1,316, followed by West Germany. Canada, Japan and Australia are the next three countries on the consumer table. High consumption of steel generally indicates a high standard of living. Much of the steel used in the U.S. for instance, goes into cars, washing machines, toasters – and girders for high-rise buildings. However, some nations with low steel consumption still have a high standard of living – Switzerland, for example. The nature and quantity of a country's exports must also be considered. The Swedes export a lot of cars and heavy machinery – not all of those 1,614 pounds stay at home. And Volvos use more steel than Swiss watches.

Source:
British Steel Corporation
+Benelux (Belgium, the Netherlands & Luxemburg)

This table compares the number of days of paid holiday enjoyed by workers in 17 different cities around the world. There's a lot of variation in the length of holidays – not only from country to country, but also from job to job within the same country.

Obviously the number of days to which you're entitled depends on a number of different factors: how long you've been employed, your qualifications, your age and – in some countries – whether or not you're married. The Union Bank of Switzerland has specified exactly comparable people in cities all over the world and then listed the length of their holidays. We've extracted this table from their data.

On the whole, London workers don't have a lot of time in which to bask on sun-drenched Mediterranean beaches. London bus drivers get only 15 days paid holiday a year – the lowest in this category.

Car mechanics don't do much better. Their 17 days – although nowhere near the shortest holiday on the table – are only half the number enjoyed by some lucky mechanics in Madrid, who can enjoy the sun in the comfort of their own gardens.

Plant managers in London don't do too badly, with 20 to 25 paid days off. But secretaries average about four weeks a year, which – although not the lowest on the table – is the minimum throughout the Common Market for this occupation.

Whatever their job, workers in New York and Toronto must feel really hard-done-by. American workers have far fewer vacation days than most others around the world. Bus drivers in New York get 20 days – five days more than their colleagues in Toronto who, with just 15 days off a year, share bottom place with drivers in London and Zürich. Qualified car mechanics in New York and Toronto come right at the bottom of the table – with only ten days paid vacation a year. And secretaries in New York are only able to escape from their typewriters for ten days. Executive plant managers don't do all that well either, with three to five weeks holiday in New York and four weeks in Toronto.

Bus drivers in London, New York and Zürich might do well to emigrate to union-minded Sydney, where they'd get 35 days paid leave in which to travel round the outback or surf on those clear, blue Australian waves. Car mechanics and secretaries should head for Madrid's sunny climes; and plant managers for industrial Dusseldorf, where they'll get over 5½ weeks paid leave.

You'd expect that holidays in Austria would be about the same length as those in neighbouring Germany – or even that the Austrians would have more time for

Paid vacation[+] days per year:	NEW YORK	TORONTO
	1976	1976
bus driver[0]	20	15
automobile mechanic[x]	10	10
plant manager[‡]	15-25	20
secretary**	10	10

	LONDON	SYDNEY	VIENNA	BRUSSELS	COPENHAGEN	PARIS	DUSSELDORF	DUBLIN	MILAN	TOKYO	AMSTERDAM	OSLO	MADRID	STOCKHOLM	ZURICH
	1976	1976	1976	1976	1976	1976	1976	1976	1976	1976	1976	1976	1976	1976	1976
	15	35	20	21	24	26	24	17	26	20	22	20	30	20	15
	17	20	18	20	24	26	19	15	15-24	14	21	20	20-30	20	15
	20-25	20	20	24	24	26	28	15	25	20	22	20	25	27	25
	20	20	15	20	24	26	23	15	20-24	20	20	20	30	24	15

Source:
Union Bank of Switzerland
+ Working days
o Works on public transportation, ten years' experience; 35 years old, married, two children
x Qualified, five years' experience; 25 years old, single
‡ 100 people work for him, long experience; 40 years old, married, no children
** Secretary to head of department, shorthand, typing and one foreign language, five years' experience; 25 years old, single

leisure. Yet, when compared with the Germans, the Austrians come off rather badly right across the board. For example, given two plant managers doing similar jobs, the German gets a 40% longer vacation than his Austrian counterpart. On the other hand, neighbouring Belgium and Holland have very similar holiday patterns.

Al Manamah in the Arab sheikhdom of Bahrain (not on the table) is the place to work if you want really long holidays – a total of eight weeks of leisure for plant managers. And a secretary in Bahrain gets the same holiday as her boss.

For most of the world, the difference between the time off given to executives and that given to their employees is fast becoming smaller, or even disappearing; as is the difference in holiday given to skilled workers and their executive bosses. However, the two-week difference in the holiday of a New York or Zürich executive and his secretary is a glaring anachronism. Even in Ireland, they've come to accept that all workers deserve roughly the same holiday time – if anyone, it's the bus drivers who come out best. And in Spain, a secretary merits a week's longer holiday than her boss.

Three of the cities on the table believe in giving their workers exactly the same amount of holiday, whatever their occupation. Workers in Copenhagen can enjoy 24 paid days off; in Paris, 26 days, and in Oslo, 20 days. On the whole this seems the fairest way of allocating paid vacation days.

In general, it's the Latin countries – France, Italy and Spain – who are well above average for paid holidays. It doesn't seem fair when they have all that sun, too. Given more time – and sun – maybe we'd all be Latin lovers.

It's worth looking at this table in conjunction with those on inflation and taxation and also the tables comparing the cost of living in various countries in the world. The figure showing how much someone is paid gives no precise indication of the value of what he or she actually earns until you've taken into account the figures from the other tables.

For instance, the Spaniards have easily the lowest average hourly earnings. Their 89p is well below half the norm. But they pay very low taxes, and many things are cheaper in Spain than elsewhere. So although Spanish workmen are still lowest in the table, these factors help to close the gap. The opposite is true when you come to the top end of the table.

Danish workers top the list with an average hourly wage of only just under £2.90. That's nearly £120 a week. And in Sweden they earn on average just five pence less per hour. But those high-earning Scandinavians pay high taxes too – well over 30% in both Denmark and Sweden. And prices are high: in 1976 Swedes had to pay over £3 for a pound of sirloin steak. Their large earnings don't buy as much as you might think.

Neither Sweden nor Denmark have been doing well economically. These high figures reflect how inflation (with high wages always chasing high prices) has hit them.

The U.S. and Canada follow Denmark

		U.S.A.	CANADA
1.	Total average earnings in manufacturing industries (£ per per hour)	1975	1975
		2.36	2.44
2.	Highest earnings (£ per hour)[o]	3.15	3.29
3.	Lowest earnings (£ per hour)[x]	1.56	1.60
4.	Ratio of: highest earnings to the average	1.33	1.35
	lowest earnings to the average	0.66	0.66
	highest earnings to the lowest	2.02	2.06

and Sweden. Surprisingly, the Canadians – with an average of £2.44 per hour – are ahead of U.S. workers by eight pence. That adds up to over £3 more a week.

Norway, the other Scandinavian country in our table, comes next. There, they earn an average £2.33 per hour, just a fraction less than in the U.S. And although they pay high taxes (an average of 39% of earnings for a single person) the North Sea oil bonanza, which is about to arrive, means things are looking up for the Norwegians.

West Germany is comparatively low on the table. Average earnings there are equal to those of the Netherlands – but lower than Switzerland and the other countries mentioned so far. The reason – once again – is inflation.

Spain is the only country with wages below £1 an hour (89p). Ireland is next lowest with an average wage of £1.20.

Italy comes next with an average hourly figure of £1.35, only two pence

	U.K.+	AUSTRALIA+	AUSTRIA+	BELGIUM	DENMARK	FRANCE	(WEST) GERMANY	IRELAND+	ITALY	JAPAN+	NETHERLANDS	NORWAY+	SPAIN+	SWEDEN+	SWITZERLAND+
	1975	1975	1975	1975	1975	1975	1975	1975	1975	1975	1975	1975	1975	1975	1975
	1.37	2.55	1.68	1.92	2.86	1.37	1.93	1.20	1.35	1.61	1.93	2.33	0.89	2.81	1.96
	1.61	2.76	2.20	2.98	3.68	1.66	2.55	1.51	2.20	2.20	2.50	2.70	1.19	3.17	2.61
	1.08	2.32	0.98	1.33	2.36	1.05	1.43	0.98	1.06	0.94	1.79	1.94	0.53	2.39	1.93
	1.18	1.08	1.31	1.55	1.29	1.21	1.32	1.26	1.63	1.37	1.30	1.15	1.33	1.13	1.33
	0.79	0.91	0.59	0.69	0.83	0.77	0.74	0.82	0.78	0.58	0.93	0.83	0.59	0.85	0.98
	1.49	1.19	2.24	2.24	1.56	1.58	1.79	1.54	2.08	2.34	1.40	1.39	2.24	1.32	1.35

lower than the average wage here in the U.K. Most Italian workers in the manufacturing industries in the north, earn more than their counterparts in the U.K. – the Italian figure takes into account wages earned in southern Italy, which are often on a par with those in Spain.

It's interesting to see that in France – where there's a minimum wage law (£1.25 per hour in 1977) – the lowest rate of pay is smaller than in the U.K. (where we have no such law). Although the U.K. has the fourth lowest average earnings, the gap between lowest and highest wages is comparatively narrow. In many other countries, there is a large difference between the highest and lowest average wage. Top wage earners get more than twice as much as the lowest in the U.S., Canada, Austria, Belgium, Italy, Japan and Spain. This invariably reflects an even greater disparity between wages and salaries. In these countries, there tend to be extremes of wealth and poverty.

Sources:
International Labour Organisation
+Heron House estimates
0Includes petroleum refineries and chemical, paper, beverage, tobacco, iron and steel, rubber products, and printing and publishing industries
xIncludes textile, clothing, footwear, leather, and wood furniture industries

	New York	Montreal
	1976	1976
Net earnings (£) of:		
an automobile mechanic[+]	5340	5718
a construction worker[o]	6641	6202
a toolmaker/lathe operator[x]	4754	7620
a female textile worker[o]	2927	3531
a department manager[‡]	11,036	10,406
a bank teller**	5309	3788[++]

If you want to earn a higher-than-average salary, go to Zurich. The Swiss score three firsts in the six occupations shown on the table, two seconds and one third. In New York, in the land of opportunity, only the department manager and the construction worker earn the highest rates. If you are a female textile worker in New York, you'd make more if you worked in Zurich, Sydney, Copenhagen, Montreal, Stockholm or Oslo.

London is well below average, right across the board – in fact, it's at the bottom of the table for department managers and bank tellers.

Overall, Dublin pays the lowest wages of the cities listed, scoring lowest for car mechanics, lathe operators and female textile workers. Irish salaries never range higher than third from the bottom of the table. Londoners come next, earning less in all but one category than their equivalent workers in Madrid.

One of Zurich's firsts is the salary paid to a car mechanic – more than £5,756 per year. Montreal comes a close second with £5,718. New York is third at £5,340. After these three cities there's a big drop of £770 a year till you get to Sydney.

In the U.K., a car mechanic earns only £2,153 a year – less than half the salary paid his counterpart in Australia.

Dublin car mechanics are the worst off, earning only £1,763. Mechanics in Milan and Madrid don't do much better. They both earn around £1,900 a year. Even these salaries are generous compared with a mechanic's pay in Manila in the Phillipines (not on the table) – just over £550 a year.

The same pattern is reflected in the construction industry. If you're a construction worker your best bet is New York, where you'll earn £6,641. A Parisian construction worker gets £1,665. He's the odd man out in the Paris figures – Parisian salaries for the other occupations are further up in the table.

For lathe operators, Montreal is tops at over £7,620. Zurich is next with an income of £6,867. Then there's a drop to Sydney (just under £4,890) and New York, fourth on the table at £4,754. The lowest salaries are paid in Dublin, Milan and London.

The next category produces at least one surprise. Women in New York are well down the scale, earning just £2,927 a

London	Sydney	Vienna	Brussels	Copenhagen	Paris	Dusseldorf	Dublin	Milan	Tokyo	Amsterdam	Oslo	Madrid	Stockholm	Zurich
1976	1976	1976	1976	1976	1976	1976	1976	1976	1976	1976	1976	1976	1976	1976
2153	4570	2997	3550	4168	2948	3198	1763	1952	2856	3073	3997	1868	3970	5756
2122	4068	2556	2921	4322	1665	2763	1837	1886	2950	3171	4688	2127	4735	4734
2523	4883	4405	4136	4437	3821	3765	2309	2497	4023	3568	4464	2699	4600	6867
1473	3615	1890	2348	3608	2232	1954	1328	1385	2373	2502	3269	1853	3478	3863
3842	7351	6533	6967	6995	9007	8479	4253	4610	10001	7078	6476	3929	5971	10475
2653	4891	4387	4878	4842	3798	4678	3395	4690	6824	4215	4540	2697	4849	7803

++Without complete bank training
00Bank employee

year – seventh of the 17 cities. Again, Zurich scores highest in this category. Women textile workers there earn more than department managers in London.

If you look at the figures for department managers, you'll see that there's an enormous disparity between the New York salaries of over £11,000 and our skimpy U.K. figure of £3,842. Management is given a greater incentive in the U.S. – the salary indicates priorities.

Bank tellers are laughing all the way to the bank in Zurich, where they earn more than £7,800 – way ahead of the rest of the field. Their New York counterparts come a poor second with £2,494 less. And banking isn't a funny business at all in London, where tellers earn a meagre £2,653, hardly enough to keep them in bowlers and pinstripe suits. Perhaps they should follow the example of their Irish colleagues whose recent strikes closed the banks for several months.

Source:
Union Bank of Switzerland
+25 years old, single; completed apprenticeship, five years' experience
025 years old, single; unskilled or semi-skilled
×35 years old, married, two children; skilled, ten years' experience
‡40 years old, married, no children; head of a production department (100 employees), many years' experience
**35 years old, married, two children; completed training, ten years' experience

This table is another myth-shatterer. For many years we've believed that the U.K. is the world's most strike-plagued country. Not true: it comes only fifth in the table, with just over six million man-working days lost through strikes. The U.K. figure is even lower proportionately. In Australia for instance, over 3.5 million working days were lost through strikes – and their work force is a quarter the size of ours.

The U.S. comes top here – by a long way: more than 31 million man days lost. Although the U.S. has a work force three times the size of the U.K.'s *five* times as many days are lost through strikes.

Italy is next on the table with over 27 million man days lost – very nearly three out of four workers are involved in disputes. This is an amazingly high figure when you consider that their work force is over twenty per cent smaller than the U.K.'s. With under a quarter of the U.S. work force, Italy is losing almost as many working days through strikes. Italy's number of disputes is 57.8% higher than that of the U.K.

The reasons for this appalling record are hotly disputed. But the extreme poverty in many areas of Italy and high unemployment, are both undeniable factors. The fact that the Communist Party is the largest political party in Italy must also play a part. You could call it Euro-communism, but nevertheless one of the main aims of communism is to

	U.S.A.	CANADA
	1975	1975
1. Number of disputes	5031	1171
% versus U.K.+	+120	51.3
2. Workers involved (000s)	1746	506.4
3. Man working days lost (000s)	31237	10909
4. Average number of workers per dispute+	347	432

overthrow the capitalist system – and one of the easiest ways to do that is to paralyse its economy.

Canada comes third on the table with about one in 20 workers walking off the job during the year. That's nearly 11 million man days lost, a very large figure when you consider that their work force is only just over a third that of the U.K.

Strong unions certainly play a large part in the high North American figures on the table. But the disparity in wages between the workers and the well-heeled executives must also be a factor, to say nothing of the North American independence of character.

Surprisingly, Japan comes fourth on the table with over eight million man days lost. Their work force is more than twice that of the U.K.'s. They suffer from 48.6% more strikes than the U.K. Big corporation paternalism – with its guaranteed employment for life and strict regimentation – obviously isn't the

	U.K.	AUSTRALIA	AUSTRIA	BELGIUM	DENMARK	FRANCE	(WEST) GERMANY	IRELAND	ITALY	JAPAN	NETHERLANDS	NORWAY	SPAIN	SWEDEN	SWITZERLAND
	1975	1975	1975	1975	1975	1975	1975	1975	1975	1975	1975	1975	1975	1975	1975
	2282	2432	NA	243	147	3888	NA	151	3601	3391	5^0	22	2807	86	6
	—	+6.6	NA	10.6	6.4	+70.4	NA	6.6	+57.8	+48.6	0.2	1.0	+23.0	3.8	0.3
	808.9	1398	3.8	85.8	59.1	3814	35.8	29.1	14110	2732	0.27^0	3.3	504.3	23.6	0.32
	6012	3510	5.5	610.2	100.1	5011	68.7	295.7	27189	8016	0.48^0	14.5	1815	365.5	1.7
	354	575	NA	353	402	980	NA	193	3918	806	54	149	180	275	54

NA not available ^{0}*In 1974 there were 17 disputes, 2,979 workers involved and 6,854 days lost. In 1976 there were 11 disputes, 15,255 workers involved and 13,984 days lost.*

answer to strikes. After all, if Japan's economy is booming, it's understandable that the workers who make it boom should want a slice of the cake too.

If you don't believe in crossing picket lines, head for Holland, Switzerland, Austria or West Germany. The peace-loving Dutch lost only 480 man days. A miraculous record by any standards, especially when you consider that they have over 4.5 million people going to work every day.

Let's break those Dutch figures down and see what they show. Throughout the year there were just five disputes involving 268 people in the whole of the Netherlands. (That's 0.2% of the U.K.'s figure where over 2,000 disputes involved over six million workers.) The Dutch are clearly too busy making money to spare the time to strike.

As for the Swiss – when did your banker last man the barricades? Switzerland lost only just over 1,500 man

Sources:
International Labour Organisation
+Heron House estimates

days. And Austria, with almost the same size work force (just under 3 million) lost 5.5 thousand man days. The average Austrian worker was out for just 1½ days – barely long enough to *schuss* down a mountain and buy a bar of chocolate.

West Germany had easily the best record of all the large economies. They lost only 68,000 man days. Their work force is around the same size as that of the U.K. and France, yet they lost less than a *thousandth* of the man days.

The answer here is that the Germans try to resolve their disputes without strikes. And their union system was completely reorganised after the war to work in co-operation with management. Perhaps we could all learn a lesson here.

How safe are you at work? Obviously this depends on the kind of job you have: you're safer working in an office or cream cheese factory than on a construction site. The table on the right shows three potentially dangerous occupations (manufacturing, construction work and mining) and compares fatal accident rates in different countries for these occupations. To show how much progress each country is making in on-the-job safety, figures for two years – 1967 and 1976 – are given.

Not surprisingly, the figures show that fatal accidents occur more often in the building trades than in manufacturing, and that mining is even more dangerous. Obviously there can be tremendous variations within each of these occupations: you'll be safer in a mattress factory than in a steel mill or digging an open pit mine, rather than working a mile-deep tunnel. Taking the figures as a whole, mining is about seven times as dangerous as factory work.

Japan has the best figures for safety in the factory: 0.01 fatal accidents for every million man hours worked in 1976. That means you'd have to work eight hours a day for over 30,000 years before you might expect to be killed.

The Japanese are very safety-conscious in their factories. However, this figure is impressive in a country with so much heavy industry. You'd be more

	U.S.A.	CANADA
	1967	1967
Fatal accident rates for:		
manufacturing	0.03^0	0.12^x
construction	0.19^0	0.96^x
mining and quarrying	0.51^0	$3.09^\ddagger$
	1976	1976
Fatal accident rates for:		
manufacturing	0.03^0	0.10^x
construction	0.16^0	0.75^x
mining and quarrying	0.31^0	$1.72^\ddagger$

than 30 times as safe working in Japanese industry than on a mining job.

On the whole, the U.K. comes off well in its safety figures. Our industrial accident rate is holding steady at 0.04 per million man hours of work in manufacturing industries – not as good as the Japanese, but better than most other countries listed. Between 1967 and 1976, we reduced on-the-job construction fatalities from 0.16 to 0.15. This is still higher than in some countries. Mining accidents have declined more dramatically – from 0.52 to 0.32 in nine years.

Canadian accident rates are high in all categories. Even though they almost cut their mining accidents by half between 1967 and 1976, they will have nearly two mining fatalities per thousand wage ear-

	U.K.	AUSTRALIA	AUSTRIA	BELGIUM	DENMARK	FRANCE	(WEST) GERMANY+	IRELAND	ITALY+	JAPAN	NETHERLANDS+	NORWAY	SPAIN+	SWEDEN	SWITZERLAND+
	1967	1967	1967	–	–	1967	1967	1967	1967	1967	1967	1967	1967	1967	1967
	0.04^x	NA	0.36^x	NA	NA	$^{x++0}$0.12	$^{+0}$.08	0.09‡	$^{+0}$0.04	0.04^0	$^{+0}$0.02	0.08‡	$^{+0}$0.02	0.04^0	$^{+0}$0.07
	^{x00}0.16	NA	NA	NA	NA	0.48^x	$^{+0}$0.20	0.14‡	$^{+0}$0.29	0.29^0	$^{+0}$0.06	0.43‡	$^{+0}$0.07	0.12^0	$^{+0}$0.24
	^{x**}0.52	0.79^x	0.50^x	NA	NA	$^{+0\ xx}$0.23	$^{+0}$0.32	1.08‡	$^{+0}$0.18	0.59^0	NA	0.47	$^{+0}$0.20	0.31^0	NA
	1976	1976	1976	NA	NA	1975	1975	1975	1974	1976	1975	1975	1975	1975	1975
	0.04^x	NA	0.18^x	NA	NA	$^{x++}$0.10	$^{+0}$0.06	0.09‡	$^{+0}$0.03	0.01^0	$^{+0}$0.02	0.06‡	$^{+0}$0.03	0.03^0	$^{+0}$0.05
	^{00}0.15^x	NA	0.62^x	NA	NA	0.46^x	$^{+0}$0.14	0.09‡	$^{+0}$0.25	0.06^0	$^{+0}$0.04	0.17‡	$^{+0}$0.09	0.08^0	$^{+0}$0.25
	^{x**}0.32	0.71^x	0.52^x	NA	NA	$^{+0xx}$0.16	$^{+0}$0.18	0.65‡	$^{+0}$0.16	0.33^0	NA	0.16	$^{+0}$0.43	0.20$^+$	NA

^{0}Per million man hours worked ^{x}Per 1,000 persons employed, excluding uranium mining
‡Per 1,000 wage earners   **Excluding Northern Ireland and quarrying
$^{++}$Including mining and quarrying ^{00}Excluding Northern Ireland
^{xx}Excluding quarrying   NA not available

ners. Part of the reason for this is that most of Canada's mines – it's the world's largest producer of zinc – are in newly opened up areas of the frozen north where conditions are treacherous and accidents correspondingly more common.

Austria is the worst country for industrial accidents among manufacturing workers. Even though the Austrians halved their rate from 1967 to 1976, they're still at the top of the table with 0.18 fatalities per thousand employees over four times the U.K. figure. However theirs is at least an improvement. French and Canadian factories come next with 0.12 fatalities per thousand.

While the U.K.'s figures for mining and quarrying have improved noticeably

Sources:
International Labour Organization
+Heron House estimates

(a drop from 0.52 per thousand employees in 1967 to 0.32 in 1976) there's been no change at all in the accident rate for manufacturing. Although our figure is below average, it would be good to see the accident rate fall even lower.

Most countries on the table have improved their safety records, with one dire exception: in Spain, mining fatalities have more than doubled – from 0.20 per million man hours worked in 1967 to 0.43 in 1975. The figures for accidents in the manufacturing and construction industries have also increased.

123

An old joke has it that, "Money may not buy happiness, but at least it allows you to be miserable in comfort." Sad to say, age often brings poverty, and while there may never be an elixir that will rejuvenate our bodies, most people will settle for enough money to live in security and perhaps with enjoyment after retiring.

Is your government pension worth waiting for? If you want the highest benefits, live in Milan or Vienna – the Italian and Austrian governments provide pensions giving you a maximum of four-fifths of your highest-earning years. Of course, you pay all your working life for this eventual nest egg. In all state pension systems, workers, employers and government contribute to the retirement funds.

When governments began taking responsibility for those who were too old to work, everyone usually received the same basic weekly or monthly amount. Now, with such vast numbers of people involved, at such a wide range of salaries, most countries have differential benefits.

In some countries, retirement benefits are designed to help mainly the lower income groups. This is true of our own Social Security system. A couple living in New York and earning £19,600 a year in 1976 would have paid Social Security contributions on only £8,568 of their income and would collect annual benefits

	NEW YORK	TORONTO
	1976	1976
1. Retirement age	65	65
2. Annual pension for a single person (UK £): +		
minimum	681	928
maximum	2446	1580
3. Annual pension for a married couple (UK £): +		
minimum	1022	1857
maximum	3669	3013

of £3,660 upon retirement. This isn't much of a pension for someone used to relative affluence. For many Americans, however, Social Security payments represent only a supplementary income which they receive in addition to a company or union pension.)

Other countries pay much larger retirement benefits than we do. In Stockholm, Sweden, for instance, a retired couple can receive up to £6,097, and in Düsseldorf, Germany, £5,571.

Many countries adjust their retirement benefits to inflation and the cost of living, but for more affluent old age you should look not to the government but the private sector. With almost all unions pressing for higher pensions, and large corporations offering lavish pensions to lure promising executives many people's prospects are improving.

	LONDON	SYDNEY	VIENNA	BRUSSELS	COPENHAGEN	PARIS	DUSSELDORF	DUBLIN	MILAN	TOKYO	AMSTERDAM	OSLO	MADRID	STOCKHOLM	ZURICH
	1976	1976	1976	1976	1976	1976	1976	1976	1976	1976	1976	1976	1976	1976	1976
	60/65	60/65	60	65	67	65	65	65	60	55/60	65	67	65	60/65	65
	702	NM	1109	965	1476	959	NM	583	568	623	1946	1550	V	1260	1351
	NMx	1479	3657	V	2318	2258	5571	641	7128	1558	V	2752	V	5359	2701
	1118	NM	1587	1328	1793	1435	NM	989	657	677	2802	2194	V	2167	2025
	NM	2457	3657	V	5957	2705	5571	1123	7213	2878	V	3334	V	6097	4052

NM no minimum   NMx no maximum   V varies (see notes below)

Source: Union Bank of Switzerland

+These figures can vary according to individual circumstances. For further details see notes below..

RETIREMENT BENEFIT QUALIFICATIONS

New York Based on the average of the 17 contributory years with the highest earnings in accordance with the maximum earnings which are creditable for social security.

Toronto Determined by financial circumstances at the time of retirement, not on previous earnings.

London The basic retirement benefit is independent of previous earnings, whereas the supplementary benefit depends on earnings-related contributions.

Sydney Determined by financial circumstances at the time of retirement, not on previous earnings. Benefits can be reduced to nothing as assets or additional income increase.

Vienna Determined by the length of insurance coverage. After 45 years, benefits reach 79.5% of the average creditable earnings during the last five years before retirement.

Brussels For a single person, about 60% of average earnings during the years contributions were made. For a married man, about 75%.

Copenhagen Determined by financial circumstances at the time of retirement, not on previous earnings.

Paris Contributing for 37.5 years gives a maximum regular retirement pension of about 50% of the creditable earnings during the ten covered years with the highest earnings. Pensions are increased by mandatory insurance for supplementary benefits.

Dusseldorf Determined by the number of years insured.

Dublin Fixed amount, not dependent on earnings.

Milan Maximum benefits equal about 80% of the average covered earnings of the three years with the highest earnings out of the last five contributory years before retirement.

Tokyo Determined by the number of contributory months and the amount of previous earnings.

Amsterdam Fixed amount, not dependent on earnings. Sometimes, supplementary payments are made from company-operated pension funds.

Oslo Determined by previous earnings and the number of years of coverage.

Madrid Determined by the amount of contributions and the number of years they were paid into the scheme.

Stockholm Minimum benefits equal the national pension received by everyone over the age of 65. The supplementary pension depends on previous earnings and the number of covered years. The maximum is 60% of creditable average earnings during the 15 contributory years with the highest income.

Zurich Determined by the number of years insured and the average earnings for the years in which contributions were paid. Benefits are also available from company-operated pension funds.

National economy (C£-000s) per person employed

10
9
8
7
6
5
4
3
2
1

The popular image of the hard-working efficient Swiss may be true. Each industrious worker in Switzerland produces £8,300 worth of goods annually. It's not all cuckoo clocks and chocolate, of course. Switzerland is banker to the world, and that generates wealth without tangible effort.

U.S. workers run the Swiss a close second when it comes to high productivity. Each one, on average, generates £8,000. Next door, Canadians produce £7,400 each.

Australia, like North America, is a young, vigorous country. This is reflected in its productivity figure of over £5,000 per head. Apparently the Aussies like to work as hard as they play.

Despite Japan's booming economy, each worker there generates only slightly more wealth (£4,260) than his counterpart in the U.K.

The average Irish worker obviously doesn't believe in overwork. But Ireland has produced some of the world's greatest writers — and talkers. The value of a tale well told can't really be measured in pounds.

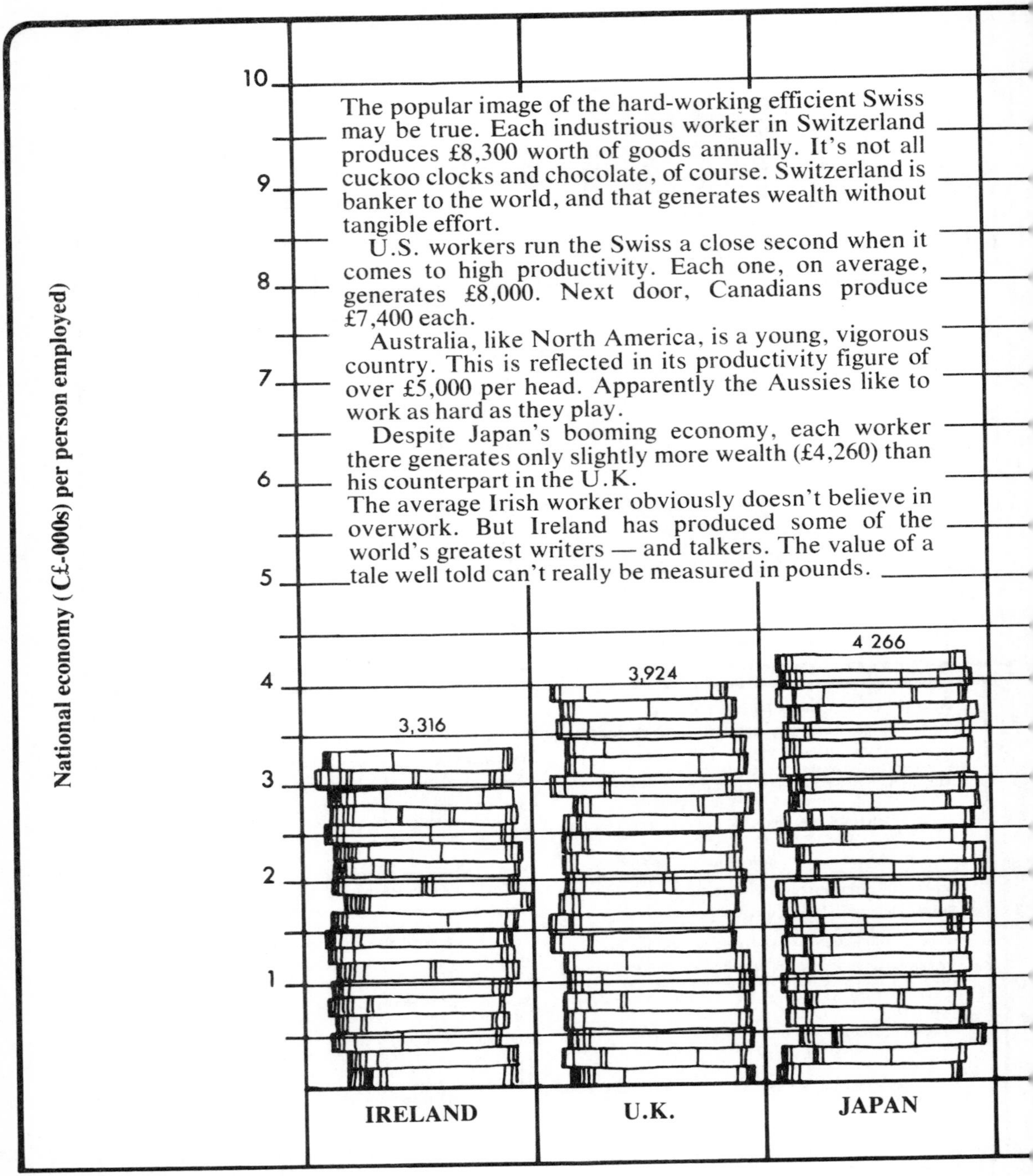

3,316

3,924

4 266

IRELAND   U.K.   JAPAN

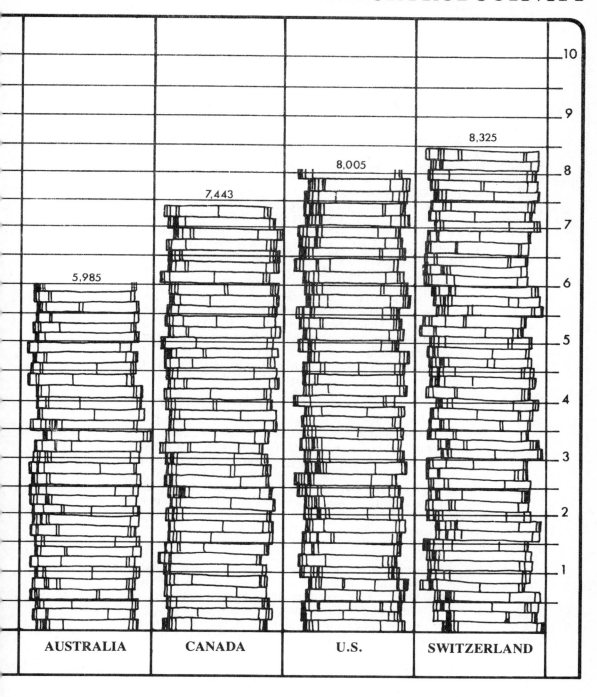

AUSTRALIA	CANADA	U.S.	SWITZERLAND
5,985	7,443	8,005	8,325

At Home

How many houses do you own? How many people on average are there in each house? What sort of appliances (if any) do you have in them? How much does all this cost you? These are just some of the questions answered in this chapter.

The first chart sheds some light on the subject of electricity. Remember the last blackout? Life changes radically when there's no electricity. That's true for a staggering 99% of house-holders in nearly every country surveyed, who all depend on electric power to light their homes. Nowadays we depend on it for a lot more than light. What we once thought of as luxuries have become commonplace labour saving devices all over the world. Think of refrigerators and washing machines. Are these two items really essential? From the look of the figures, the answer is an overwhelming "yes". That's true in a lot of countries, not just North America and the U.K. as you might expect. More than 50% of the households in the table have their own washing machine. Most households can live quite well without a telephone, however. Only in the U.S., Canada, Switzerland and Sweden do half or more of the households have telephone service. We're about average here in the U.K.

So now we know what the homes in various countries are like and have an idea of how they're equipped. But how many people live in each one? The next table shows the average number of people per household. What's most surprising here is how *little* the figure varies from country to country – despite difference in culture and living standards. The old saying tells us "two's company, three's a crowd", but not many people seem to agree. The average for the countries we've covered is about three per household. Only in Ireland is the average marginally over four, and no country has an average of less than 2.5. What conclusion can we draw from this? Is there something about human beings which makes them naturally tend to live together in groups of three? That's the great thing about figures – they're just facts. You can always draw your own conclusions – so long as they fit the figures, of course. This section also points out some fascinating cultural differences. For example, over twice as many people live alone in fast-paced industrialized West Germany, while the gregarious Italians seem to feel "the more the merrier".

Next we have a table on home ownership. How many people own their own homes? Where do you think seven out of ten people own their own homes – in tiny, economically-troubled Ireland or in large, affluent France? How many rent and how many live in homes provided by their government or employers? Where do you think more people own their own homes – the U.K. or West Germany? You may be in for some big surprises when you look at the table. Home ownership used to be for the rich – but from the look of the figures that's no longer true. In more than half the countries we've covered, at least half the population live in houses of their own. There's a pleasant alternative to buying or renting your house in countries where the government or your employer provides free accommodation. In affluent Sweden, for example, only a third of the population owns their own home. Home ownership is obviously becoming a matter of life-style and convenience, rather than simple affluence.

There are other aspects to housing also. Rent, for instance. You're always meeting people who say how cheap it is to live in various cities. Now you'll be able to check up on their statements – we've got the facts for you. We've taken some of the largest cities in all of the countries listed and given the figures for renting different sizes of flats, both furnished and unfurnished. Here in the U.K., we hear a lot of complaints about how expensive a London flat is. Now you can see how London renters fare compared to other big city dwellers. You may be pleasantly surprised, especially when you look at the figures for that other small nation, Japan.

But even if it appears comparatively expensive to rent a flat, it may still turn out to be worth your while to stay in the city of your choice – especially if local wages are higher. There's a lot more to keeping a roof over your head than just paying the rent. Fuel and power costs probably take a big chunk out of your household budget. The next table shows what percentage of household expenditure goes on rent, power and fuel. So have a look to see how your costs in this department compare with other countries. This gives you more of an idea of how much it actually costs people to live in their homes. After those basic necessities have been paid for, some people like to put their surplus cash into a second or vacation home. This table shows just who likes to get away from it all by heading for a holiday home.

Electricity, refrigerators, washing machines and telephones – now we're getting down to basics when it comes to home requirements. Running water, bathtubs and flush toilets are all right as far as they go. But in a modern western society accustomed to electric toothbrushes, frozen gourmet foods and drip-dry clothes, we need another standard of household comparison.

It's no surprise to find that electricity is almost universal in the households in the countries on the list. In fact more households have electricity than running water – which just goes to show that perhaps more people would rather watch T.V. than bathe.

Ninety-nine out of every hundred households in almost all the countries in the table have electricity. There are only three exceptions. In Spain, 96% of the households have electricity. And in Ireland and Japan, the figure is 98%.

Once you've got electricity, you might as well have a place to chill your beer. Life is so much more enjoyable if you've got a fridge. Ice cubes, ice cream, chilled *Chablis,* crisper lettuce, unsour milk – these are just some of the reasons to have one. Ninety-nine per cent of the households in the U.S. and Canada wouldn't be able to cope without a refrigerator. In Australia, almost as many consider it vital to have somewhere to keep those cans of lager cool.

		U.S.A.	CANADA
		1975	1975
1.	Homes equipped with electric lighting (%)	99	99
		1976	1976
2.	Households owning refrigerators (%)	99	99+
3.	Households owning washing machines (%)	83	82
		1976	1976
4.	People with telephones as % of population	70	57

In most of the other countries on the list, around 90% of the households have refrigerators. But in the U.K., only 85% of us have fridges and in Norway the figure is 84%. Perhaps it's so cold there electricity isn't needed to chill food.

In Japan, more than one in four households does *not* have a refrigerator.

But the lowest country on the list is Ireland. Less than half the Irish households have refrigerators. And in Spain, one out of every three households drinks warm *sangria* and keeps its left-over *paella* in the cupboard.

After the fridge, the next home essential is the washing machine. It's not the U.S. that heads this list, but the spick and span Dutch. In 85% of Dutch households there's a washing machine. The West Germans run a close second; there, 84% of all households have washing machines. And in the U.S., where so many house-

ELECTRICITY

	U.K.	AUSTRALIA	AUSTRIA	BELGIUM	DENMARK	FRANCE	(WEST) GERMANY	IRELAND	ITALY	JAPAN	NETHERLANDS	NORWAY	SPAIN	SWEDEN	SWITZERLAND
	1975	1975	1975	1975	1975	1975	1975	1975	1975	1975	1975	1975	1975	1975	1975
	99+	99	99	99	99+	99	99+	98	99+	98	99+	99+	96	99+	99+
	1975	1976	1975	1975	1975	1975	1975	1975	1975	1976	1975	1975	1975	1975	1975
	85	96	89	86	92	88	91	49	91	71	95	84	69	91	91
	71	68	70	75	60	69	84	62	79	61	85	75	68	80	51
	1975	1975	1975	1975	1975	1976	1975	1976	1976	1976	1975	1975	1975	1975	1976
	38	39	28	29	45	26	32	14	26	41	37	35	22	66	61

Sources:
1, 2 & 3 Euromonitor
4 United Nations
+ Heron House estimates

wives say they'd leave home if they didn't have a washer, only 83% have one.

The U.K. is about average on this list. Seventy-one per cent of all British households have washing machines. Lowest of all on the table is Switzerland. Half of all Swiss housewives send their laundry out or wash it all by hand. The Danes are second lowest at 60%. And the third lowest is Japan with 61%.

There are people who say they'd go mad without a telephone in the house. Yet it's surprising how many people still don't have them in some of the most communications-sophisticated countries in the world. Well below half the households in many countries on the table have telephones. Even in the U.S., where Alexander Graham Bell started the black wire fever, only 70% of all households have a telephone.

Canada is third, with telephones in over half of all Canadian households. In most other countries on the table, less than half the households have a phone. The U.K. is about average with almost 40%.

Ireland is the lowest on our table – only one household in every seven has a phone. Even Spain has a higher figure than Ireland. But the loquacious French are surprisingly low at a mere 26%. The same figure maintains in Italy – only about one household in four has a phone. Perhaps people with Gallic and Latin temperaments feel restricted when they can't express themselves with hands.

131

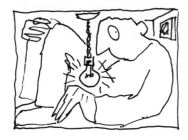

Household size depends upon where you come from. The U.K. and the U.S. are the lands of couples, with 31% of us living in twos. Germany tops the table for singles, with 28% of its people living alone – as many as live in two-person households.

The largest households are in Ireland, where 31% of people live in households of five or more. Ireland also has almost the lowest number of single-person households. Spain comes after Ireland, with an average of four per dwelling.

Why do people in these countries live together in larger units? For one thing, in Catholic countries families are usually larger and they stay together longer. Children in more affluent countries are likely to take on jobs or training that involve leaving home. In agricultural countries such as Ireland and Spain, they're more likely to stay and help with the work at home. The same is true of older people: in poorer countries their families care for them at home. The extended family of aunts, uncles and cousins is a tradition which lingers. In more affluent societies, the elderly often spend their latter years in nursing homes.

The U.S., the U.K. and Switzerland are in the lead when it comes to small-sized households and space to accommodate them. All three average fewer than three people in a household and just 0.6 people per room. The Canadians, with an average household size of 3.1 people – and almost two rooms per person, obviously have space to spare. Among the affluent countries, Japan must be feeling the squeeze with an average household of 3.4 people and 0.9 people per room.

The overall impression from this table is that, as countries get richer in terms of G.N.P. per capita (see G.N.P. table) households get smaller. Technological advances breakdown both tradition and the extended family. The smaller, nuclear family becomes more prevalent. And, of course, as people get richer, they buy more – including more breathing space.

		U.S.A.	CANADA
		1977	1977
1.	Average number of people per household	2.9	3.1
		1977	1977
2.	Households (%) consisting of:		
	one person	21	17
	two people	31	28
	three people	17	18
	four people	16	19
	five or more people	15	18
		1970	1971
3.	Average number of people per room	0.6	0.6

	U.K.	AUSTRALIA	AUSTRIA	BELGIUM	DENMARK	FRANCE	(WEST) GERMANY	IRELAND	ITALY	JAPAN	NETHERLANDS	NORWAY	SPAIN	SWEDEN	SWITZERLAND
	1977	1971	1977	1977	1977	1977	1977	1977	1977	1975	1977	1977	1977	1977	1977
	2.9	3.5	2.9	3.1	2.7	3.0	2.6	4.1	3.1	3.4	3.3	2.9	4.0	2.6	2.8
	1977	–	1977	1977	1977	1977	1977	1977	1977	1977	1977	1977	–	1977	1977
	20	NA	25	17	24	21	28	15	12	14	15	18	NA	26	21
	31	NA	27	30	30	28	28	22	20	17	27	24	NA	30	29
	19	NA	18	21	19	19	18	17	23	20	19	21	NA	19	19
	16	NA	15	15	17	15	15	15	21	27	20	19	NA	16	15
	14	NA	16	16	10	17	11	31	24	22	20	18	NA	9	16
	1971	1971	1972	1970	1970	1973	1972	1971	1971	1973	1970	1970	1970	1970	1970
	0.6+	0.7	0.9+	0.6+	0.8+	0.8+	0.7+	0.9+	0.9+	0.9	0.7+	0.7+	0.9+	0.7+	0.6+

NA *not available*

Sources:
Euromonitor
+National statistical offices

Owning their own home was once the dream of nearly every young couple. Today, values have begun to change. There's been an enormous increase in the price of property, and mortgage rates have soared.

In the U.K. the population is almost evenly divided between people who own their homes and those who rent accommodation. However, our extensive council housing may mean that the number of tenants in the country is artificially high.

When it comes to home-ownership, rural Ireland heads the list. Centuries of British rule and absentee landlords prevented most people owning their land or homes. With the Republic came a slump in land prices which made it possible for more people to own their own land and homes. Now about seven in every ten homes are owned by their occupants.

Historically, this passion for land also applies – in a different sense – to the colonial settlers of Australia and Canada, where about two-thirds of the people live in their own homes.

In the U.S. nearly 65% of the population own their own homes – which puts them fourth in the table.

Do the millions of tenants get what they pay for? The normal local rent (estimates based on middle to high rents) is as low as £14 a month in Vienna. The normal local rent in London – the next lowest city in the table – is three times as high, but the rents in both cities are still comparatively low. Vienna, like London, has many government-owned flats.

The highest, normal, local rents are in Toronto, with New York, Paris and Oslo running a close second. In crowded New York, tenants pay a premium for space. A medium-priced, furnished, four-room flat will cost £688 a month compared to a mere £108 in Ireland. Citizens of Tokyo pay an even greater premium for space – £819 a month for an average-priced, furnished four room flat.

		U.S.A.	CANADA
		1974	1977
1.	Home owners as % of population	64.6[+]	65.4[+]
2.	Tenants as % of population	35.4[+]	34.6[+]
3.	Free housing provided by state or company	*	*
		1976	1976
4.	Average monthly rent[‡] in big cities (£)[x]:	a	b
	for average-priced furnished 4-room apartment	688	473
	for average-priced unfurnished 3-room apartment	402	265
	normal local** rent	129	133

	U.K.	AUSTRALIA	AUSTRIA	BELGIUM	DENMARK	FRANCE	(WEST) GERMANY	IRELAND	ITALY	JAPAN	NETHERLANDS	NORWAY	SPAIN	SWEDEN	SWITZERLAND
	1976	1971	1972	1970	1970	1973	1972	1971	1971	1973	1971	1970	1970	1970	1970
	53.3^+	67.3^0	49.4^0	55.9^0	47.0^0	45.5^+	33.5^0	70.8^+	50.8^+	59.2^+	35.4^+	52.6^+	57.2^+	35.2^0	28.5^0
	46.7^+	27.3^0	45.1^0	44.1^0	48.6^0	44.3^+	66.5^0	26.8^+	44.2^+	34.4^+	64.6^+	42.4^+	24.7^+	51.6^0	68.4
	*	5.4^0	5.5^0	*	4.4^0	10.2^+	*	2.4^+	5.0^+	6.4^+	*	5.0^+	18.1^+	13.2^0	3.1^0
	1976	1976	1976	1976	1976	1976	1976	1976	1976	1976	1976	1976	1976	1976	1976
	c	d	e	f	g	h	i	j	k	l	m	n	o	p	q
	263	303	216	193	139	191	195	108	122	819	166	191	159	162	304
	142	126	95	65	83	131	124	92	72	410	68	122	79	86	144
	56	101	14	91	83	105	94	73	72	62	64	107	85	81	76

a = New York b = Toronto c = London d = Sydney e = Vienna f = Brussels g = Copenhagen
h = Paris i = Dusseldorf j = Dublin = k = Milan l = Tokyo = m = Amsterdam n = Oslo
o = Madrid p = Stockhom q = Zurich *less than 0.5%

Sources:
[+] National statistical offices
[0] United Nations
[x] Union Bank of Switzerland
[‡] Rent is defined as payments for rented dwellings, implied rental value of rent-free or subsidised housing and owner-occupied dwellings.
[**] General estimates based on middle to high rents.

135

The song tells us that those old bare necessities will come to us. It's a pity, but that's just wishful thinking. Those necessities don't come to you – you have to go out and get them. So how much today do we have to set aside from our household incomes to cover the cost of rent (or mortgage repayments), fuel and power?

The Swedes are at the very top of the table – 20.9% of their income is swallowed up by these essential items. The Canadians are very close behind with 19% – just under one-fifth of their family budget. Both Sweden and Canada are northerly countries and although their summers are hot they are also short. They have to endure long cold, dark winters, and so need to spend more on keeping their dwellings, and themselves, warm and comfortable.

We're always being told that we have a mild climate in the U.K. Well, we might not have Siberian blizzards but you don't have to live through many of our endless, grey, damp, chilly winters to quickly realise the need to stoke up the fire for as much as six months of the year. So it's not surprising that we're third on the list when it comes to paying out to keep a roof over our heads and Jack Frost outside the door.

However, the 18.4% that we spend on rent, fuel and power really puts us in the luxury league compared with the Irish. With virtually the same climate to contend with, the Irish spend only ten per cent of their money on providing them-selves with a warm place to live. Either rents are very much cheaper or they must all be huddling together for warmth round peat fires.

In the U.S., they pay their rent and the bills for the fuel and power needed to keep their houses comfortable and still have 89.9% left of their budgets for household expenses. This low expen-diture on these items reflects the coun-try's wealth in natural resources, but it must also have something to do with their climate. While in some parts of the U.S. it's possible to spend as much on keeping cool in summer as on keeping warm in winter, many areas have ideal outdoor weather for much of the year.

The Norwegians, whose country extends up to the edge of the Arctic, have to spend 14.2% of their family income on the basic necessities of shelter, warmth and lighting which places them eighth on the table with the French. It's not for nothing that Norway has been called the "Saudi Arabia of the North", and their careful development and exploitation of their North Sea oil and gas resources certainly seems to be paying dividends.

	U.S.A.	CANADA
	1974	1976
1. Households (%) who own a second or vacation home	4.1[+]	6.8
	1976	1976
2. Expenditure on rent[X], fuel and power as % of household expenses	11.1	19.0

LIVING EXPENSES

	U.K.	AUSTRALIA	AUSTRIA	BELGIUM	DENMARK	FRANCE	(WEST) GERMANY	IRELAND	ITALY	JAPAN	NETHERLANDS	NORWAY	SPAIN	SWEDEN	SWITZERLAND
	–	–	1976/7	1970	1974	1975	1977	–	–	–	1971	1973	1975	1974	–
	NA	NA	6.0^0	3.3	8.3	9.6	3.0^0	NA	NA	NA	0.2	18	7.4	14	NA
	1976	1976	1976	1976	1976	1976	1976	1976	1976	1976	1976	1976	1976	1976	1976
	18.4	15.0	17.9	14.1	13.1	14.2	16.0	10.0	13.9	13.7	13.5	14.2	14.1	20.9	17.5

NA not available

Not only are Norwegian families better off than most others in the family income groupings, but more of them (18%) own a second or holiday house, which places them in first place on that table.

The French are not nearly so well-off, but nor are they so poor as they would often have us believe. They spend 14.2% of the household budget on rent (or mortgage), fuel and power, but only 9.6% of French families believe in putting what surplus cash they have into a second or holiday home.

The Netherlands, though among the most prosperous countries in Europe, manages to keep the cost of rent, fuel and power very low indeed. The Dutch, for some years, have suffered acutely from a housing shortage but recent government policy, channelling more resources into solving the housing problem, has brought the possibility of adequate accommodation within the reach of all. Still, the Dutch are not particularly interested in owning second or holiday homes. As the table indicates only a minute 0.2% spend their surplus cash in this way, whereas in Spain, one of the poorer countries in

Sources:
1 National statistical offices
2 Euromonitor
+US Bureau of Census
0 Institut für Markt und Sozialanalysen
xRent is defined as payments for rented dwellings, implied rental value of rent free or subsidized housing and owner-occupied dwellings, including mortgage repayments

Europe, 7.4% of Spanish families own a second or holiday house according to the 1975 figures.

The Americans, who appear to have more money left over than anybody else to spend in pursuit of the "good life", don't seem to bother that much with a second or holiday home. In 1974, only 4.1% of American families spent their surplus income in this way.

Everyone these days has complaints about the high cost of living and the great difficulties in making ends meet. According to this table, though, people in most countries spend well under one-fifth of their household budget on those essentials – rent, fuel and power.

Average monthly rent for furnished four-room apartment £

The cost of keeping a roof over your head isn't a big worry in Dublin. At least, not if you live in a furnished four-room apartment. The rent is only £108. It's almost double that in London, where the cost of comparable accommodation is £263.

The cost of a four-room apartment in Sydney leaps to £303. That's close to the rent in Zurich (£304). Toronto's in about the same bracket, with rent taking a bite of £473 out of your income.

If you cross the border into the U.S., there's an another 50% increase in the cost of an apartment. The rent in New York sky rockets to £688 a month. But the Japanese rent-payer is even more hard-pressed. After coming up with nearly £819 a month, there can't be much left over for *sukiyaki*.

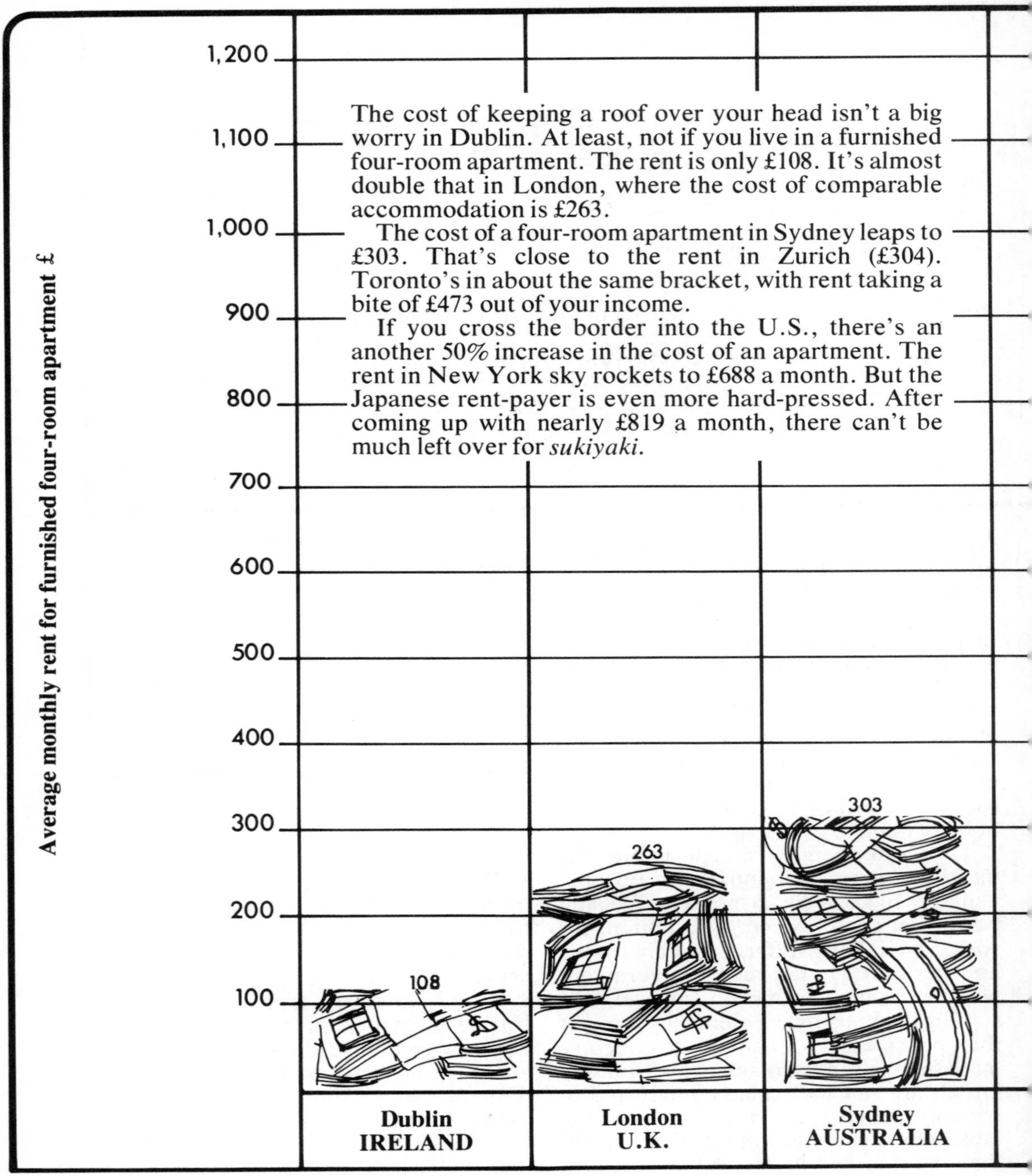

| 1,200 |
| 1,100 |
| 1,000 |
| 900 |
| 800 |
| 700 |
| 600 |
| 500 |
| 400 |
| 300 |
| 200 |
| 100 |

108

263

303

| Dublin
IRELAND | London
U.K. | Sydney
AUSTRALIA |

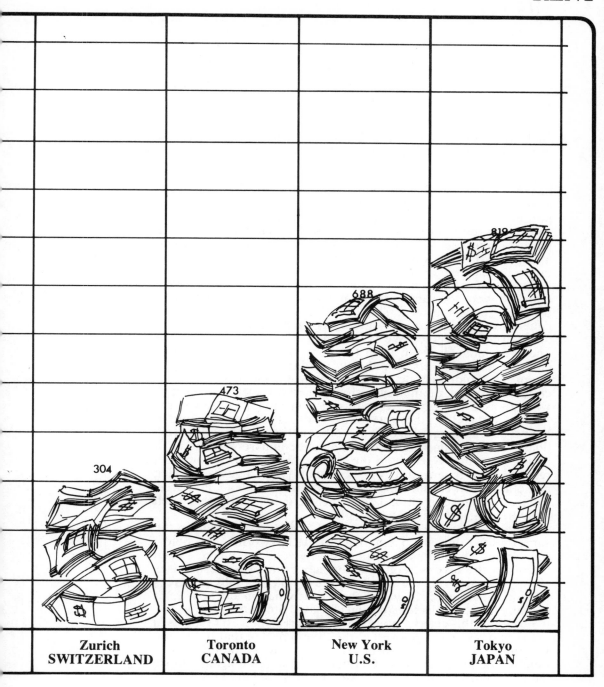

Zurich SWITZERLAND	Toronto CANADA	New York U.S.	Tokyo JAPAN
304	473	688	819

At Play

In this chapter, we'll be taking a look at the ways we find to amuse ourselves. They vary surprisingly from country to country, as you'll see.

First there's the subject of leisure activities, right across the board. That takes in a wide field of possibilities – everything from propping up those tired feet and unwinding in front of the T.V. to putting on your dancing shoes and going to a party. And we got some very different responses according to nationality. Like to hazard a guess as to what *your* fellow countrymen prefer to do after a hard day's work? Check it out against the section on favourite leisure pursuits.

As you may have guessed, a lot of the responses to the above question were the same. And that brings us to the next figures, those on television ownership. No doubt about it – in just one generation that electronic miracle has changed our life-styles to an enormous degree. We've taken a long hard look at the one-eyed monster, both from the point of view of programme content and also its popularity in different countries. An important topic, and one that crops up again and again. Are our children growing up in a television-addicted society? Is most of what you see just drivel? Read this section and take a viewpoint.

Over in the U.S., they own more T.V. sets and spend more of their time watching them than any other nation in the world. How does our desire for instant entertainment compare with the U.S. habit? How many people in the warm Mediterranean countries feel the need to own a T.V.? Is it mainly an English speaking addiction? Don't think America is a cultural wasteland, however. Figures on museum attendance, theatre-going, and other such activities suggest quite the opposite.

Of course, the mention of the U.S. and entertainment in the same breath brings us to America's gift to the world in this field – the movies. As befits the home of Hollywood, they take a keen interest in the silver screen. But you may be in for some surprises about the rest of the world. Who, for example, actually produces the most movies each year? And which nation do you think actually has the most theatre seats for avid film buffs?

One thing this chapter points out is the great interest in sports in many nations. So we looked at the "big time" in that field – the Olympics. You may be surprised to see how certain nations – notably the U.K. and East Germany – have fared against those giants of international competition, the U.S. and the U.S.S.R.

Then we turned our attentions to the sport of golf. Golf has come a long way since its origin in ancient Scotland. The whole world has taken to the sport in a big way.

Next we take a look at the distinctly more glamorous – you might even say decadent! – leisure pursuits of gambling and champagne drinking. You can probably guess why we've linked the tables of those two high-living activities. After all, what better way to celebrate a successful bet than with a bottle of bubbly?

On a less frivolous note, we examined the reading habits of many nations – and exploded some popular myths in the process. It's not true that Americans seldom read, nor is it true that when they *do* read they prefer books of fact to books of fiction. In this section, we also examine the current state of the struggling newspaper industry.

Then there's another aspect – book production. Our table here provides a basis for some thought-provoking speculation on the state of a nation's culture. We look not just at the number of books produced, but also the main types of books each country publishes – another indicator of cultural patterns.

And where do a lot of books end up? In libraries, of course. Our table on these worthy institutions offers some real shocks. For instance, which European nation has a mere 31 libraries to serve its entire population? Do we fulfill our national obligation to provide all forms of reading materials for our citizens?

Another popular form of entertainment is listening to music. That's especially true in both the U.K. and in the U.S., homes of the Beatles and the Beach Boys, Benjamin Britten and Aaron Copland. In fact, a look at the relevant section in this chapter will show you just where record sales are booming. You can also tell from the tables which countries prefer classical music to more modern strains. How do cassette sales compare in different countries?

We then turn to the important subjects of holidays and tourism. Do we all need to get away from it all and where do we go when we do?

What do you like to do after a hard day's work? Stretch out in front of the television set, play a few sets of tennis or maybe relax over a drink with some friends?

When Americans were asked how they like to spend an evening, a staggering 46% opted for – you guessed it – the television set. Only half as many of us in the U.K. choose T.V. The 23% who choose the goggle-box, though, far outstrip the 16% who choose reading as their favourite pastime.

The Italians at 25% are Europe's most dedicated television addicts. They rate television high on their list of favourite leisure activities – day or evenings. They're also Europe's most gregarious nation – 18% gave socialising as their preferred way of enjoying themselves.

Only seven per cent of us choose parties and visiting friends as our favourite activity. Nearly as many prefer to potter around in our gardens or do those odd jobs on the car or in the house.

Compared with the other figures, those for sports are also low. Charging around a football field or swinging a tennis racquet just aren't many people's ideas of fun.

The French, Austrians and West Germans are the most active people on the table. One out of ten people in these countries lists a participant sport as his or her favourite diversion.

The Germans are the bookworms of the world. Twenty-one per cent of them prefer reading to any other pastime – more than in any other country. They're avid for culture of all kinds – music, the performing arts, going to museums and art galleries. This shows on the table giving the amounts that various nations spend on culture. West Germany is in a class of its own. It spends a staggering £328.4 million a year on cultural activities. Germany's postwar economic growth has enabled her to spend vast sums on cultural activities while hardly cutting into the national budget – the money spent on culture is only one per cent of West Germany's total public expenditure. Tiny Belgium allots twice that percentage for similar activities – but this only amounts to £76.5 million. The

	U.S.A.	U.K.
	1974	1975
1. Favourite leisure pursuit (%):[+]		
watching television	46	23
reading	14	16
sport (as participant)	5	8
house and vehicle maintenance	3	6
parties; entertaining; visiting friends	16	7
	1975	1974
2. Government expenditure on culture[x] (£ — millions)	57.7[0]	29.1
as % of total expenditure	0.03[0]	0.2

RELAXING

	AUSTRIA	BELGIUM	DENMARK	FRANCE	(WEST) GERMANY	ITALY	NETHERLANDS	NORWAY	SPAIN	SWEDEN	SWITZERLAND
	1975	1975	–	1975	1975	1975	1975	–	1975	1975	1975
	13	18	NA	14	15	25	18	NA	19	21	18
	14	13	NA	13	21	11	17	NA	12	15	16
	10	7	NA	10	10	7	8	NA	6	8	7
	4	6	NA	7	4	3	6	NA	3	6	6
	16	14	NA	14	13	18	13	NA	17	13	13
	1974	1974	1974	1974	1974	1974	1974	1974	–	1974	–
	22.3	76.5	43.5	81.3	328.4	60.7	26.1	9.9	NA	35.0	NA
	1.0	2.2	1.4	0.5	1.0	0.7	0.5	0.6	NA	0.7	NA

NA not available

Sources:
Europe: Euromonitor
US: The Gallup Organisation Inc.
[+]The question asked (Europe): "What is your favourite leisure pursuit?" In the US the question was: "What is your favourite way of spending an evening?" The US figure for sports participation is therefore probably comparatively lower than the one for Europe, and those for television watching and entertaining probably higher.
[0]US Information Office, London
[x]Creative and performing arts, museums, art galleries, etc.
Totals do not add up to 100 as categories have been extracted from a more extensive questionnaire.

Germans may be world champion art subsidisers, but the Belgians are there trying.

At the other end of the scale, we British are hardly in the public expenditure race. We spend a tiny 0.2% on culture. That's the second smallest percentage on the table. It's hardly surprising that our theatres are closing or that unique works of art are increasingly disappearing abroad to countries that place a greater emphasis on culture than we do.

T.V. has changed our lives more than anything else in the past 30 years. Gone are the days when a family would sit around the fireplace reading, talking or playing games. Today, we sit in front of the T.V. set and only talk during advertisements. We spend more time with Angela Rippon and Michael Parkinson than we do with our own friends. For many of us, life without T.V. would be unimaginable.

Where has T.V. taken the firmest hold? In North America, of course. There they have 571 sets for every thousand people. Placed end to end those sets would stretch twice around the world. There are 1.7 people to every T.V. set. That's a lot of two – and three – set families. They have more colour sets, more channels to choose from, and more hours of programmes.

The Canadians come second with one set for every 2.7 people, closely followed by the Swedes with 2.8. The U.K. is third – about 315 sets per thousand, or one for every 3.1 people.

Even in Spain – which has the lowest number of sets of any of the countries on the table – you're never far from a T.V. There are about 5.7 people to a set. Ireland and Italy are also low in the T.V. watching stakes. It's tempting to see these figures simply reflecting the relative lack of affluence or larger households in those countries. More likely, they've got better things to do with their spare time. Think of all the bustling Mediterranean restaurants and cafés. And it could just be that the Irish prefer to spend their evenings in the pub.

When the T.V. viewer turns on his set, what sort of programmes does he have to choose from? You might think there would be more programmes devoted to entertainment than to anything else, but that's not the case. In most countries, fewer than 20% of broadcasting hours are devoted to entertainment. U.K. figures are about average – 15.5% – and the U.S. is much higher with 34.8%. Except for Canada and Italy, all countries give more broadcasting time to education than to either information (news, documentaries and so on) or entertainment programmes. Of course, few educational broadcasts

	U.S.A.	CANADA
	1974	1974
1. Television sets per 1,000 inhabitants	571	366
2. People per television set	1.7	2.7
	1977	1974
3. Type of program as %[+] of all programs:		
information	15.7[x]	27.5
education	43.3[x]	25.8
entertainment	34.8[x]	44.0
programs for special audiences[0]	6.2[x]	2.8

	U.K.	AUSTRALIA	AUSTRIA	BELGIUM	DENMARK	FRANCE	(WEST) GERMANY	IRELAND	ITALY	JAPAN	NETHERLANDS	NORWAY	SPAIN	SWEDEN	SWITZERLAND
	1974	1973	1974	1974	1974	1974	1974	1974	1974	1974	1974	1974	1974	1974	1974
	315	226	247	252	308	235	305	178	213	233	259	256	174	348	264
	3.1	4.4	4.0	3.9	3.2	4.2	3.3	5.6	4.7	4.3	3.8	3.9	5.7	2.8	3.8
	1975	–	1975	1974	1974	1972	–	1974	1974	1975	–	1974	1974	1975	1974
	20.0	NA	27.4	23.4	19.6	18.5	NA	30.7	32.9	16.2	NA	26.7	21.7	23.1	21.7
	56.4	NA	51.1	57.2	40.9	53.3	NA	33.6	21.9	60.1	NA	30.8	25.6	36.4	31.3
	15.5	NA	8.6	8.3	10.7	19.6	NA	11.0	10.4	23.4	NA	19.9	21.8	17.1	19.7
	8.1	NA	8.1	8.8	14.1	6.4	NA	24.8	15.0	–	NA	12.1	15.0	14.8	14.3

NA not available

take place during peak viewing times. In the U.K., the figure is 56.4% of total broadcasting time. This puts us in second place to the Japanese who have allocated more than 60% of broadcasting time to communicating ideas and knowledge.

The Italians have fewer educational programmes than anyone else. They don't go in for entertainment either. Only about ten per cent of viewing time is devoted to dramas and serials, quiz shows, music, sports, etc. You'll find more news information programmes on Italian T.V. than anywhere else. That's understandable in a country undergoing social and political turmoil. Italians look to T.V. to tell them what's going on – and events in Italy are happening almost too fast for anyone to follow.

The percentage of time the U.S. devotes to news and documentary programmes is much smaller. After education, most T.V. time is given to entertainment. Many of these programmes are shown around the world. There are Lucy-lovers in Antwerp and Kojak viewers in Athens.

The U.K. score is quite low when it comes to programmes for special audiences.

Sources:
1 Organization for Economic Co-operation and Development
2 & 3 Heron House estimates
+ Totals do not always add up as some categories have been excluded
0 Women, children, religious groups, etc
x Industry estimates

The U.S. is known as the home of Hollywood. For the better half of a century, the Hollywood Dream Machine has churned out celluloid icons for the whole world to adore and emulate.

Of course they are the most film-fixated people in the world. They have got nearly 17,000 cinemas, with a total seating capacity of over seven million. That means they could take the entire populations of Norway and Ireland to the cinema on a rainy day. And that's not even counting their 4,000 drive-in cinemas.

	U.S.A.	CANADA
	1977	1973
1. Number of movie theaters	16,829	1135
2. Seating capacity (000s)	7167	650.9
per 1,000 inhabitants	29.8	28.7
3. Annual attendance (millions)	1066	89.0
per capita	5.0	3.9
	1974	1975
4. Number of long+ films produced	156^{x0}	41‡

We, on the other hand, have just over 1,500 cinemas, with a capacity of nearly a million. That's about average on the table.

But look at the figures for seating capacity per thousand population. Here the U.S. is actually deprived. Spain is at the top of the list with 74 cinema seats per thousand population. Next comes Belgium with half that amount, and Australia, France and Norway all with about 35 seats per thousand. The Japanese surprisingly, have just over ten seats per thousand population. We in Britain have a meagre 17. But the U.S., despite their 17,000 cinemas has just under 30 seats per thousand. That's the same number as Canada. Perhaps that says they're all staying home to watch their favourite films on T.V.

If Americans are the most film-fixated people in the world, it follows that they go to the cinema the most often. Wrong. The Italians go twice as often as the Americans. And the Spanish go three times for their every two.

In Canada the ordinary person sees an average of only four films each year. That puts them sixth in the table. The Japanese have the distinction of seeing the least number of films a year – just under two. This is peculiar because they make twice as many films annually as the Italians, and over twice as many as the U.S. If the Japanese built more cinemas they'd have the seats to sell, so that people would pay to see the 405 films made there every year – and everyone would go home enriched and happy after enjoying home-produced entertainment.

France, not the U.S., comes after Japan when it comes to film-making. The

	U.K.	AUSTRALIA	AUSTRIA	BELGIUM	DENMARK	FRANCE	(WEST) GERMANY	IRELAND	ITALY	JAPAN	NETHERLANDS	NORWAY	SPAIN	SWEDEN	SWITZERLAND
	1974	1972	1974	1972	1974	1974	1974	1974	1974	1974	1973	1974	1974	1974	1975
	1535	976	645	728	367	5844	3114	240	5924	2468	380	448	5178	1199	521
	973.0	478.4	NA	359.0	128.0	1817	1189	NA	NA	1107	NA	142.0	2600	NA	195.8
	17.4	36.9	NA	37.0	25.4	34.6	19.2	NA	NA	10.1	NA	35.5	73.8	NA	30.1
	138.5	NA	23.7	32.8	19.9	178.5	136.2	NA	546.1	185.7	26.5	17.9	262.9	25.4	27.0
	2.5	NA	3.2	3.4	3.9	3.4	2.2	NA	9.8	1.7	2.0	4.5	7.5	3.1	4.2
	1975	1970	1975	1973	1974	1974	1975	1975	1975	1974	1975	1975	1975	1975	1975
	70‡	11‡	6ˣ	17‡	22ˣ	234‡	81‡	2**	203**	405‡	16ˣ	14ˣ	105**	14**	15**

ˣ Released or shown commercially for the first time by censor   ‡ Production completed   ** Approved

French turned out 234 films in 1974. In Italy, in 1975, they made over 200. The reason the U.S. ranks a poor fourth may be because of the frugal budgets and elaborate procedures of the Hollywood film-makers.

In the U.K., where the film industry is said to be in very bad straits, we still made 70 full-length films in 1975. All the figures for numbers of films produced don't include the full-length films which were made solely for T.V.

In Europe and Japan cineastes turn out first-rate films with skeleton crews – not multi-million epics with casts of thousands. FrancoisTruffaut, the French film director who starred in 'Close Encounters of the Third Kind', was astounded by the number of people on the sets. He normally works with no more than 20 people on his film sets.

Sources:
1, 2 & 3 United Nations
4 United Nations Educational, Scientific, and Cultural Organization
0 Confidential industry source
+ For commercial exhibition in cinemas, duration 37 minutes and over.

It's a long time since 776 B.C., the year when the ancient Greeks held the first athletic contests on the plains of Olympia. The Roman Emperor Theodosius abolished the Games in A.D. 393 after Greece lost its independence, and it was 1500 years before they were revived in 1896. Except for lapses during the two World Wars, they've been held at four year intervals ever since.

The tables on the right show how different countries have fared in the Games. However, not all the countries listed have been competing for the same length of time. Only four – the U.S., the U.K., Australia and Switzerland – have never missed an Olympic Games. They've participated in all 19 – something which deserves a medal in itself. At the other end of the scale, East Germany has only competed three times as a separate nation. Considering the number of medals they take home, they've certainly made up for lost time.

The U.S. has won by far the most medals in the Summer Olympics: 628 gold, 473½ silver and 413 bronze – more than 1,500 altogether. (If you're wondering about that half medal, it means they tied with another country.) The other giant of the sports world the

	U.S.A.	CANADA
1. Number of Olympic Games[+] in which country has taken part	19	18
2. Medals won at Summer[0] Olympic Games		
gold	628	26
silver	473½	43
bronze	413	53
total	1514½	122
3. Medals won at Winter[x] Olympic Games:		
gold	30	12
silver	38	7
bronze	27	14
total	95	33

U.S.S.R. (not on the table), is in second place with a total of 670, including 238 gold medals. This is still an impressive record: Russia came into the Olympics later than other countries and has competed only 11 times against our 19. However, if you compare the average number of medals won in each Olympics, the U.S. comes out the overall winner. Their average figure is 79, compared with 62 for Russia.

The U.K. comes next. With only a quarter the population of the U.S. or

	U.K.	AUSTRALIA	AUSTRIA	BELGIUM	DENMARK	FRANCE	(WEST) GERMANY‡	IRELAND	ITALY	JAPAN	NETHERLANDS	NORWAY	SPAIN	SWEDEN	SWITZERLAND
	19	19	18	17	18	18	16	10	17	13	16	18	12	18	19
	160½	62	18	35	28½	140	135½	4	124	73	38	40	1	127½	40
	198	51	27	48	58	152	181	3	115	64	42	30	6	122	63
	168	65	34	43	51	157	178	6	106	63	53	31	4	152	53
	526½	178	79	126	137½	449	494½	13	345	200	133	101	11	401½	156
	5	0	22	1	0	12	25	0	10	1	9	50	1	25	15
	4	0	31	1	0	9	20	0	7	2	13	52	0	23	17
	10	0	27	3	0	12	23	0	7	1	9	43	1	26	16
	19	0	80	5	0	33	68	0	24	4	31	145	1	74	48

‡*East and West Germany competed as one country up to 1964 (as an 'All German' team 1956–64). From 1968 they competed as separate teams and East Germany has won a total of 19 medals in the Winter Games, 181 in the Summer Games: 7 and 69 gold; 5 and 57 silver; 7 and 55 bronze.*

Russia, we have won a total of 526½ medals, including 160½ gold. Our record in the early years of the Games was even better, but competition has recently become much more intense – and we've sometimes lacked the resources to provide adequate training for our athletes.

The Germans have 494½ medals to their credit for the 16 years they've been competing in the Summer Games. This includes 135½ gold medals – about the same average per contest as the U.K. East Germany (not on the table) has competed separately since 1968 – and what a record! In just three Games, the East Germans won a total of 181 medals. That's an average of 60 for each Games – close to the Russian figure. It's a staggering performance for a country with such a small population.

Source:
International Olympic Committee archives
+Summer and Winter Games
⁰19 Games, 1896-1976
ˣ12 Games, 1924-76

	U.S.A.	CANADA
	1974	1974
Number of:		
public and private courses	11,956	1,127
golfers (000s)	11,000	1,750
teaching professionals	11,000	1,200

The origins of golf are obscure. The Scots claim to have invented it, but they have their rivals. A game called *kolf* was played in Holland in medieval times.

Be that as it may, golf was well established in Scotland by the 15th century, when James II issued a decree that "fute-ball and golfe be utterly cryed down". The reason? They encouraged people to miss their archery practice. That ancient regal decree seems to have had little effect on either sport. And archery is hardly flourishing today.

James II also considered golf an "unprofitable sportis". Times have certainly changed. The top pros in the game win hundreds of thousands of pounds annually and millions of pounds are spent every year on equipment and green fees.

By the end of the 15th century, James IV became addicted to the sport and changed royal policy. A bill for "golf clubbis and ballis" was found among his equipment. Scottish royalty also provided the first recorded woman golfer – Mary Queen of Scots. She was charged with playing golf in the fields around her castle just a few days after her husband's murder. It looks as though hitting a golf ball has always been an escape from matrimonial difficulties, as many golf widows, or widowers, will confirm.

Golf spread from Scotland as soon as the Scots laid down their swords and taught the game to the Sassenachs, south of the border in England. Today, the major golf-playing country is the U.S. By 1974 there were nearly 12,000 public and private courses. The U.S. even boasts a thriving association of professional golf-course architects.

The U.K. comes second, with nearly 2,000 courses. Australia is next with 1,345. If you look at the number of golfers in the U.K., there's less than 350 of them on each course. In Australia there's one course for every 261 players. In the U.S., over 900 players have to share the average course. But in Italy, there are only 181 players per course. Your best bet is Austria, with one course for every 150 golfers.

Both Canada and Japan (more than 1,000 courses each) take golf seriously. The Japanese figure is impressive – 40 years ago you could count Japan's golf courses on the fingers of one hand. Today golf is the big boom sport in Japan. The only trouble is that there are over 4,300 golfers for every course – and there isn't the available space to allow for enough extra courses to meet the demand.

The Norwegians evidently have better things to do than play golf. There are only seven courses in the entire country.

	U.K.	AUSTRALIA	AUSTRIA	BELGIUM	DENMARK	FRANCE	(WEST) GERMANY	IRELAND	ITALY	JAPAN	NETHERLANDS	NORWAY	SPAIN	SWEDEN	SWITZERLAND
	1974	1974	1974	1974	1974	1973	1974	1973	1974	1974	1973	1974	1974	1974	1974
	1,982	1,345	16	13	44	111	103	238	49	1,045	25	7	47	132	27
	682.2	350.6	2.4	5.0	14.5	154.1	25.6	70.0	8.9	4,500	7.6	3.4	25.0	55.7	7.0
	NA	486	7	16	21	182	125	65	80	924	32	7	212	101	60

NA not available

Many golfers would consider this a criminal waste – so much beautiful countryside without a fairway in sight.

Belgium, Austria and the Netherlands come after Norway. The Dutch may have been playing *kolf* for centuries, but they lack staying power. Maybe too many golf balls were lost in the canals. Or, possibly, golf just doesn't appeal to the Flemish temperament: Belgium, with many Flemish citizens, has only 13 courses.

As you'd expect, the U.S. has more golf players than any other country – more than 11 million of them. That's more than the populations of Norway and Denmark added together. Over one in every 20 Americans plays golf.

Japan comes next with 4.5 million golfers. Nearly one in every 25 Japanese spends a large proportion of his leisure time swinging a club and worrying about his handicap. Canada comes third, with 1.75 million golfers. One in every 15 Canadians play golf – an even higher proportion of golfers than across the border in the U.S.

The U.S., Japan and Canada are way

Source:
"Census of World Golf", *Golf Digest*, February 1975

ahead of the rest of the field. The U.K. is fourth, with nearly 700,000 golfers. In the U.K., only one person in 85 plays golf.

It's interesting to see the number of pros per country. They're the expert professionals who teach golf at various courses and enter for professional tournaments. According to Jack Nicklaus, one of the greatest golfers playing today, there are at least a hundred professional golfers on the golf tournament circuit who, ten years ago, would all have been better than anyone else in the world.

In the U.S. there are 11,000 professionals, one for every thousand golfers. The Canadians are second on the table, with 1,200 pros. The U.K. figure isn't available, but there's at least one professional to every course – which puts the U.K. figure around the 2,000 mark.

Professional golfers in Austria and Norway must be pretty lonely – there are only seven of them in each country.

	U.S.A.	CANADA
	1974	–
1. Money spent on gambling (£) per person per year	76.1[0]	NA
	1977	1977
2. Champagne (bottles) imported[+] (millions)	4.8	1.4

Almost everyone likes to gamble from time to time, but for most of us this doesn't go much beyond the Saturday night bingo game or occasional bet on the horses. That's a long way from the casinos of Las Vegas or Monte Carlo.

Some people take gambling more seriously than others. Over in the U.S., there's a lot of flirting with lady luck. Americans spend a lot more on gambling than anyone else in the world. They each gamble to the tune of about £76 a year, or nearly £1.50 a week. Since this figure includes children, we can assume that the real figure for adults is even higher. Or else there's a lot of school kids squandering their lunch money in poker games.

The U.K. comes second in the table. Each Briton gambles away about £45 a year. There's no equivalent of Reno or Las Vegas in Britain, but we make up for that with horse racing. Even the Royal family are devotees of the thoroughbreds.

"Going to the dogs" is a popular pastime in Britain – bets on greyhound racing account for a large percentage of that £45. It's a pretty safe bet that we are just as interested in gambling as they are in the U.S. In fact, gambling is legal throughout Britain, unlike the U.S. But we have a much lower average income, and so the total amount of money spent on betting per person is less.

Here in Britain there's a bookie's shop on almost every street corner. In addition to this, there are millions of people who have a flutter on the pools every week.

Then there are the countless bingo halls where millions of us get our eyes down to look hopefully for the lucky combination of numbers that will give us a prize.

Lotteries are on the increase too, tempting us to spend a little to win that elusive fortune.

As in most areas of life, the Swedes have a liberal approach to this sometimes controversial pastime. The average Swede spends approximately £29 on gambling, just a little more than his French counterpart. But there's more temptation in France, from the National Lottery to the Longchamps races.

Members of Gamblers Anonymous might try moving to Austria, where they squander a mere £3.10 each per year. Or perhaps Switzerland, where they channel their gambling instincts in other directions – such as banking.

Even if you never manage to break the bank at Monte Carlo, you'll still feel like celebrating from time to time. That's when a bottle of champagne comes to mind. No prizes for guessing where they knock back the most – France. They go through a staggering 124.5 million bottles

152

U.K.	AUSTRALIA	AUSTRIA	BELGIUM	DENMARK	FRANCE	(WEST) GERMANY	IRELAND	ITALY	JAPAN	NETHERLANDS	NORWAY	SPAIN	SWEDEN	SWITZERLAND
1975	–	1975	1975	1975	1975	1975	–	1975	–	1975	1975	1975	1975	1975
44.9	NA	3.1	7.9	8.2	26.7	19.3	NA	7.9	NA	9.5	21.9	5.6	29.3	7.3
1977	1977	1977	1977	1977	1977	1977	1977	1977	1977	1977	1977	1977	1977	1977
7.3	0.4	0.2	6.8	0.4	124.5 x	4.0	0.1	7.3	0.2 ‡	1.4	0.07	0.3	0.3	2.3

x *Home consumption figures*   ‡ *Includes sparkling wines*

a year. That's about 2½ bottles for each man, woman and child.

Of course, France is the producer of the world's finest champagne. And at the rate they're downing it, we're lucky any gets out the country at all. On this front, no other country approaches France. The next largest number of champagne consumers is in the U.K. where we drink about 7.3 million bottles. That might sound paltry, compared with those effervescent Frenchmen, but look at it another way, corks pop all over Britain about 20,000 times a day.

In the U.S., they get through 4.8 million bottles of imported champagne a year, rather a low figure compared to the size of the population. But then they drink a lot of domestic bubbly from California and New York State.

Norway is the land where champagne is least popular. One big factor here is the fact that large parts of Norway are dry, so no one drinks alcohol. The lucky Norwegians who *are* allowed to crack open a bottle of something prefer to drink their native *aquavit*.

In Ireland, they only import about

Sources:
1 Euromonitor
2 Government trade offices
+ Champagne imported from France
0 Gambling in America, Commission on the Review of the National Policy Toward Gambling

100,000 bottles of champagne each year. Per head of population, that's far more than it seems at first. The Irishman is fond of his Guinness stout, but the Irish like to blend the best of both worlds in the land of misty mountains. They mix Guinness with champagne and call it Black Velvet.

It looks as if the Japanese prefer to stick to *sake*: they drink only 200,000 bottles of imported champagne a year. The Austrians drink the same amount. If you enjoy the high life, think twice before heading there, because Austria also ranked lowest on our table on gambling. Those rugged Austrians are probably too busy scaling Alpine peaks to be bothered with such frivolous pastimes.

153

In the U.K., 21% of us claim to read books every day, the highest figure for the countries listed. The other side of the coin, however, is that 29% admit to never reading books at all. That's not as bad as it looks; only four countries have lower figures, Switzerland (19%), Denmark (25%), the Netherlands (27%) and Sweden (28%). It seems as if our excellent free public library system allows those who wish to read to do so unhindered by the cost of buying books. Those who choose not to read don't appear to be tempted anyway.

Americans are the worst-read people of the nationalities on our table.

	U.S.A.	CANADA
	1971	1975
1. People who, in a year, read books:		
never (%)	74$^+$	45^0
daily (%)	NA	NA
2. Average number of books read per person per year	NA	NA
	1975	—
3. Retail sales of books (£ UK — millions)	2249.6	NA
4. Expenditure on books (£ UK — millions)	10.52	NA

In the U.S. three-quarters of all adults don't seem to read a book from one year to the next.

Their nearest rivals as non-readers are the Italians. Only a third of them ever read a book from cover to cover. There's a bit more interest in Spain – 48% of the Spanish read books, 52% don't. The U.S., Italy and Spain form a kind of anti-literacy league. More than half the people in these places never read a book.

But more than one British adult in five does pick up a book every day. We are closely followed by the Danes and Swedes – one in five of whom reads a book daily. The daily reading norm for all the countries listed is 15% – so about one adult in seven reads a book every day, except in Spain and Italy. Only one Italian in 11 reads daily and in Spain the figure is one in 14.

The most revealing figure on this table is the number of books a person reads each year. The Danes are the most enthusiastic readers: they average 5.7 books a year – roughly one book every two months. Britain, with 5.5 books a year, is next, and Sweden, with 5.2, is in third place. The Dutch, who read 5.1 books a year, come next. The French and Swiss are also reasonably avid readers. They turn the pages of just under five books a year. Italians are right at the bottom of the ladder– they read less than two books a year. The Austrians (2.6) and Spanish (2.8) are just above.

	U.K.	AUSTRALIA	AUSTRIA	BELGIUM	DENMARK	FRANCE	(WEST) GERMANY	IRELAND	ITALY	JAPAN	NETHERLANDS	NORWAY	SPAIN	SWEDEN	SWITZERLAND
	1974	1974	1974	1974	1974	1974	1974	1974	1974	1974	1974	1974	1974	1974	1974
	29	48ˣ	33	45	25	38	34	40	68	NA	27	30	52	28	19
	21	NA	12	13	20	14	18	NA	9	NA	15	13	7	20	14
	5.5	NA	2.6	3.4	5.7	4.9	4.8	NA	1.9	NA	5.1	4.3	2.8	5.2	4.5
	1975	–	1975	1975	1975	1975	1975	–	1975	–	1975	1975	1975	1975	1975
	264.8	NA	30.4	40.7	20.8	156.8	196.2	NA	105.6	NA	58.8	12.1	41.4	54.2	36.8
	4.73	NA	4.03	4.00	2.13	2.98	3.17	NA	1.86	NA	4.34	3.03	1.17	6.62	6.85

NA not available

Americans may not read books, but they certainly buy them. On average they spend around £10.50 a year on books – more than twice as much as readers spend in any country except Switzerland and Sweden. In Switzerland, the figure is nearly £7.00 and in Sweden, about £6.60. The British certainly read more books than the Americans or the Swiss and Swedes, but we spend less than half as much on them as the average American – we have an excellent public library system (see libraries table) and books are cheaper in the U.K. than they are in the U.S. It's not really clear what the Americans do with all these books. Maybe they give them to other non-readers to decorate their coffee tables.

Sources:
Euromonitor
+Gallup Organization Inc
0Leisure Survey of the Secretary of State
ˣAustralian Embassy
‡Confidential source

		U.S.A.	CANADA
		1974	1973
1.	Number of daily newspapers	1798	121
2.	Estimated circulation (000s)	62,156	5207
	copies per 1,000 inhabitants	293	235

There's a shock here for us in the U.K. We have long believed that we read more newspapers than anyone else. If that is true it must be that we share them round because we only buy 443 newspapers per thousand – or one paper for every 2.3 people. That places us third in the league. Reading someone's paper over their shoulder on the train obviously doesn't count.

John F. Kennedy was famous for speed reading his way through ten or fifteen newspapers a day. The average American is less well informed. On average, there are only 293 daily newspaper copies printed for every thousand people. That means about one for every three readers.

Sweden has the highest rate of copies per thousand on the table. Over half the population buys a newspaper every day. Here the rate is 536 newspapers purchased for every thousand people – one paper for every two people. Japan is next on the list with 526 newspapers purchased for every thousand people. If newspaper sales are an indication of being well-informed, the Swedes and the Japanese are the most up to the minute people on the table.

In some countries readers share newspapers as an established procedure. In the Netherlands, Austria, Spain and other countries, many cafes have newspaper racks with a selection of the daily papers hanging over poles for the customers to read over their daily refreshments. These countries have slightly lower rates of purchases per thousand people. But the fact that their figures aren't so high doesn't mean that they're not keeping up with the news.

The lowest figures are for Spain. Only 96 Spaniards buy a newspaper each day for every thousand inhabitants. That's a rate of only one in ten people who feel it's important to know what's going on. But the Spanish figures are four or five years old – from a time when the press was firmly controlled by Franco's government. Since then, Spain has undergone dramatic changes – Franco's death, the coronation of the king and the restitution of democracy. Presumably the Spanish are buying the newspapers more often nowadays.

Today people want the news fast – and that's how we get it on T.V. and radio. With satellite communications, videotape, and instant replay, we see the news almost as it happens – even when it's taking place on the other side of the globe. This must be worrying newspaper journalists and circulation managers. The result is an almost universal decline in newspaper readership. Over the last decade, newspaper sales in the U.K., as elsewhere, have fallen dramatically. Many British families used to take two national morning newspapers, one for the home and the other would go to the office with the breadwinner. Around London, a

	U.K.	AUSTRALIA	AUSTRIA	BELGIUM	DENMARK	FRANCE	(WEST) GERMANY	IRELAND	ITALY	JAPAN	NETHERLANDS	NORWAY	SPAIN	SWEDEN	SWITZERLAND
	1974	1973	1974	1974	1974	1973	1974	1974	1974	1974	1973	1974	1974	1974	1974
	109	58	30	31	51	103	320	7	79	180	93	75	115	111	92
	24800	5126	2316	2416	1792	11458	17872	729	6963	57820	4175	1567	3396	4362	2535
	443	386	308	247	355	220	289	236	126	526	311	391	96	536	391

Source:
United Nations Educational,
Scientific, and Cultural Organization

lot of people bought one of the three evening newspapers as well. And a Sunday newspaper brought the total up to 19 a week. Today, people buy less than half that number in a week.

The number of daily newspapers printed also indicates readership trends. France, West Germany and the U.K. all have about the same populations – but in West Germany there are three times as many newspapers. Perhaps this is explained by the fact that in France and the U.K., the journalistic action tends to focus on the capitals of Paris and London. Whereas in West Germany the press tends to be more evenly distributed among the large cities. Instead of reading just *The Times* or *Le Monde*, West Germans, depending on where they live, may subscribe to *Die Frankfurter Allgemeine, Die Zeit, Die Welt* or others. And this variety seems to be good for the newspaper market.

In general, things don't look too good for the newspaper industry. With ever more sophisticated T.V. techniques, video cassettes that tape programmes for replay even when you're not at home, and other new technology, some experts believe that in 20 years time newspapers as we know them may have become a thing of the past.

Data on book production can be read as an informal profile of a nation's culture. We all know, for instance, about China and Chairman Mao's Little Red Book. In the countries studied, there's a bit more variety. This table shows what kind of books people like to read and which subjects are studied.

It should be kept in mind, however, that our figures are for 1974. Book production is a growing industry and up-to-date figures would be somewhat higher.

At first glance, the U.S., with a massive total of 68,600 works published, looks as if it is well in the lead in book production. This figure, however, includes government publications, university theses

Number of titles[+] published	U.S.A. 1974	CANADA 1974
generalities	908	428
philosophy	1033	150
religion	1612	220
social sciences	7014	1656
philology	319	291
pure sciences	2523	338
applied sciences	4833	827
arts	2519	937
literature	4992	1131
geography/history	2390	605
Total	68,600[0]	6583

and juvenile fiction, without which the total would be 28,143. This means that West Germany is the leading book producer with a total of 39,612. The U.S., with almost four times as many people as are in West Germany, actually publishes 10,000 fewer books a year.

The U.K. comes in fourth place with 24,310 titles published a year. For a relatively small country, that figure certainly justifies Caxton's invention. The highest output of books in the U.K. is in the 5,149 titles which come under the vague heading of "literature". Presumably this includes everything from literary classics to thrillers and por-

nography. In the literature category, as in most other categories, the West Germans produce far more titles than we do. In Social Sciences, for example, they publish 7,000 more titles than the U.K..

The one area where we publish more volumes than they do, is in the field of Pure Sciences where our total was 2,744 – nearly five hundred more than theirs. Since the U.K. is the home of more Nobel Prize-winning scientists than any other country, it's no surprise that we lead the world in books on such subjects. If only we could get together with the Japanese – who top the list in applied science – we might work to each other's

BOOKS

	U.K.	AUSTRALIA	AUSTRIA	BELGIUM	DENMARK	FRANCE	(WEST) GERMANY	IRELAND	ITALY	JAPAN	NETHERLANDS	NORWAY	SPAIN	SWEDEN	SWITZERLAND
	1974		1974		1974		1974	1974	1974	1974	1974	1974	1974	1976	1974
	714	NA	81	NA	187	NA	3176	29	279	634	151	139	2170	NA	127
	762	NA	177	NA	181	NA	1017	12	355	648	293	93	711	NA	238
	784	NA	165	NA	189	NA	1894	31	434	613	343	226	1102	NA	553
	4891	NA	1054	NA	1110	NA	11583	115	1802	5465	1239	1557	4347	NA	1766
	518	NA	68	NA	177	NA	1366	4	171	514	685	302	487	NA	236
	2744	NA	598	NA	550	NA	2284	23	409	1148	600	531	1059	NA	904
	3770	NA	819	NA	1091	NA	4998	110	741	6362	701	799	2041	NA	1658
	2623	NA	407	NA	396	NA	2715	59	629	1828	452	237	1082	NA	916
	5149	NA	599	NA	1470	NA	7318	198	2449	6007	2197	1281	5355	NA	1480
	2355	NA	455	NA	608	NA	3261	41	644	1833	538	378	1670	NA	679
	24310	NA	4423	NA	5959	NA	39612	622	7913	25052	7199	5543	20024	6990	8557

0 Includes government publications, university theses and juvenile fiction   NA not available

benefit. The Japanese would have a fresh source of scientific ideas and theory. We, in return, could take some of their efficiency in putting such theories into practice. Their emphasis on practicality is seen in the fact that they publish twice as many books on applied science as we do.

Overall, our total publication figure is about 700 less than the Japanese, and they come third in the book publication table behind West Germany and the U.S.. Other categories in which their publishing performance is particularly strong are literature and social sciences.

Spain ranks fifth overall – publishing

Source:
United Nations Educational, Scientific, and Cultural Organization
+First editions only

over 20,000 titles annually. They come second in that undefined category of "general" works, and third in both religion and literature. The Spanish have always had a rich literary heritage. They're deeply religious, too. In the field of religious works, they produce over 300 more books a year than we do.

	U.S.A.	CANADA
	1974	1971
1. Number of public libraries (000s):	8.3	0.7
per 100,000 inhabitants[+]	4	4
2. Number of volumes in public libraries (millions)	387.6	26.2
3. Average number of books per public library (000s)	46.5	37.7
4. Number of registered borrowers (000s)[+]	NA	NA

Whether your interest is gardening or politics, football or philosophy, there are books on your favourite subject, which you can borrow from your local library.

From the figures, it would appear that the U.K. possesses the most comprehensive library system. But the figures are somewhat misleading. In Britain we have an unusual public library system that makes comparisons practically impossible. Many of the 239 libraries per hundred thousand population are small vans carrying a rotating stock from a larger central library. The advantage of this system is that readers can get practically any book through an inter-library loan.

The ratio of libraries to population in the U.S. is surprisingly low. Every U.S. public library serves 25,000 people – which is perhaps the reason why many people may stay at home to watch T.V. Nonetheless, virtually every town and city in the country has a public library, with almost 50,000 volumes to browse through. The average Canadian library has a respectable 38,000 volumes.

Italy claims almost 9,000 public libraries – more than the U.S. But each has an average of only 2,000 volumes. Even with 16 libraries per hundred thousand people there are fewer than 32,000 books per hundred thousand, or roughly one book for every 30 people.

Norway claims 11 libraries per hundred thousand, each with about 21,000 books. West Germany and Spain equal the U.S. in the proportion of libraries to population. But West Germany has about 17,000 books per library, less than half as many as the U.S..

Sweden and Denmark have excellent libraries. Sweden has five for every hundred thousand people – or one for every 20,000. The average size of a Swedish library is over 73,000 volumes. Denmark, with almost as many libraries, averages more than 130,000 volumes.

Japan with Ireland has the lowest proportion of libraries to population – less than one for every 100,000 Japanese and the average number of books is 4,300. France has about the same number of libraries as Japan.

The most interesting figures on the table are for Ireland. The Republic has only 30 libraries. But each library has a remarkable average of 165,000 volumes.

	U.K.°	AUSTRALIA	AUSTRIA	BELGIUM^x	DENMARK	FRANCE	(WEST) GERMANY	IRELAND	ITALY	JAPAN	NETHERLANDS	NORWAY	SPAIN	SWEDEN	SWITZERLAND
	1977	1974	1974	1974	1974	1973	1974	1974	1974	1974	1974	1973	1974	1974	1974
	135.0	0.9	0.4	1.6	0.3	0.8	2.5	0.03	8.7	0.9	0.4	0.5	1.4	0.4	1.9
	239	6	6	16	5	2	4	1	16	1	3	11	4	5	29
	125.5	11.5	4.4	20.5	32.7	37.4	43.0	5.1	17.0	38.8	17.9	9.4	8.7	30.5	NA
	0.93	13.5	10.3	12.7	130.3	44.7	17.2	165.3	2.0	4.3	43.7	21.0	6.1	73.4	NA
	16,928	NA	635.3	NA	NA	NA	90.8	720.8	2944	3756	2292	621.2	943.4	NA	NA

NA not available

Sources:
United Nations Educational, Scientific, and Cultural Organization
[+]Heron House estimates
[0]Library Association. Figure includes mobile libraries, prison libraries, etc.
[x]National statistical office

The Americans are the biggest buyers of records – 276 million albums a year. They also top the table for pre-recorded tapes, with 114 million sales. Yet sales per head of the population give a different picture. The U.S. average is one L.P. and half a tape every year, not including single records or E.P.'s. In the U.K. the figure is much higher. We buy 106 million L.P.'s annually – that works out at almost two for each person in the country.

Third in this list is Japan, where over 90 million L.P.s are sold every year, closely followed by the West Germans (80.5 million). Who buys the fewest records? Norway, where they purchase only about two million followed by the Austrians (2.6 million), The Danes and the Swiss (three million each). It seems Scandinavians don't go for sounds.

The real story, however, isn't in total sales but in L.P. records sold per person.

That's where the U.K. comes out top – almost 2,000 records are sold each year per thousand inhabitants. Next comes Canada, where the figure is 1,800. Australia – home of the Beachcombers – is third. The U.S. is fourth in this category, with 1,300 records bought per thousand people, about the same as Sweden and West Germany.

The Italians – well-known for bursting into song at the least provocation – buy the fewest records of all – 165 per thousand inhabitants. In the land of opera they obviously prefer to do their own singing in the bath or shower.

Austria, home of Wolfgang Amadeus Mozart, is second lowest. They sell just 300 L.P.s per thousand people.

The Spanish figure (500) includes singles and E.P.s as well as L.P.s, so it may be the Spanish who are the worst customers of all for long-playing records.

One reason why the U.S. comes off poorly in record sales is that it is heavily into tapes. They buy 540 pre-recorded tapes per thousand inhabitants – by far the highest figure. We come next at 320 per thousand. Combine the record-buying figures with the ones for tape purchases, and we are still world champion recorded music buyers. The U.S. comes in second with Canada close behind.

Austria, second from the bottom in total record sales, comes top in the

		U.S.A.	CANADA
		1974	1974
1.	Total sales of records (millions):	276.0[0]	40.8[x]
	per 1,000 inhabitants[+]	1300	1800
2.	Number of classical records manufactured as % of all records	5.2	NA
3.	Total sales of pre-recorded tapes (millions)	114.0[0]	NA
	per 1,000 inhabitants[+]	540	NA

RECORD AND CASSETTE SALES

	U.K.	AUSTRALIA	AUSTRIA	BELGIUM	DENMARK	FRANCE	(WEST) GERMANY	IRELAND	ITALY	JAPAN	NETHERLANDS	NORWAY	SPAIN	SWEDEN	SWITZERLAND
	1974	1973	1973	1973	1972	1974	1973	–	1974	1974	1973	1973	1974	1974	1972
	106.0^x	22.0^x	2.6^0	10.6^0	3.0^0	54.6^0	80.5^0	NA	9.1^0	90.5^x	NA	2.1^0	$16.0^{x‡}$	10.7^0	3.0^0
	1900	1600	300	1100	600	1000	1300	NA	165	800	NA	500	500	1300	500
	10.0	9.0	25.0	NA	7.0	NA	13.0	NA	NA	9.0	16.0	NA	NA	NA	NA
	18.0^x	NA	1.2^0	1.0^0	1.2^0	7.5^0	9.7^0	NA	7.7^0	NA	NA	0.8^0	7.7^x	1.8^0	1.5^0
	320	NA	160	100	240	140	160	NA	140	NA	NA	200	220	220	230

NA not available ^{0}Sold ^{x}Produced ‡Includes singles and EPs

classical league: twenty-five per cent of all records manufactured are classical. That's a lot of Mozart and Strauss. The Netherlands is second in the classical sweepstakes, 16% of their production is classical. West Germany, home of von Karajan and the Berlin Philharmonic, comes in third with 13%. In the U.K. about ten per cent of records manufactured are classical and, in Australia, nine per cent. It seems that Americans, too, are rock freaks rather than classical enthusiasts: only 5.2 of all the records they manufacture are classical.

The newest trend in the recording industry is the shift to tapes instead of records. It's well under way here and in the U.S. No doubt other countries will soon follow.

Source:
EMI Ltd
+Heron House estimates

	U.S.A.	CANADA
	1977	–
1. People (%) who take a vacation	58	NA
2. Percentage of people taking a vacation who:		
go abroad	30	NA
stay in their own country	70	NA

This table is about holiday habits in different countries. Where do the most people go abroad? Where do they prefer to stay at home? Where do they simply do without any holiday?

More Swedes take holidays than any other nationality on our table – 84% of them. That's over ten per cent more than their Scandinavian neighbours in Norway. Here 73% spend the long winter nights browsing through coloured brochures and planning their holidays – ready to get up and go as soon as the summer arrives. The Norwegian figure is closely followed by that of Italy, 72% of whom take holidays.

Denmark has the next highest figure which completes the Scandinavian trio. If you want to set up a tourist agency, go to Scandinavia.

Only 63% of us British take holidays. This is about average for the countries in our table – exactly the same percentage as in Belgium and West Germany. But it's a big surprise to find that only 58% of Americans go on holiday.

The Spanish are at the bottom of the holiday table – only 32% take one. That's understandable, sunny Spain is a major tourist attraction, and most of us think that every day there must seem like a holiday.

Holland follows Spain. Only 40% of Dutch take holidays. It's hard to believe. Over-crowded Holland is one of the wettest countries in Europe.

Australia has the third lowest figure (50%). Of course, if you're living in Sydney you can go to Bondi Beach any weekend.

The other figures on this table show where people go when they do go on holiday. Who prefers to go on holiday in their own country? Who likes to get up and go abroad? The Swiss top the table when it comes to taking holidays abroad –74% of them leave their home country at least once a year. Switzerland isn't exactly large – and it's also near lots of interesting places. Paris, Brussels, Vienna or Rome are all just 300 miles across the border. So they're never more than 50 miles from a foreign country.

The West Germans, famous for their wanderlust, are the next most dedicated foreign travellers. Sixty-eight per cent of West Germans who take a holiday prefer to take it abroad. They usually head for the sun – that's one reason why beaches in Italy, Spain and Greece always have so many well-tanned West Germans in residence. And hard-currency German Deutschmarks go a long way in Mediterranean countries.

Two Scandinavian countries come next. In Norway, 53% of holidaymakers go abroad; in Denmark the figure is 50%.

U.K.	AUSTRALIA⁰	AUSTRIA	BELGIUM	DENMARK	FRANCE	(WEST) GERMANY	IRELAND	ITALY	JAPAN	NETHERLANDS	NORWAY	SPAIN	SWEDEN	SWITZERLAND
1975	1976	1975	1975	1975	1975	1975	—	1975	—	1975	1975	1975	1975	1975
63	50	59	63	71	53	63	NA	72	NA	40	73	32	84	69
33	15	47	49	50	36	68	NA	28	NA	49	53	28	38	74
67	85	53	51	50	64	32	NA	72	NA	51	47	72	62	26

The Swedish figure is comparatively low. Only 38% of Swedes who take holidays go abroad. There are two main reasons. Sweden is a large and beautiful country and many Swedes prefer to spend their holidays in their own lakeside cottages, or pottering about the lakes in boats. Also, Sweden is a long way from European holiday areas.

The Swedes aren't the most dedicated stay-at-homes on the table – not by a long way. Eighty-five per cent of holiday-making Australians don't go abroad. The reasons are obvious. The Swedes may be a thousand miles from the Mediterranean – but many Australians have to travel that far before they leave their native shores.

The Spanish and Italians come second in the stay-at-home stakes. Who can blame them? Many people travel hundreds of miles for the sunshine and scenery that they have on their doorsteps.

We in the U.K. don't have the same inducements to stay at home as the Spanish and Italians yet 67% of us prefer to stay in Britain. Perhaps we don't mind our weather as much as we say we do. Or could it be the poor currency exchange rates when we go abroad?

Source:
Euromonitor
+Industry estimates
⁰Australian Tourist Board

	U.S.A.	CANADA
	1976	1976
1. Foreign arrivals at frontiers (millions)	17.5	14.1
comparison with 1975 (%)	+11.6	−5.1
2. International tourist receipts millions)	3282[+]	919[+]
comparison with 1975 (%)	+20	+7

Once a year they pack their suitcases, stop the milk and the papers, board the cat and – where *do* all the tourists go?

You'll find a lot of them gazing at the Colosseum, floating on Venice's canals in gondolas and studying the old masters in Florence museums. Italy gets more foreign visitors than any other country. There were nearly 38 million of them in 1976. These tourists outnumbered the entire population of Spain, Egypt or Poland. They left just under £1½ thousand million behind them, or about £38 apiece.

The second biggest crowds were to be found exploring the Prado, climbing the steep cobblestoned streets of Toledo or lounging in the sun on the Costa del Sol. Thirty million tourists opted for Spain that year – not as many as went to Italy, perhaps, but then they did spend more money – that's £1726 million, or about £51 each. When the price is right, tourists shell out more money. In Spain prices and sales taxes are low.

The third most visited country was the U.S. They had 17.5 million foreign visitors in 1976 – more than the combined populations of Ohio, Indiana, Illinois and Wisconsin. When they weren't gazing out at the Grand Canyon, riding up in elevators to the top of the Empire State Building or taking a tour of the Universal Studios in Hollywood, all these visitors were busy spending money: £3282 million of it, to be precise, or £192 apiece. That's a lot more travellers' cheques than in any other country. America's a big place, and lots of tourist money went to airlines and car rental companies. Nonetheless, the amounts spent show that America – like Spain – gives good value for the tourist's money. They may be plagued by inflation, but New York is less expensive for visitors than London, Paris or Tokyo. The new cheap air fares make it even better value for the European tourist.

An astounding 14.4 million tourists chose tiny Denmark as their vacation spot in 1976 – that's almost three times the entire Danish population. Denmark's visitors spent £450 million – that's about £32 each. Denmark's quiet charm and pleasant scenery are what draws the crowds: there's still a real fairy-tale charm about the land of Hans Christian Andersen.

But all those tourists didn't take the opportunity to visit nearby Sweden or

	U.K.	AUSTRALIA	AUSTRIA	BELGIUM[x]	DENMARK	FRANCE	(WEST) GERMANY	IRELAND	ITALY	JAPAN	NETHERLANDS	NORWAY‡	SPAIN	SWEDEN	SWITZERLAND
	1976	1976	1976	1976	1976	1976	1976	1976	1976	1976	1976	1976	1976	1976	1976
	10.1	0.5	11.6	NA	$14.4^{‡}$	13.4	NA	1.2	37.7	0.8	NA	$0.4^{‡}$	30.0	0.8	NA
	+14.1	+3.1	+4.1	NA	−3.6	+3.1	NA	+0.1	+4.5	+12.4	NA	+13.2	−0.4	+4.3	NA
	1618	171^{0}	1753^{0}	537^{0}	450^{0}	2023^{+0}	1798^{0}	112	1414	175^{0}	594^{0}	221^{0}	1726^{0}	198^{0}	940
	+17.3	+9.7	+3.2	+9	+7.7	+4.1	+12.8	−0.5	−2.0	+23.8	−4.2	+7.7	−9.4	+3.2	+4.4

$^{+}$OECD estimates ^{0}Figures refer to receipts registered in foreign currency grouped regionally according to the denomination of the currency. These figures do not correspond exactly to the amount of receipts from visitors coming from certain countries. ^{x}Includes Luxembourg
‡Excludes Scandinavian citizens

Norway. Those two countries welcomed fewer tourists than almost any other nation in our table. Apparently the legendary beautiful blondes don't make up for the very high prices.

Japan didn't attract many foreigners either. This is probably accounted for by a combination of high prices and the language barrier. But the Australians may be wondering why they're so unpopular. Let's face it – they just live in the wrong neighbourhood. They've got a California-style climate and great surfing beaches, but after paying the air fare you can't even afford swimming trunks.

When it comes to counting up tourist money, the Irish leprechauns must be trying to figure out what happened to the pot of gold. They made a measly £112 million. But that does work out at £95 spent by each visitor. That kind of money buys an awful lot of Guinness.

Canada entertained almost as many

Source:
Organisation for Economic Co-operation and Development

tourists as the U.S. Canada has a lot of attractions – great fishing and hunting, and spectacular scenery. There's no language problems for their neighbours in the U.S. either. It's a safe bet that a lot of Canada's 14 million tourists are American, and a lot of the 17.5 million tourists in the U.S. are Canadian.

Things were looking up for beleagured Britain in 1976 – a record ten million visitors showed up to watch the Changing of the Guard, trek to the Tower of London and to wave at the Queen. That meant a shot in the arm for our ailing economy – to the tune of £1618 million. Our reputation for bad weather is obviously counter-balanced by a lot of very splendid British features.

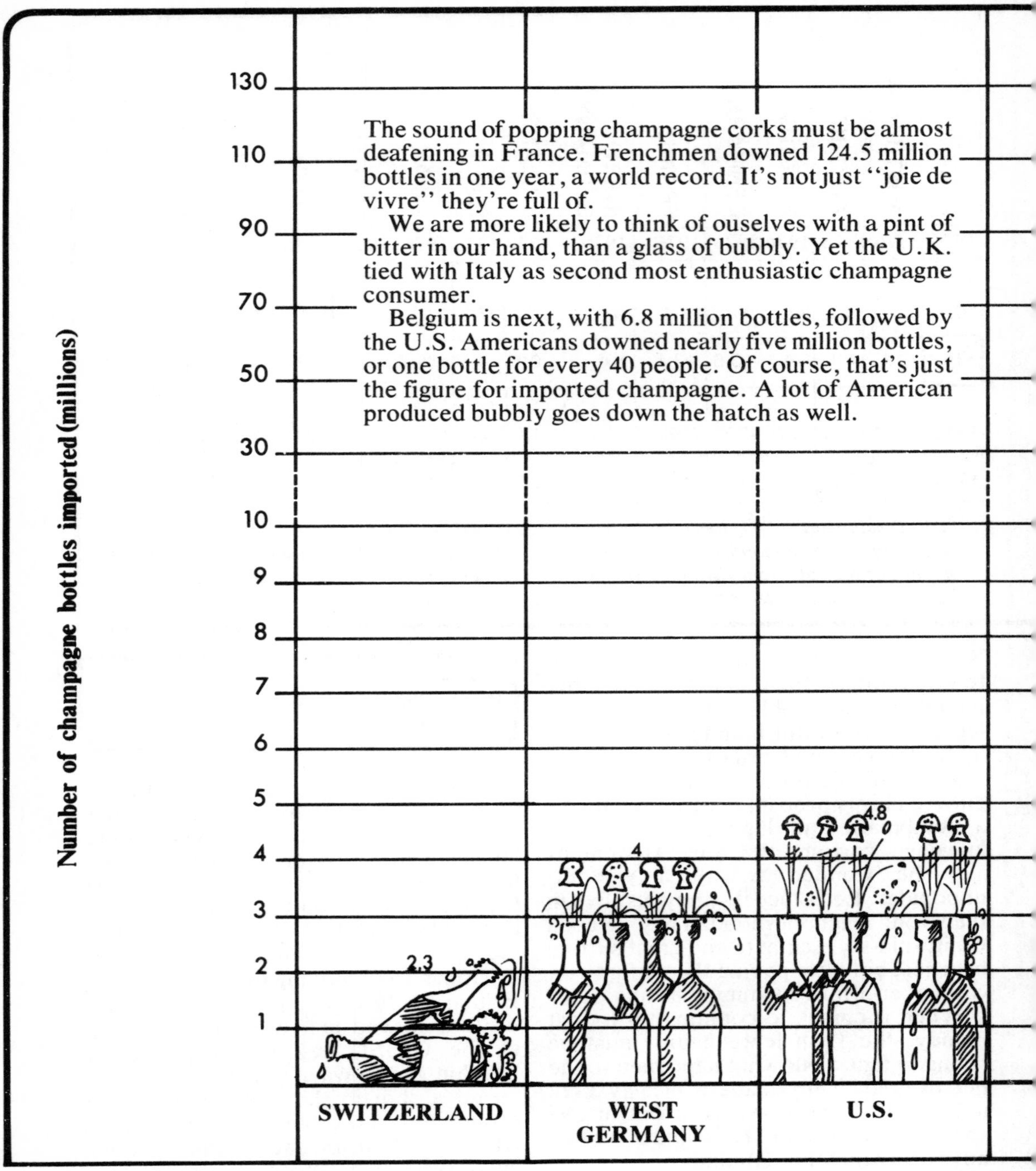

Number of champagne bottles imported (millions)

130 —
110 —

The sound of popping champagne corks must be almost deafening in France. Frenchmen downed 124.5 million bottles in one year, a world record. It's not just "joie de vivre" they're full of.

90 —

We are more likely to think of ouselves with a pint of bitter in our hand, than a glass of bubbly. Yet the U.K. tied with Italy as second most enthusiastic champagne consumer.

70 —

Belgium is next, with 6.8 million bottles, followed by the U.S. Americans downed nearly five million bottles, or one bottle for every 40 people. Of course, that's just the figure for imported champagne. A lot of American produced bubbly goes down the hatch as well.

50 —

30 —
10 —
9 —
8 —
7 —
6 —
5 —
4 —
3 —
2 —
1 —

2.3 (SWITZERLAND)

4 (WEST GERMANY)

4.8 (U.S.)

SWITZERLAND

WEST GERMANY

U.S.

CHAMPAGNE

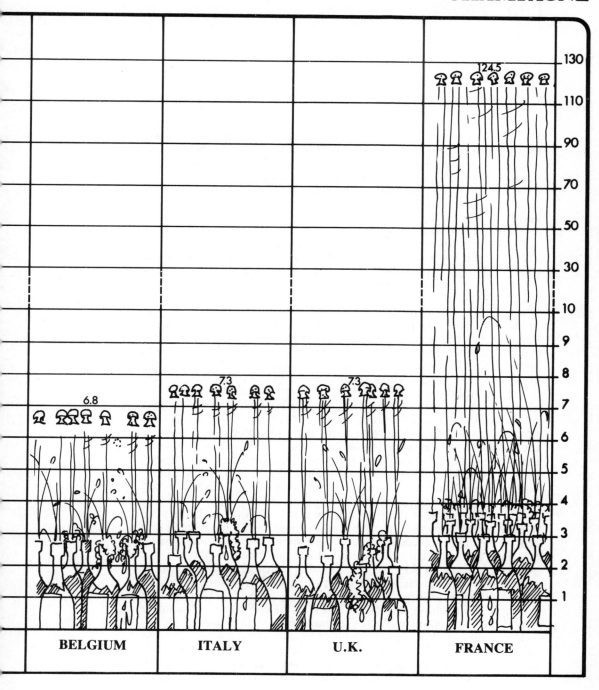

| BELGIUM | ITALY | U.K. | FRANCE |

Underway

We all like to travel. Right from our first faltering footsteps, there's that urge to get somewhere. But nowadays, for most of us, what comes to mind when we think of travel isn't a pair of hiking boots and a rucksack, or even a bicycle. It's a car.

In this chapter, you'll find some interesting facts and figures on the use and abuse of the car. Some of these may come as a surprise. The sheer number of vehicles on the road, for instance, as shown in the first table.

But before you even get your vehicle on the road, you'll want to know what it's going to cost you. Well, we've got some facts here too. You might think that people buy more cars in countries where they're cheap and not so many in areas where the prices are high. It just doesn't work out that way. In some countries, people just refuse to do without the convenience of the family car, no matter how high the price. And, of course, buying one is only the beginning. You've then got the insurance to think about. In most countries this is compulsory. Then you need petrol. We've got some telling figures on the comparative cost of these in different countries. You might well look at these tables before taking off on your next holiday.

Now you've got your car and a tankful of petrol. There's just one other thing you're going to need to get very far – a road. Our next table shows you where to go for the wide open motorways of the world. You'll be able to see just how driving conditions vary from one country to another. In some countries there's a superb system of roads – but hardly anyone on them. And some of the countries with the most extensive network of roads in the world have one big drawback – you can expect a bumpy ride.

Of course, most of the time it's not pot-holes that make us take it easy on the throttle. It's the dreaded sight of police cars in our rear-view mirror. Every country's got speed limits of one kind or another, but they can vary quite a bit from place to place. So if you're planning to do a lot of travelling it could well pay you to take a look at our table which illustrates this. If you intend to take your 140 mile-an-hour Lamborghini on a tour of Japan, for example, you'll have a slow time of it. And you wouldn't want to travel the *autobahns* of West Germany on a motor-bike.

If you do any driving at all, you're bound to have some interest in road safety. All the topics we've dealt with so far – number of cars on the road, driving conditions, speed limits – have a great deal to do with that. But if you want to discover the plain facts about the road accident records of different countries, our table will provide you with the information you want.

If the figures on road accidents make you think twice about going on a drive – take heart. There are other ways of getting around. If it's more than a short trip you have in mind, maybe you'll want to fly. If so, take a look at our next set of tables. We've brought together a lot of material here, on all aspects of flying. Where better to start than at the airport. Just how much traffic do the major airports handle? Our table has the answers.

Then we turn to the airlines themselves. From the hundreds of airlines throughout the world, we've chosen a few of the larger ones from each of the major nations. If you want to know just how these different airlines compare with one another – in terms of distances covered and number of flights – our table will help you out. The accompanying text adds quite a lot of interesting information also. You might expect to find the American airlines leading the world's air traffic – and you'd be right. But are you so sure who comes next? Most of the leading countries have just one major national airline. But how about the lesser airlines?

Our tables show that some of these smaller lines do a surprising amount of business – in some cases more than the well known international giants.

There's another aspect of flying that you're going to want to know something about – and that's air safety. The next table in the chapter takes ten different airlines and gives the number of deaths per billion passenger miles. Over the page, there are figures showing at what stage of your plane journey you're most at risk.

You can also find out which airports have the worst safety records – something to bear in mind when you're thinking about your next holiday.

Finally, we have a table of figures for one of the more frightening phenomena of modern times – air transport terrorism. The figures show the number of incidents and deaths there were between 1968 and 1976. One encouraging fact is that successful terrorist acts are getting fewer and fewer.

Whether you're more used to travelling on a bicycle or a Jumbo, there's something for you in this chapter.

Fast, reliable and cheap transport is the keystone of our culture. Yet when we're caught in yet another traffic jam, we in the U.K. may feel that we are a nation of slaves to the car. The truth is that, despite Rolls—Royce and other famous names of British motoring history, we are not in the same league as other countries when it comes to producing, owning or driving cars. British Leyland, even without all its industrial strife, has always been a minor competitor compared with the American giants. Today the Ford Motor Company's annual revenues are greater than the national budgets of many small countires, it is not surprising that the U.S. has many many more cars on the road than any other country.

No one else comes near the American figure of nearly 110 million. However, they do have more people. And they have the biggest road system in the world. But the proportion of cars to people is still almost ridiculously high. With 500 cars to every thousand of the population, there's one to every two people. This includes infants, children and bed-ridden grannies. Cars, in fact, are almost as common as legs – and for the vast majority of Americans, they have come to do the service once performed by those humble limbs. With about 3.1 million miles of paved roads in the U.S., there are approximately 36 cars per mile of road.

	U.S.A.	CANADA
	1976	1975
1. Passenger automobiles on the road (millions)	109.7	8.9
per 1,000 inhabitants[+]	500	386
2. People per automobile[0]	2.0	2.6
	1975	1976
3. Motorcycles/scooters on the road (000s)	4967.0	355.5
per 1,000 inhabitants[+]	2^x	15

Canada comes second in the ratio of cars to population with 386 cars for every thousand inhabitants. But that's only marginally ahead of Australia with 380. Canada and Australia are both large countries with comparatively small, young, populations. But Australia's paved road network is nearly twice as big as that of Canada. Add in all the unpaved roads in the Australian outback and Australians are far better off for roads than Canadians. Australia has 38 cars for every mile of paved road. Even Britain, with relatively few cars, has 70 per mile. But Canada has a crowded 111 cars for every mile of paved roads.

Sweden has 350 cars per thousand inhabitants – the highest ratio in Europe. In fact, Sweden is the only comparatively small country that produces enough cars to cater for an export market as well as a home market. It's the amount of pro-

	GREAT BRITAIN	AUSTRALIA	AUSTRIA	BELGIUM	DENMARK	FRANCE	(WEST) GERMANY	IRELAND	ITALY	JAPAN	NETHERLANDS	NORWAY	SPAIN	SWEDEN	SWITZERLAND
	1976	1976	1976	1976	1976	1976	1976	1976	1976	1976	1976	1976	1976	1976	1976
	14.4	5.3	1.8	2.7	11.3	16.2	19.2	0.55	15.9	18.5	3.8	0.97	5.3	2.9	1.9
	263	380	243	274	264	306	312	174	285	163	275	240	147	350	291
	3.9	2.6	4.1	3.6	3.8	3.2	3.2	5.7	3.5	6.1	3.7	4.2	6.8	2.9	3.4
	1976	1976	1976	1976	1976	1976	1976	1976	1976	1976	1976	1976	1976	1976	1976
	682.0	304.0	82.2	100.0	36.5	500.0	300.3	36.0	1221.0	750.4	68.0	22.0	1142.0	28.4	93.7
	12	22	11	10	7	9	5	11	22	7	5	5	32	3	15

x Motorcycles only

duction the domestic industry can get out that counts. This seems to determine how readily available cars are. It's always too expensive to provide transport for the masses from imports. Germany and France, with over 300 cars per thousand population, are the other densely-wheeled countries.

Great Britain is down the table, with 263 cars per thousand population – less even than Italy with 285. In both these places production is plagued by labour disputes and wildly inflationary costs. Anyway, we've both traditionally been producers of luxury or specialist cars. Rolls and Maserati are not really in the same league as Ford and General Motors.

The lowest country on this table is Spain, with only 147 cars per thousand population. Surprisingly, Japan is next lowest with only 163 per thousand.

Sources:
International Road Federation
+Heron House estimates
0Society for Motor Manufacturers and Traders, London

There aren't many hard and fast conclusions as far as motorcycles and scooters are concerned. In general, poorer countries tend to use more motorcycles than richer ones. They cost less to buy, use less petrol, and aren't so expensive to repair as cars are. Spain tops the table with 32 motorcycles per thousand people. This balances its low car figures. Italy is traditionally the home of the little motor scooter. It comes next with 22 per thousand – the same as Australia.

The U.S. has the fewest motorcycles of all, with only two per thousand.

The table on the right compares the cost of buying, and running, a car in 17 cities. Data has been extracted from information provided by the Union Bank of Switzerland, who've compared purchase and maintenance costs for the most popular, medium-sized (over 1500cc) cars in 41 major cities.

A car is obviously a luxury in Scandinavia. It costs more to buy one there than anywhere else.

In Norway, the average car costs over £6,400 – easily the highest price in the table, and more than double the price of our 17-country average. Once a Norwegian has a car, though, he only pays £148 a year to maintain it, which is about the norm for all the countries we've listed.

Denmark comes next: an average car costs almost £4,500. Sweden follows at just over £4,000. Despite these high figures, the Scandinavians are above average in car ownership. But the real surprises are at the bottom end of the price scale. Spain and Japan are lowest – the average car costs just under £2,000 in each country. Both have their own motor car manufacturing industries and plentiful cheap labour. They're also easily the lowest countries in car ownership – at around 150 per thousand people.

The U.K. is next in low-price car production. A Ford Cortina costs just £2,130 – despite all those strikes at Dagenham. The next country is U.S. It costs £2,320 to buy a Chevy Nova. British cars may be slightly less expensive but our lower average earnings prevent some of us owning a car.

Once you've bought your car and paid the taxes, you have to pay for insurance. "Third party liability insurance" covers the damage you do to someone or something, but not the damage you do to yourself or your car.

The highest insurance rates are in Dublin (£172). The Irish are famous for their drinking, and maybe this accounts for the cost of their insurance. Next is Paris; where the French have also been known to favour a glass of wine or two. Madrid (£16) and Tokyo (£32) are cheapest on the table.

London (£71) and New York (£84) come roughly in the middle of the

	NEW YORK	TORONTO
	1976	1976
1. Average cost of automobile (£)	a 2,320	b 3,056
2. Maintenance cost (£)[+]	113	171
3. Annual insurance premium[0] (£)	84	137
4. Average price (£) of 1 gallon of regular gasoline[x]	.41	.41

COST OF DRIVING

	LONDON	SYDNEY	VIENNA	BRUSSELS	COPENHAGEN	PARIS	DUSSELDORF	DUBLIN	MILAN	TOKYO	AMSTERDAM	OSLO	MADRID	STOCKHOLM	ZURICH
	1976	1976	1976	1976	1976	1976	1976	1976	1976	1976	1976	1976	1976	1976	1976
	c 2130	d 3549	e 2819	f 2849	g 4451	h 2938	i 2950	j 2591	k 2850	l 1929	m 3700	n 6421	o 1915	p 4034	q 3100
	127	97	149	222	170	211	198	260	103	118	283	148	53	193	251
	71	57	101	151	85	161	135	172	81	32	115	89	16	101	150
	.68	.45	.90	.99	.95	.95	.90	.86	1.13	1.01	.99	.99	.77	.90	101

a = Chevy Nova Standard   b = Chevy Nova   c = Ford Cortina   d = Holden Kingswood
e = Ford Taunus 1600L   f = Renault 16TL   g = VW Golf   h = Renault 16L
i = VW Passat TS   j = Hillman Hunter DL   k = Fiat 132, 1600   l = Toyota Corona Mark II
m = Peugeot 504L   n = Volvo 244   o = Seat 1430   p = Ford Taunus 1600L
q = Opel Ascona 16S

Source:
Union Bank of Switzerland
+Annual tax, insurance and check up
0For third party automobile liability insurance written for an unlimited amount. In some countries, third party liability is not compulsory and in others, insurance companies grant discounts according to age and the policy holder's accident record.
×Heron House calculations based on UBS data

insurance cost table, along with Copenhagen (£85) and Milan (£81).

In Milan, petrol cost more than £1.13 a gallon in 1976. Fuel prices, like everything else in Italy, have been badly hit by inflation, so you have to update that price.

Tokyo and Zürich come next on this list, with petrol at £1.01 a gallon.

The Scandinavian countries all have high petrol prices, and Norway is top at 99 pence a gallon. But this will surely change with the development of North Sea oil. The U.K., as the table shows, fares well compared with all other European countries. At 68 pence per gallon (1976 price), petrol is nine pence cheaper than Spain. Both Belgium and the Netherlands are high at 99 pence a gallon.

And who's the lowest? Canada and the U.S. are kings of the road on this score, at 41 pence per gallon. Of course, the U.S. has oil of its own, but they aren't self-sufficient like the Canadians. Short of new discoveries, the U.S. price will soon begin to rise.

	U.S.A.	CANADA
	1975	1975
1. **Miles of paved roads (000s)**	3100	80.2
per 1,000 inhabitants[+]	14.1	3.4
2. **Total road network (miles — 000s)**	3837	182.2
3. **Paved roads as % all roads**	80.8	44.0

If you drove every mile of paved road in the U.S., you'd travel a distance equal to six round-trip journeys to the moon, or 120 circumnavigations of the earth. No matter how you picture it, Americans have satisfied their love of the car with a truly monumental system of roads. With just over 200,000 miles of paved roads, we in the U.K. have a very low ratio of just over 3.5 miles of paved roads to every thousand population. This is a tiny figure compared to the U.S. and, in fact, only Spain, Japan, Italy and Canada are lower. But the mitigating factor is the unusual density of Britain's population – among the highest in the world. With less distance between towns and villages – on average, less distance between people – we simply don't have as far to go.

When it comes to highways, trunk roads, by-passes and motorways Britain is one of the best-served countries in Europe. On the other hand our service station restaurants on the motorway network have international – and national – notoriety as a gourmet's nightmare.

With more than three million miles of roads, the U.S. has nearly three times as much paved roadway as all the paved roads in France, Germany, the U.K. and Italy put together. That's more than 14 miles of paved roads for every thousand inhabitants. And it's not counting thousands of miles of paved driveways. Only Ireland has a higher ratio of paved roads to people – which says more about the small size of the Irish population than about the roads! In fact, more Irish have left Ireland for the U.S. than the current population of the island. Not a few of these immigrants were originally employed laying out the roadways that now serve their descendants in the U.S.

The U.S. has vast open spaces that are still reached by more primitive routes. Only 80.8% of the entire network of roads in the U.S. is paved. Roughly one road in five is unpaved. In countries where road systems have been laid out for centuries, the progress toward paving them all is considerably more advanced compared to those countries settled more recently.

Naturally, this means that Europeans have a much larger percentage of paved roads than Australia. At 80.8% paved, the U.S. is not doing too badly. But the Dutch and the Austrians claim to have no unpaved roads at all. This must take a lot of the adventure out of camping trips. Except for Norway and Sweden, most of the rest of Europe has virtually eliminated dirt roads.

France has the second highest number of miles of paved roads – nearly 440,000 miles. But Australia's spindly network of

	U.K.	AUSTRALIA	AUSTRIA	BELGIUM	DENMARK	FRANCE	(WEST) GERMANY	IRELAND	ITALY	JAPAN	NETHERLANDS	NORWAY	SPAIN	SWEDEN	SWITZERLAND°
	1976	1976	1976	1976	1976	1976	1976	1976	1976	1976	1976	1976	1976	1976	1976
	207.0	139.2	84.7	66.3	39.7	438.8	277.2	50.3ˣ	168.2	234.5	53.7	23.0	78.8	41.1	38.2
	3.6	9.5	11.2	6.7	7.8	8.1	4.5	16.0ˣ	3.0**	2.0	3.9	5.6	2.2	4.9	5.8
	214.4	525.0	84.7	71.3	41.3	494.4	291.7	55.3	180.8	670.0	53.7	48.5	90.3	77.6	38.6
	96.5	26.5	100.0	93.0	96.0‡	89.0	95.0	91.0	93.0+	35.0	100.0	47.3	87.3	53.0	99.0

ˣ1975 figures   ‡Average estimate   **Non-official figure

outback track gives her a larger total of road miles. Germany – with just over half France's total paved miles – comes third.

But the French and German figures obscure the fact that for decades Germany had the finest and most extensive highway system in Europe – one of Hitler's few praiseworthy legacies. (He built the *autobahns* for military purposes.) Although the *autoroute du soleil* – the turnpike from Paris to the south of France – is one of the best long stretches of highway in Europe, France is one of the few important European countries that is still waiting for a comprehensive highway system.

With over 234,000 paved miles, Japan has an impressive road system. But to serve their massive population, the Japanese rely primarily on a more primitive network of unpaved roads. In this, as in many other ways Japan is a tantalising mix of the traditional and the modern. In spite of its relatively small size, it has a proportion of unpaved roads exceeded only by the Aussies. And their number of paved miles per inhabitant is the lowest shown – even below that of Spain (whose

Source:
International Road Federation
+Heron House estimates
°National statistical office

industrial development is considerably less advanced).

Australia, with massive distances between its territories and cities, obviously needs an extensive road system. Not surprisingly, almost three-quarters of the country's roadways are unpaved, the highest figure in the table. You have only to think of the pot-holed dirt tracks that connect the widely scattered sheep farms of the outback to understand why. Australia's paved and unpaved roads make up the most extensive network of all the countries surveyed with the exception of the U.S. and Japan. The U.S. has over seven times Australia's mileage. Japan has 670,000 miles and Australia weighs in with 525,000.

If you think that better roads mean fewer accidents you would, unfortunately, be wrong – as our table on dangerous driving indicates.

At one time, a man had to walk in front of a car carrying a red flag. That was one way of limiting the speed of cars. When the red flag went, the number of road accidents soared.

Today speed limits are generally high. Cars are built to go fast. Most new cars cruise comfortably at over 70 miles per hour – the average speed limit for the countries listed. Most drivers are tempted to speed on a stretch of clear road.

	U.S.A.	CANADA
	1978	1978
Maximum speed (mph):		
on a motorway	55+	70
on a secondary road	25-40	50
in a built-up area	0	30

The speed limits in many countries were reduced during the oil crisis. In the U.K., speed limits were temporarily reduced from 70 to 50 m.p.h. There are so many driving laws in this country that the Metropolitan Police estimate that a driver breaks the law every other time he takes his car out on the road. It's surprising that the figures for road deaths in the U.K. are only just overtaking the number killed 40 years ago. In those days, there were fewer regulations and fewer cars – but people weren't as aware of the danger of traffic as they are today.

In the U.S., too, speed limits were reduced to 50 m.p.h. to restrict petrol consumption. A 55 m.p.h. limit is still in effect on inter-state motorways, but it's no secret that this is widely ignored.

In the wide, open spaces of Canada, the speed limit is a more realistic 70 m.p.h. But on secondary roads, it's a low 50 m.p.h. In town, it's 30 m.p.h.

The Australians zip along at a reasonable 60 to 68 m.p.h. on inter-state motorways. But the best place for speed demons to roam free is New South Wales, where there's no maximum speed limit at all.

Italy has the highest speed limits on the table. There are limits of 56 to 90 m.p.h., "depending upon engine c.c.". Italy has excellent *autostradas*, and it's also the home of a first-rate sports car manufacturing industry – with cars like Ferrari, Maserati and Lamborghini. Not every Italian can afford an expensive sports car. But in his heart, every Fiat driver sees himself in an Alfa Romeo.

West Germany comes next, with a speed limit of 81 m.p.h. on its *autobahns*. Germany has an excellent motorway system – the most extensive in Europe. It also has Europe's largest car manufacturing industry. As in Italy, there's a strong lobby against any reduction in speed. People tend not to buy the fastest, most expensive, cars when they're not legally allowed to use them near capacity.

Austria, France and Switzerland rank after Italy and Germany. All three have a limit of 80 m.p.h. on motorways.

SPEED LIMITS

	U.K.	AUSTRALIA	AUSTRIA	BELGIUM	DENMARK	FRANCE	(WEST) GERMANY	IRELAND	ITALY	JAPAN	NETHERLANDS	NORWAY	SPAIN	SWEDEN	SWITZERLAND
	1978	1978	1978	1978	1978	1978	1978	1978	1978	1978	1978	1978	1978	1978	1978
	70	60-68[x]	80	74	68	80	81[++]	60	56-90[xx]	49	62	56	62	56-68	80
	70	‡	62	56	56	68[**]	62-75[00]	40	50-68[00]	37	49	43	56	43	62
	30	37	31	37	37	37	31[++]	30	31	24	31	31	37	31	37

[+]Interstate   [0]Varies from state to state   [x]New South Wales has no maximum speed limit
[‡]Criteria for categorising secondary roads and designating special speed limits vary from state to state   [**]Dual carriageway   [++]Recommended   [00]As signposted   [xx]Varies according to engine cc

The Japanese figure of 49 m.p.h. is the lowest on the table. From this you'd think that Japan has very safe roads filled with careful drivers. Not so. The car accident table shows that Japan heads the list in accidents per number of cars on the road, with 25 per thousand cars. The Japanese limit is unrealistically low, and many motorists simply ignore it – which defeats the point of having a speed limit.

The Austrians, also high on the accident table, have generous speed limits. The figure of 62 m.p.h. for secondary roads is higher than for any other country except Germany, Switzerland and the U.K. In general, speed limits on secondary roads have an effect on the accident rate. They're also easier to enforce. In Ireland, which has the lowest limit for secondary roads (except for the U.S. and Japan), the accident record is good. Only safety-conscious Sweden has a better record.

As far as accidents are concerned, there are undoubtedly many other factors involved besides speed limits.

Sources:
Automobile Association
Department of Motor Transport/Australia
Embassies

The U.S. heads the list of people killed in car accidents with over 45,000 deaths each year. That's as many people as attend a football match on an average Saturday afternoon. In other words, over 120 people are killed every day of the year. Even Great Britain's total, which is comparatively low, means that nearly 18 people are killed every day throughout the country.

But these gross figures say more about the number of drivers than about the safety of the roads. To see the countries with the most dangerous drivers, look at the line showing deaths per hundred thousand inhabitants.

Canada, Australia and France head this list, each with 26 per hundred thousand. These figures include people who died within 24 hours of an accident as a direct result of injuries received. The high figure in France is undoubtedly related to their high rate of alcoholism. Belgium and Austria come next on the death toll, with 25 deaths per hundred thousand, followed by Germany with 24.

Great Britain has the comparatively good figure of 12 deaths per hundred thousand, the result of our persistent road safety campaigns. Only Norway (11 per hundred thousand) and Japan (eight per hundred thousand) are lower. At first glance Japan's seems to be an excellent figure, but if you look at their record in non-fatal accidents you'll see that they have the highest figure for accidents: 25 per thousand cars.

Sweden has easily the lowest rate of accidents – 5.9 per thousand cars. This is certainly because of their strict laws on drink and driving.

There's no apparent reason for the low figures shown for Ireland – 8.4 accidents per thousand cars. The Irish authorities

	U.S.A.	CANADA
	1975	1975
1. Deaths from automobile accidents:[o]		
people killed	45,500	6,061
people killed per 100,000 inhabitants[+]	21	26
2. Injuries from automobile accidents:		
people injured (000s)	2,802.3	220.9
people injured per 1,000 inhabitants[+]	13.1	9.6
3. Automobile accidents:		
accidents (000s)	1,861.0	154.8
accidents in built-up areas as % of number of accidents	69.3	NA
accidents at night as % of number of accidents	NA	31.6
accidents per 1,000 cars[+]	17.4	17.4

DANGEROUS DRIVERS

	GREAT BRITAIN	AUSTRALIA	AUSTRIA	BELGIUM	DENMARK	FRANCE	(WEST) GERMANY	IRELAND	ITALY	JAPAN	NETHERLANDS	NORWAY	SPAIN	SWEDEN	SWITZERLAND
	1976	1975	1976	1976	1976	1976	1976	1976	1975	1976	1976	1976	1976	1976	1976
	6,520	3,694	1,903	2,486	873.0	13,787	14,804	477.0	9,511	9,734	2,440	470.0	4,500	1,168	1,188
	12	26	25	25	17	26	24	15	17	8^x	18	11	12	14	18
	331.6	89.4	62.7	84.0	19.5	357.4	480.5	7.6	229.8	613.9	63.5	10.2	94.0	21.8	28.7
	6.1	6.0	8.3	8.5	3.8	6.7	7.8	2.4	4.1	5.4	4.6	2.5	2.6	2.6	4.5
	253.0	65.7	45.0	62.5	15.9	261.2	359.6	4.6	279.8	471.0	55.4	9.9	62.5	17.0	23.5
	NA	NA	82.1	NA	NA	NA	70.0	NA	NA	NA	72.0	NA	48.0	57.6	67.7
	NA	40.6	42.8	NA	NA	NA	NA	NA	NA	NA	NA	NA	30.0	38.0	28.2
	18.0	12.0	24.6	23.1	11.8	16.0	18.7	8.4	18.5	25.4	14.7	9.7	11.6	5.9	12.6

NA not available ^{x}*Deaths within 24 hours of accident*

introduced breathalyser tests, but they were fought so strongly, they didn't work. Ireland also has the best figure for injuries: 2.4 per thousand people.

Great Britain's injury rate is 6.1 per thousand, followed by Australia. But it's by no means the worst figure on our table. The U.S. tops the list with the alarmingly high rate of 13.1 people per thousand injured on its roads. And Canada isn't far behind, with 9.6. Belgium and Austria are both high: 8.5 and 8.3 respectively.

Source:
International Road Federation

+   Heron House estimates
o   This includes people who died 24 hours after the accident or as a consequence of the accident

	U.S.A.	CANADA
	1[0] 1976	2 1976
Total number of aircraft movements[+] (000s)	714.0	244.9
Scheduled airline traffic[+] (000s)	661.8	160.7
Scheduled international airline traffic[+] (000s)	26.2	64.1

Chicago's O'Hare is by far the busiest airport with 714,000 flights in and out every year. That's nearly 2,000 flights a day, and approximately one flight every 45 seconds throughout the entire 24 hours. (That's every day of the year, so don't buy a house near the flight path to O'Hare!) Yet only a tiny 3.5% of O'Hare's traffic is international.

London's Heathrow Airport is a distant second, with 278,000 flights per year. Unlike O'Hare, nearly 75% of these flights are international. This doesn't mean that flights to and from Heathrow are necessarily covering greater mileage. A Chicago to New York plane ride covers about three times the mileage of a London to Paris flight.

Ranking after these two comes Canada's Toronto International, which handles nearly a quarter of a million flights each year; and Germany's Frankfurt Main Airport which handles over 200,000. Here again the difference in the number of international flights is on the side of the smaller European airport, where distances between countries are so much shorter. Of Frankfurt's flights, 62% are international compared to a mere 26% of Toronto's. But then Frankfurt is within 400 miles of 15 other countries.

After Frankfurt, the main airports in Europe are Amsterdam's Schiphol (176,000 flights per year), Rome's Ciampino/Fiumicino (just under 172,000) and Orly, just outside Paris, with nearly 152,000. Most of these European airports – except for Orly – deal mainly with international flights.

The surprising European figure is the low number of flights out of Madrid's Barajas. With just 122,000 flights each year, they deal with fewer flights than Switzerland's Zürich.

Australia's largest (Sydney Kingsford International) and Japan's largest (Tokyo Haneda International) both handle as much traffic as the major European airports. Yet they both have a very low percentage of international flights.

U.K.	AUSTRALIA	AUSTRIA	BELGIUM	DENMARK	FRANCE	(WEST) GERMANY	IRELAND	ITALY	JAPAN	NETHERLANDS	NORWAY	SPAIN	SWEDEN	SWITZERLAND
3	4	5	6	7	8×	9	10	11	12	13	14	15	16	17
1976	1976	1976	1976	1976	1976	1976	1976	1976	1976	1976	1976	1976	1976	1976
278.1	132.8	66.3	102.2	163.7	151.9	208.2	76.1	171.9	168.4	176.4	83.3	122.0	105.1	139.8
256.1	101.6	46.3	74.4	147.7	144.5	193.2	35.8	162.8	165.8	132.2	52.5	117.9	66.1	105.1
206.4	20.5	46.2	74.3	115.9	76.4	129.4	32.1	91.7	55.3	128.8	24.6	49.2	47.6	101.4

0 *John F Kennedy has more international commercial air traffic (95.1) than O'Hare*
× *Charles de Gaulle has more international commercial air traffic (82.0) than Orly*

1 Chicago O'Hare
2 Toronto International
3 London Heathrow
4 Sydney Kingsford International
5 Vienna Schwechat
6 Brussels National
7 Copenhagen Kastrup
8 Paris Orly
9 Frankfurt Main
10 Dublin
11 Rome Ciampino/ Fiumicino
12 Tokyo International
13 Amsterdam Schiphol
14 Oslo Fornebu
15 Madrid Barajas
16 Gothenburg Torslanda
17 Zürich

The airport on our table with the lowest volume of traffic is Vienna's Schwechat. It only has 66,000 flights a year, of which just over two-thirds are international. In some ways, Vienna is badly placed for international flights – on the way to nowhere but Eastern Europe. Unless the international situation changes, it doesn't look as if there's much room for expansion in Vienna.

The second lowest on the table is Ireland's Dublin Airport. But then Ireland has steadily built up quite a large international traffic via Shannon Airport.

There are several other things to keep in mind when studying the table. Paris, for instance, has two main airports: Paris Orly, and the brand new Charles de Gaulle, which in 1976 had almost 100,000 flights. Similarly, London has Gatwick as well as Heathrow – and Gatwick is as busy as some of the smaller European airports. And while Chicago's O'Hare may have the largest volume of air traffic in the U.S., New York's Kennedy Airport handles most of the international traffic. In addition New York has another large terminal at La Guardia.

Source:
International Civil Aviation Organisation
+Take-offs and landings

183

A national airline is a prestige symbol. Every country that can afford it (along with several that can't) has an airline. It may fly a very limited selection of routes – often only between the national capital and one of the major European capitals or New York. Many of these airlines run at a loss and have to be subsidised by their national government.

Even large airlines like our British Airways chalk up losses which have to be subsidised out of government funds. This is principally because of rising fuel prices, unrealistic international price agreements and commitments to maintain an uneconomical number of flights on certain routes. For instance, many airlines that fly the Atlantic are half empty.

U.S. airlines are by far the leaders in total number of passenger miles travelled. The biggest *single* airline in the world is Russia's Aeroflot – not represented in our statistics. Aeroflot is a state monopoly and carries all of the U.S.S.R.'s air traffic, both domestic and international.

Each of the four largest U.S. airlines carries more passengers than the combined airlines of any other country. While they have a large international network, their domestic route structure is even larger. United Airlines, which has mainly domestic flights, makes up almost 30 thousand million passenger miles a year. The second largest carrier, American Airlines (23 thousand million passenger miles) and the fourth, Eastern (almost 20 thousand million) are also principally domestic carriers.

The U.S. has heavy internal air traffic making O'Hare Airport in Chicago (not Kennedy, Dulles, Heathrow or Orly) the world's busiest airport in terms of aircraft departures and arrivals. Americans hop on a plane almost as casually as people in other countries take a bus.

After the American airlines, the next figure is for British Airways Overseas. Our national airline covers over 13.5 thousand million passenger miles – nearly 1.5 thousand million miles more than their neighbouring competitor – Air France – which covers nearly 12 thousand million passenger miles. J.A.L. of Japan follows closely with a figure of over 11.5 billion passenger miles. Air Canada is eighth in our table (nearly 11 thousand million passenger miles), and West Germany's Lufthansa is ninth (over nine thousand million passenger miles).

PASSENGER MILES

Passenger miles per airline (scheduled[+] flights)			
Country	Airline	Passenger miles (millions)	Date
U.S.A.	United	29,784	1976
	American	23,069	1976
	Trans World	22,292	1976
	Eastern	19,479	1976
CANADA	Air Canada	10,703	1976
	CP Air	4207	1976
	Pacific Western	530	1976
U.K.	British Airways Overseas Division	13,588	1976
	British Airways European Division	3773	1976
	British Airways Caledonian	920	1976
AUSTRALIA	Qantas	7083	1976
AUSTRIA	AUA	512	1976
BELGIUM	Sabena	2418	1976
DENMARK	SAS	1500	1976
FRANCE	Air France	11,869	1976
	UTA	2230	1976
GERMANY (WEST)	Lufthansa	9308	1976
IRELAND	Aer Lingus	949	1976
ITALY	Alitalia	6003	1976
	Alisarda	64	1976
JAPAN	JAL	11,601	1976
	Japan Asia Airways	275	1976
NETHERLANDS	KLM	6407	1976
	NLM	45	1976
NORWAY	SAS	1695	1976
SPAIN	Iberia	6599	1976
	Aviaco	315	1976
SWEDEN	SAS	2204	1976
SWITZERLAND	Swissair	5276	1976

Source:
ICAO
[+]Scheduled flights chosen according to the greatest passenger-mile records

The last year or so hasn't been a good time for airlines, economically. However, this is almost certainly a temporary state of affairs. Once fares are more rationalised – and more competitive – the airlines that are still in business will almost certainly go on expanding.

Which airlines are the biggest in the world? Which one, for instance, has the most planes in the air? That's what this table is all about.

One glance will tell you that the U.S. airlines are the highest fliers.

There are more scheduled flights in the U.S. than in all the other countries in the table put together. Eastern fastened its seatbelts 534,500 times in 1976. That's nearly three times more than the U.K.'s total aircraft departures. It's also a staggering 17,000 more flights than its nearest American rival– United.

The U.S. is not only one of the world's richest countries, it's also one of the largest. And as people become more wealthy they get itchy feet. From the Texas oil mogul in his private jet to the California-bound super-saver, the U.S. is the land of the long-distance traveller. And much of their travelling is done by air.

For instance, in 1974, U.S. airlines flew the equivalent of 770 miles for every single U.S. citizen.

Restless jet-setters would also do all right in Canada. Their airlines total was 283,300 flights in 1976. Canada's population is only an eighth of America's, so you can see that Canadians like to have their heads in the clouds.

If you look at the low figures in the departure league, you'll find that the Australians are surprisingly earthbound.

Qantas, the main airline "down under", has a mere 19,200 flights – peanuts compared with Canada. However, most Qantas flights are international. And when you consider that most Australians have to fly about 1,000 miles before they reach any other country you can begin to see why these figures are so low. Obviously when you fly Qantas you're really flying somewhere – and there's no question of it being just a short shuttle flight to the big city in the next-door state or adjacent country – as it often is in the States and Europe.

After Canada and America comes the U.K. – not bad considering our size. British Airways, our nationalised airline, got 183,600 planes off the ground in 1976. And that was before we went supersonic and put Concorde into service.

Those efficient West Germans come next with Lufthansa (179,300), followed by Iberia with 162,400. It's rather surprising that Iberia should come so high in the departure league. The answer probably lies partly in their trans-Atlantic flights to South America. And there's all that tourist traffic which keeps Spanish skies buzzing – French and Spanish air traffic controllers permitting.

Air France (160,900) follows hot on Iberia's tail; and if you combine the Scandinavian countries – Norway, Denmark and Sweden – S.A.S. takes fifth place with 152,400.

Austria's A.U.A. (23,000) and Ireland's Aer Lingus (27,100) follow Qantas at the bottom of the table for national airlines. However, the population of Ireland is only around four million, so there's a large proportion of globe trotting men living in the Emerald Isle.

TAKE OFF

Scheduled aircraft departures by airline

Country	Airline	Takeoffs[+] (000s)	Date
USA	Eastern	534.5	1976
	United	517.4	1976
	Delta	500.2	1976
	American	369.8	1976
CANADA	Air Canada	194.0	1976
	Pacific Western	48.2	1976
	Canadian Pacific Airlines	41.1	1976
UK	British Airways European Division	111.6	1976
	British Airways Overseas Division	42.6	1976
	British Caledonian	29.4	1976
AUSTRALIA	Qantas	19.2	1976
AUSTRIA	AUA	23.0	1976
BELGIUM	Sabena	39.8	1976
DENMARK	SAS	39.9	1976
FRANCE	Air France	160.9	1976
	UTA	14.7	1976
GERMANY (WEST)	Lufthansa	179.3	1976
IRELAND	Aer Lingus	27.1	1976
ITALY	Alitalia	106.0	1976
	Alisarda	5.5	1976
JAPAN	JAL	68.3	1976
	Japan Asia Airways	2.5	1976
NETHERLANDS	KLM	56.9	1976
	NLM	20.1	1976
NORWAY	SAS	56.6	1976
SPAIN	Iberia	162.4	1976
	Aviaco	22.0	1976
SWEDEN	SAS	55.9	1976
SWITZERLAND	Swissair	89.3	1976

Source:
International Civil Aviation Organization
[+]Scheduled flights only. Includes both domestic and international flights.
Figures rounded up or down

During the period from 1950–74, over 30,000 people throughout the world were killed in airplane accidents. For anyone involved in air travel, from management to maintenance, it can be a matter of great importance to look closely at these statistics. They can be of interest to the ticket-buying passenger as well. But there's just one thing to remember – air-safety statistics are notoriously difficult to interpret. So don't go flying off to hasty conclusions.

1. Deaths[+] per billion passenger miles, 1950-74	
Country	Airline[0]
CUBA	Cubana
CZECHOSLOVAKIA	CSA (Czechoslovak Airlines)
EGYPT	Egyptair[‡]
JORDAN	ALIA (Royal Jordanian Airline Corporation)**
NIGERIA	Nigeria Airways
PHILIPPINES	PAL (Philippine Air Lines)
RUMANIA	TAROM (Rumanian Airline)
SPAIN	AVIACO (Aviacion y Comercio SA)
TURKEY	THY (Turkish Airline)[++]
VENEZUELA	VIASA (Venezolana Internacional de Aviacion SA)[00]

For one thing, there are many different causes of air accidents. In an extreme case, an airline could be entirely innocent of any kind of neglect, and yet be involved in a number of fatal accidents. In the preparation of these figures the causes of accidents have been ignored. In the end, though, it's the responsibility of each airline to take precautions against all hazards and eventualities. And some lines are indeed better at doing just that, as the figures on our table indicate.

Sometimes the full figures on air accidents just aren't available. Some airlines are much more willing than others to provide this data and give full details and circumstances. Other airlines are particularly unco-operative in this respect, so they've been left out of these tables.

There's another question facing the statisticians. How do you express the accident rate so as to give the fairest comparison? Some airlines carry far more passengers and clock up many more miles than others. Obviously, a straight comparison of the number of fatalities per airline would not take these factors into account. To cover this the number of deaths per billion passenger miles has been calculated. Even this way, though, is not perfect. Some airlines are predominately involved with short-haul traffic. This involves more takeoffs and landings for miles flown and these are the most hazardous parts of any flight (see table on air deaths – when and how.)

In the second part of the table, we've looked at the very latest available figures in order to see whether there are any changing patterns that might redeem the airlines with the worst records. It's noticeable that while some airlines had recorded deaths, four of the airlines had completely clean records. This should reassure you the next time you have to put out your cigarette and strap yourself in. As you ascend into blue yonder, the music that you hear will almost certainly be pre-recorded and not the angels playing on their harps.

DEATHS IN THE AIR

Deaths[+] per billion passenger miles	Order[x]	2. Subsequent Deaths[‡‡] 1975-77***		
		1975	1976	1977
24.77	10	0	78	0
27.39	7	126	71	4
48.66	5	0	52	0
191.30	1	188	0	0
38.64	6	0	0	0
25.00	9	32	14	0
60.70	4	0	0	0
66.93	3	0	0	0
77.41	2	42	155	0
26.28	8	0	0	0

Source:

Destination Disaster, by Paul Eddy, Elaine Potter and Bruce Page
© Times Newspapers Ltd

[+]Adjusted to the nearest decimal point. The responsibility for deaths on planes owned by one airline but chartered or leased by another was determined by the authors of *Destination Disaster*. One billion is one thousand million.

[0]Only airlines in operation at the time of the writing of *Destination Disaster* are included. Figures are not restricted exclusively to scheduled flights. Aeroflot has been omitted because of lack of comprehensive data.

[x]According to the number of deaths per billion passenger miles.

[‡]Egyptair was known as United Arab Airlines and before that, Misrair.

**ALIA began operations in 1963.

[++]THY was formerly known as DHY.

[00]VIASA began operations in 1961.

[‡‡]Crew and passenger deaths; absolute figures. Figures include scheduled, non-scheduled and freight flights, and accidents due to hostile action.

***Based on *Flight International* data.

Between 1970 and 1977 more than 12,000 people were killed on scheduled and non-scheduled airline flights. That's more than the entire student population at Cambridge in 1977.

Is this an appallingly large figure? Or an astonishingly small one? Opinions differ. Yet some experts say we're safer flying in an aeroplane than we are crossing the road. Others say that's as it should be – there are comparatively few flights and endless safety precautions are taken before an aircraft can even taxi out onto the runway.

But most people, as they are strapped in their seats when the plane accelerates down the runway, feel a slight quickening of apprehension at some stage. They only relax when they're finally airborne and are allowed to unstrap their seatbelts while they rise through the clouds. But is an aeroplane most likely to crash during take-off? Is it safer once it's airborne? This table answers these questions.

The total number of fatalities has been broken down into deaths during different phases of flight: take off and climb; en route; approach and landing. Deaths as a result of terrorism, and as a result of collisions, are also included.

The figures show that you're right to be apprehensive during take-off and climb. That's when 20% of all airline fatalities happened during the last seven years. However, it's still not the most dangerous time of a flight. That comes during the approach – at the end of the flight when the aeroplane lines up with the runway and descends. Between 1970 and 1977, nearly a third (31%) of all airline fatalities occured during this stage. Nineteen per cent happened during landing.

Another figure is for deaths as a result of collisions. With more and more aircraft flying, and airports getting busier all the time, it's surprising that there aren't more of these. However, safety precautions are extremely thorough and all major airports are equipped with sophisticated equipment designed to avoid this particular possibility.

In the past decade, a whole new hazard has cropped up in flying. That's terrorism, a subject that's dealt with more fully in another section in this chapter. Airlines are particularly vulnerable to the criminal or fanatic with a gun or bomb, and a hijacking virtually guarantees world-wide publicity for a cause. Seven per cent of the total airline fatalities between 1970 and 1977 were a result of terrorist activities - that's over 800 deaths. However airline security systems are getting more effective every year.

The only other category – deaths en route – accounts for just 14% of all airborne fatalities between 1970 and 1977. However, 14% means 1,716 deaths – that's a lot of mourning families.

All the news in this field isn't gloomy – in 1977 only 440 passengers were killed on scheduled airline flights (outside Russia). That's one less than the previous record low and each year more and more flights cover more and more miles. Life in the air seems to be getting safer.

Mention must be made, however, of the 578 deaths in the 1977 collision between two 747s at Tenerife. Both aircraft involved were primarily scheduled carriers but in the particular circumstances of the accident were on non-scheduled flights. As such, they are not included in the 1977 figure for

AIR DEATHS: WHEN AND HOW

Deaths[+] on passenger flights[0], 1970-1977	
1. Total deaths 1970-1977	12,004
2. Deaths during take-off and climb	2,357
as % of total	20
3. Deaths en route	1,716
as % of total	14
4. Deaths during approach	3,682
as % of total	31
5. Deaths during landing	2,242
as % of total	19
6. Deaths as a result of terrorism	804
as % of total	7
7. Deaths as a result of collisions	1,203
as % of total	10

Sources:
Heron House Associates compilation, based on *Flight International* data.
[+]Passengers and crew   [0]Scheduled and non-scheduled

scheduled flights mentioned earlier in this section. If it had been, the figures would not look so encouraging.

Some words of comfort for tremulous travellers. According to the experts – and the statistics – more and more airline crashes these days involve elderly aircraft. And hijacking is almost always confined to scheduled flights – many charter flights require a passengers' manifest to be made up well in advance, for security checks by the authorities.

In fact, the answer to the passenger's question "what are my chances of getting killed before I get to my destination?" is about three in a million. That answer takes into account the number of fatal crashes that occur and the number of passengers who survive. Those odds are something to remember the next time you fasten your seat-belt and wave good-bye to *terra firma*.

Each year, the International Federation of Airline Pilots Association (IFALPA) draws up a list of "black star" airports. These are airports which the association considers to be "difficult". Or, by their own criteria, airports which are considered to be "critically deficient" in safety features. The list is sent annually to all members of this airline association, which considers it confidential and not for publication. For this reason, IFALPA officials were unwilling to release the 1978 list for our use.

However, at the time of the Tenerife accident involving a KLM and a Pan Am plane in April 1977, the "black star" airports for that year were leaked to the press. We've based our table on this information. The complete 1977 list cites 26 "black star" airports – but the names of two of them weren't released by the press because of negotiations between IFALPA and the authorities concerned. They aren't included in our table.

Which country has the most "black star" airports? IFALPA considers Colombia to be the world's most undesirable airline destination. Seven airports there qualify for the "black star". They're all international airports with modern navigational devices, but there's a shortage of fire and safety equipment. According to a top IFALPA executive: "In the event of a bad accident we don't give very much chance for the pilots', or the passengers', chances of survival".

The U.S. is the next worst country in this table – but there are mitigating circumstances. Two of the airports out of the four listed aren't in the U.S. mainland at all, but in dependencies or outlying territories.

There's no excuse for the other two in the American list. Logan Airport, Boston, and Los Angeles Airport are both big and busy – and carry a lot of traffic. They're considered "difficult" by IFALPA because they both have strict anti-noise rules. These require many incoming and outgoing planes to take a flight path over the ocean – which isn't safe when the wind is high, or blowing from certain directions.

Other major airports on the "black star" list are Osaka Airport in Japan and Tehran Airport in Iran. The latter has been described as an accident waiting to happen – especially during the winter months which can be very severe. Osaka Airport is also notorious – mainly because of its slippery runways.

However, all the news on the airport front isn't bad. The general opinion amongst airline pilots is that most international airports are safe. And two of the best – London's Heathrow and Dulles Airport in Washington D.C. – are also among the biggest.

In Europe, particularly in France, the U.K., West Germany and the other industrialised countries, air line technology is fully as sophisticated as in the U.S. and in some aspects the Europeans are more advanced. France, for example, uses a battery of jet engines to blast away fog from Paris' two international airports – De Gaulle and Orly. That technique has not been adopted in the U.S. largely because of the noise and pollution it creates.

No system of airport rating can be foolproof. Before the Tenerife disaster in April 1977 the airport wasn't even on IFALPA's "black star" list.

'BLACK STAR' AIRPORTS

Black Star[+] Airports: 1977

Country	Number	Name
U.S.A.	2 + 2[0]	Boston — Logan; Los Angeles; St Thomas, US Virgin Islands[0]; Pago Pago, American Samoa[0]
CHILE	1[0]	Mataveri, Easter Island[0]
COLOMBIA	7	Bogota; Barrangvilla; Cali; Cartagena; Leticia; San Andres; Medelling
FIJI	1	Suva/Nausori
GREECE	2	Corfu; Rhodes
INDONESIA	3	Denpasar, Bali; Medan; Ujung, Pandang
IRAN	1	Teheran — Mehrabad
ITALY	2	Alghero, Sardinia; Rimini
JAPAN	1	Osaka
MALAYSIA	1	Penang
TONGA	1	Fua'/Amotu

[0]*Dependency or outlying territory*

Source:
Sunday Times files
[+]The International Federation of Air Line Pilots Association Black Star Airport list was leaked to the press in 1977. A Black Star airport is one that is considered to be "critically deficient" in safety features. The complete 1977 list cites 26 Black Star airports; however, the names of two of the airports were not released by the press due to delicate negotiations between IFALPA and the authorities concerned.

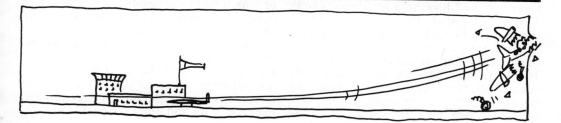

Urban terrorism is one of our modern age's most sensational and unpleasant developments. Anyone can be affected at any time. Aircraft hijacking is only one aspect of this form of terrorism, but for many years it has been the most prevalent.

An aircraft is uniquely vulnerable to this form of terrorism. Hijacking is liable to involve nationalities and airports of many different countries, and so attract the greatest possible amount of publicity. And this, when all's said and done, is what the hijackers are after. Terrorism is usually perpetrated by extremists who're desperate to publicise their cause.

From the figures on the table it looks as though hijacking is on the wane. There are many obvious reasons why. All airlines and airports take extremely sophisticated precautions against hijacking – most of which are secret. Anyone who's travelled on an aeroplane recently will be familiar with security checks like the magnetometer scanning device which each passenger has to pass through to determine if he's carrying metal objects. There are x-ray scanners to check baggage for explosives. And many aircraft carry armed plain-clothes security guards.

The table shows that hijacking was on the increase at the end of the 60s. In 1969 there were 87 hijacking incidents – only 17 of which failed. That means over 80% of attempted hijackings were successful.

From then on the number of incidents decreased – and so did the success rate. In 1970, over 68% of hijackings were successful, but after that the figure barely touched 50%. However, from 1972 onwards, hijacking turned into a fatal business. In 1972 there were 60 attempted hijacks, only half of which were successful. But there were 117 deaths as a result, whereas there had been no fatalities during the preceding four years. As if this wasn't bad enough, the deaths total almost doubled in 1973 – though the number of incidents fell to 22.

Airline security had started to get tough – and so had the terrorists. Hijacking became a very dangerous business – for hijackers, security men and passengers.

During the four years from 1973 to 1976 the number of hijacking incidents for each year stayed in the 20s. In 1973, 50% of these attempts were successful. In 1974 that figure fell to 30.8%. In 1975 it was down to 25%. And by 1976 only just under 17% of all hijacking attempts were successful.

In other words, the odds on a successful hijacking were down – from evens to six to one against – in just four years.

There were only four successful hijackings in 1976. This looks like a success story – but it has a fatal flaw. In 1973, 210 people lost their lives during 22 or so incidents. In 1974, when there were 26 attempts, fatalities dropped to 159. In 1975, with 20 hijacking attempts, there were only 29 fatalities. However, in 1976, with just 24 hijacking attempts (only four of which were successful) a colossal 173 people were killed. That's second only to the horrifying 1973 figure.

This table shows that hijacking seems to have dwindled to very small proportions. World-wide, in 1976, there was one attempt every 15 days. Fewer and fewer attempts succeed, as precautions become more and more stringent.

Terrorist incidents and number of deaths caused by air transport terrorism, 1968-1976

Year	Total incidents	Failed attempts	Successful attempts *(total)*	Successful attempts %	Deaths[+]
1968	35	5	30	85.7	0
1969	87	17	70	80.5	0
1970	82	26	56	68.3	0
1971	59	36	23	39.0	0
1972	60	30	30	50.0	117
1973	22	11	11	50.0	210
1974	26	18	8	30.8	159
1975	20	15	5	25.0	29
1976	24	20	4	16.7	173

Sources:

Heron House Associates compilation based on data from "Security in the Air", Chris Eliot, *Aerospace International*, February/March, 1978 and *Flight International* data.
[+]*Passengers and crew.*

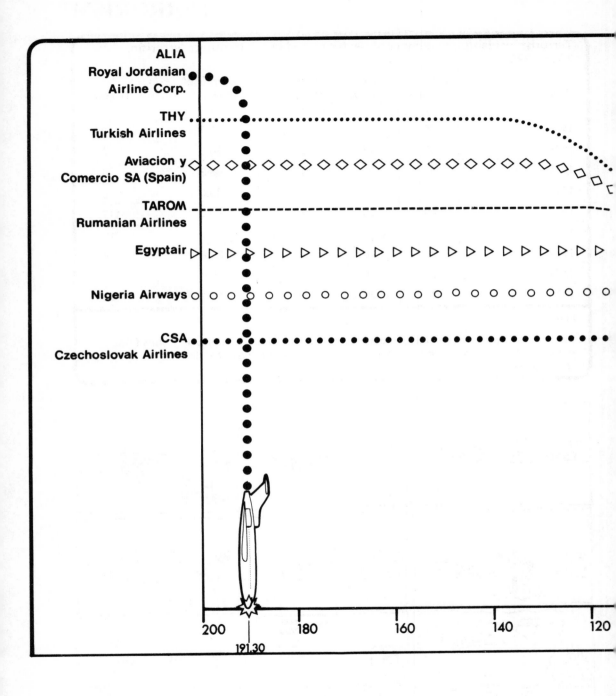

ALIA
Royal Jordanian
Airline Corp.

THY
Turkish Airlines

Aviacion y
Comercio SA (Spain)

TAROM
Rumanian Airlines

Egyptair

Nigeria Airways

CSA
Czechoslovak Airlines

200 180 160 140 120

191.30

If you're a nervous flier, you'll be especially interested in these figures. It's often said that travelling by aeroplane is as safe as crossing the street. True, but some streets are safer than others. For instance, the national Czechoslavakian airline had only around 27 fatalities per thousand million miles flown. But ALIA lost over seven times as many passengers.

The Turkish airline THY is the next most hazardous, but their rate of 77 deaths per thousand million miles is still well under half the Jordanian figure. Spain's AVIACO is next, with nearly 67 fatalities for the same number of miles flown. The airline TAROM records just over 60 deaths — not a terribly high figure, but something to bear in mind the next time you feel tempted to hop on a plane for Rumania.

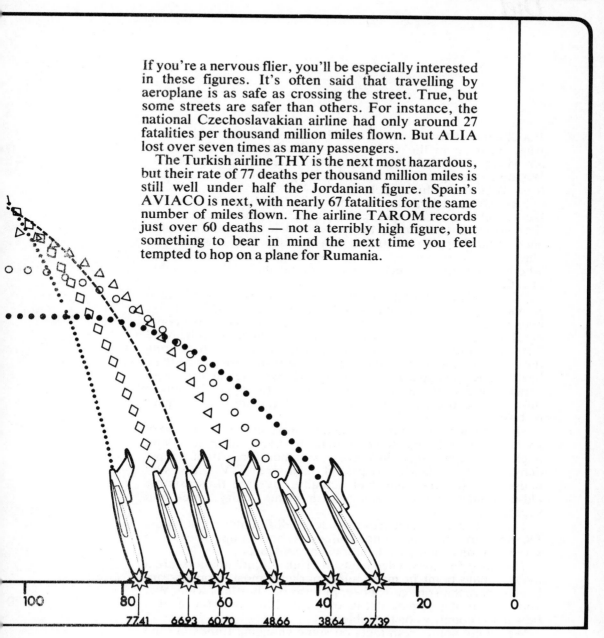

100 80 60 40 20 0

77.41 66.93 60.70 48.66 38.64 27.39

Deaths per billion passenger miles

Diet

Food, glorious food – where would we be without it? Skinnier is what we'd be in the U.K., but you'll find very few people anywhere who don't take a ravenous interest in this subject. Whatever your particular taste, this chapter should stimulate, at least, your intellectual appetite.

First, feast the brain on the Americans, and bite off some food facts from your local supermarket. The cash register checks out the calories as it gobbles up the money and, in cold cash, the table shows that shopping habits boil down to table habits. Whether a chicken in every pot dictates how we dispose of our money, or whether the money decides the national diet, is a debatable question.

What we *don't* wheel out past the supermarket counter also provides food for thought. The back-to-nature movement has been progressively gaining in the U.K. and more and more people are buying organically grown food, unpolluted by chemical fertilisers. Some people go one step further than this – they grow their own, saving some money while presumably preserving their health. There are specialist foodstores all over the country, from vegetarian delicatessens to macrobiotic restaurants. But this is only an eccentric expenditure in the large amount that we spent on food.

Sinking your teeth into the statistics on American supermarket spending, you'll find some difficult-to-digest surprises. You might not expect that, in a country famed for instant soups, dehydrated ham flakes, crisps and 'man-size' T.V. dinners ad nauseam, what the consumer actually buys are those simple old-fashioned, hard-to-prepare fresh meats, dairy foods and vegetables.

The first section compares the sales of different sorts of foods. Do they carry out more canned juice or canned soup? The article serves up some canny explanations for cold facts.

The thirst for knowledge leads to drink. Liquid libations are a lively selling point in many supermarkets; most shoppers are having more than one can of beer plus the odd bottle of wine. Wet your whistle at a running stream of home truths about the alcohol flow in the countries on the table.

We've got some fresh facts on some changing trends in food buying as well. Some foods are currently out of vogue, but bad taste one year is good taste the next. Don't miss this section if you're opening a food store. You'll notice there's been a striking drop in sugar sales even though prices are now almost normal, perhaps because sugar has suffered lately from a bad press. It seems we can't stomach extra calories these days.

How do our eating habits – and purchasing power – compare with other peoples'? The table provides you with some answers. Meat goes down very well with most Americans, but their meat-eating habits are modest compared with the Australians. They eat more than 25 times as much as the Japanese, which is small wonder, since prime beef (no bones, please) costs over £10 a pound there. We don't know, but perhaps the Japanese compensate by eating 25 times more fish than the Australians.

We've got some tasty tit-bits for dinner-time conversation. Where do they eat the most eggs? The most margarine? Where do we find the sweet-tooths of the world? Did you know that we eat more than four times as many chickens per capita as they do in the United States?

Maybe you're interested in seeing just what the secret is to that French *haute cuisine*. Well, we can't give you the recipes, but we can certainly tell you just what they're putting into the pot. And the same goes for those other Mediterranean gourmets – the Italians. They top the list twice. They eat more flour than anyone else, and with all that pasta that's probably just what you'd expect.

And after we've eaten, our thoughts naturally turn to other forms of refreshment. If you're interested in knowing who are the world's boozers, our next table has got all the information you can swallow. With all these facts you could be the star of the show at your next cocktail party. You'll see that the favourite tipples vary quite a bit from country to country. And you might find that some popular notions just aren't true. Where do they drink more beer than anywhere else? And are the Australians bigger beer swillers than the Americans? We all know about the French taste for wine, but who actually drinks more wine, the French or the Italians?

No figures on drinking would be complete without the Irish. Do they really drink Guinness for breakfast? You'll find that their position in the red-nosed league is surprisingly modest. Perhaps it would look different if we had some figures on Irish poteen.

DEEP FREEZE

This table shows you how much people consume of three very 20th century foods: breakfast cereals, frozen foods and ice cream. It also leads to some interesting conclusions about changing eating habits in the U.K. and Europe.

	U.K.	AUSTRIA
	1975	1975
Consumption per annum of:		
breakfast cereals[+] (per household)	23.6	16.7
frozen food[+] (per capita)	19.6	8.8
ice cream[o] (per capita)	1.2	0.8

The traditional English breakfast used to start with porridge, followed up with eggs and bacon, kippers or kedgeree, and finally topped off with toast, butter and marmalade. But as our table for breakfast cereal consumption shows, that's all changed. Now each U.K. household munches its way through nearly 24 lbs of cereal a year. That's about 2 lbs a month, and puts us way out front of nearly every other nation in the table. A bowl of cornflakes may not be as appetizing as a cooked breakfast, but it's cheaper and quicker, which counts for a lot these days.

The Netherlands (20.9 lbs per year) is the only other place that approaches our enthusiastic consumption. Those hard-working Dutch must really appreciate the time they save by simply pouring breakfast out of a box. Where do the most people shun so-called convenience foods when they begin their day? In Spain. The average Spanish household gets through 4.2 lbs. of breakfast cereal a year – a mere one-sixth of our consumption. The Spanish have decided to stick to their traditional breakfast of *churros* and strong coffee. (No wonder – it often includes a brandy or two).

Maybe snap, crackle and pop first thing in the morning just doesn't agree with the Latin temperament. The Italians prefer the gentle hiss of an expresso machine – each household eats less than half a pound of breakfast cereal a month.

In Denmark too, cereals haven't lured many people away from their traditional hearty meat and cheese breakfast. Like the Italians, a Danish household consumes only 5.07 lbs a year.

Do you like to eat raspberries in December, or sweet corn in January? If so, you're one of the many people in the U.K. who use frozen foods. We each eat nearly 20 lbs of it a year. But we're way behind Sweden. There, the average person consumes a whopping 37 lbs a year. With those long, cold winters you'd think Scandinavians would shudder at the mere sight of anything with frost on it. But frozen food is almost as popular in Norway and Denmark as in Sweden. Of course, that northern climate with its short summers means a limited supply of fresh home-grown produce. The Scandinavians import a lot of fresh food and that often means freezing it.

Travel south and you'll have a hard time finding anything edible that comes from a freezer. For that matter, you'll have a hard time finding a freezer. The

SPECIAL FOODS

	BELGIUM	DENMARK	FRANCE	(WEST) GERMANY	ITALY	NETHERLANDS	NORWAY	SPAIN	SWEDEN	SWITZERLAND
	1975	1975	1975	1975	1975	1975	1975	1975	1975	1975
	6.8	5.07	8.4	18.7	5.07	20.9	8.1	4.2	8.6	8.1
	9.2	25.8	5.9	11.5	3.3	18.5	22.5	3.08	37.0	15.6
	0.8	1.5	0.5	0.9	0.8	0.8	1.9	0.1	2.02	1.4

Source:
Euromonitor
+Pounds
0Gallons

Spanish and Italians both eat less than four pounds of frozen food a year. There are several reasons. The sunny climate means a year-round abundance of fresh fruits and vegetables. Then too, the lower standard of living means few people can afford fridges and freezers.

France, on the other hand, is an affluent country. But the average Frenchman eats only 5.9 lbs of frozen food a year. The abundance of agricultural produce there means the French don't have to import much; they just head for the nearest shop and buy fresh, home-grown *petits pois* and *fraises*. Of course, France is the home of haute cuisine. It'll probably be a while before those haughty French chefs are convinced that frozen vegetables can be almost as good as the straight-from-the-garden variety.

The ultimate frozen food is ice cream. And the shocking fact our figures reveal is that Sweden laps up more of it than any other nation in the table. The mere sight of all that snow isn't enough – the Swedes seem to want to eat it too. Over two gallons a year disappears down eager Swedish throats. Northerly Norway (1.9 gallons) and Denmark (1.5 gallons) are

close behind. Of course, the Scandinavians are famous for their glowing good health, and the main ingredient in ice cream is milk. So they've really just found a way of staying healthy and having fun at the same time.

The Swiss are next fondest of ice cream, while in the U.K., the average person consumes 1.2 gallons a year. Almost every town and village has a shop selling one brand or another and in recent years, British housewives have been taught to think of ice cream as an easy dessert.

Recently, American ice cream has begun to compete with the traditional three flavours of vanilla, strawberry and chocolate. Our conservative British palates have been confronted with such exotic flavours as mango, peach and even peanut butter.

With all that blazing sun and hot weather, it's surprising to find Spain at the bottom of the ice cream table.

"What is food to one, is to others bitter poison," said Lucretius many years ago. Each country has its own national dishes; yet, just as many of us don't like snails or octopus, many people elsewhere would hate fish and chips.

Yet many national dishes are different ways of cooking the same foods. The main ingredient of spaghetti bolognese, keema curry, Mexican tacos, and shepherd's pie is the same – ground beef.

But apart from the different methods of preparation, what do people actually eat in different countries?

For most people (with the notable exception of fish-eating Japan) meat is

Annual per capita consumption (lbs) of:	U.S.A. 1975	CANADA 1975
butter	4	13
eggs	39	29
cheese	15+	NA
margarine	27	11
sugar	92	100
poultry	6	50
fresh fruit	82+	NA
fresh vegetables	179+0	NA
flour	186	183
meat	123	116

the main ingredient of any meal – if they can afford it. And that's no small consideration these days.

In conjunction with this table, it's worth looking at the foodbasket table showing prices of various foods in the different countries.

While steak is very expensive in Japan, it's comparatively cheap in Australia; so it's not surprising to find the Australians topping the figures for meat consumption. The average Australian chomps his way through over 250 pounds of meat each year. That's 11 ounces of meat for every man, woman and child every day of the year.

This is in surprising contrast to the figures for steak-loving U.S.. They eat very little meat compared with some other countries on our list – 123 pounds a year per person (less than half the Australian figure).

West Germany is second – the average German consumes 220 pounds of sauerbraten, sausages and other meat each year. Australia and West Germany are the only countries to top the 200 mark, though the French come close at 198. The home of *haute cuisine*, and the originator of the word "*gourmet*" really cares about its food, as you'll see in the table on household expenditure.

With the price of food as high as it is in France, the average Frenchman has to

U.K.	AUSTRALIA	AUSTRIA	BELGIUM	DENMARK	FRANCE	(WEST) GERMANY	IRELAND	ITALY	JAPAN	NETHERLANDS	NORWAY	SPAIN	SWEDEN	SWITZERLAND
1975	1975	1975	1975	1975	1975	1975	1975	1975	1975	1975	1975	1975	1975	1975
20	20	11	22	17	20	14	25	18	0.8	9	10	0.6	9	15
31	31	31	24	25	28	37	29	23	35	25	21	35	27	23
13	NA	14	21	21	33	24	7	26	NA	26	23	3	25	23
11	12	17	27	42	7	18	9	2	7	30	44	5	41	15
100	126	97	76	114	90	81	124	71	55	129	55	67	99	74
24	29	19	21	16	31	20	22	36	15	15	4	39	10	13
106	NA	171	164	118	168	278	104^x	239	NA	188	136	213	111	96
149	NA	161	209	112	242	159	147	341	NA	193	77	210	62	184
135	211	117	163	84	159	98	178	276	80	117	138	161	99	145
135	255	172	198	153	198	220	143	159	10	161	108	132	134	165

NA not available ^{0}*Includes 77lb potatoes* ^{x}*Excludes citrus fruits*

earn a lot of money to buy the meat he converts into *Boeuf Bourguignonne* and *Steak au Poivre*.

Equal to the French are their neighbours, the Belgians, who also take great delight in their food. A meal in Belgium can often last more than three hours.

The roast beef of old England is world famous, but we can't compare with the really big meat eaters. However, each of us eats ten per cent more than the Americans and we are about average on the table.

The Japanese eat less meat than anyone else: ten pounds per head of population every year. That's less than half a pound a month. If we had the

Sources:
Euromonitor
+US Department of Agriculture

figures for fish consumption, Japan would certainly top the table. They're also fifth lowest when it comes to eating poultry – after Sweden, Norway, Switzerland and America. America's poultry consumption is surprisingly low for the home of Colonel Sanders and Kentucky Fried chicken.

Bottoms up! Skol! Santé! Here's mud in your eye! If you're a drinker or a drinks salesman, this table will tell you where to find the most convivial company.

Surprisingly the countries with the largest number of alcoholics (see table on alcoholism) don't follow the same order as those which consume the most alcohol. Obviously drinkers in some countries can hold their booze better.

	U.S.A.	CANADA
	1975	1975
Quantities of alcoholic beverages (pints) per person per year		
beer+	151.0	151.2
wine+	11.5	11.8
spirits0	9.1	12.6

If beer's your tipple, West Germany is the place to go. Seasoned experts claim that German *bier kellers* serve the best beers in the world. The Germans like it anyway. The average German consumes just under 260 pints every year and that's counting all the children and non-drinkers in the picture. When you consider that there are nearly 62 million people in West Germany, it means that more than 1,500 million *gallons* of beer go down the hatch each year.

The lager-loving Aussies are second in this field. They get through just over 250 pints per year per person. No wonder the Aussies have a reputation for drinking the place dry.

The Belgians come next in the beer-drinking stakes. They get through just over 237 pints per head each year. Fourth at 230 pints are the Irish – whose taste for Guinness is legendary.

The Danes are after the Irish. They consume more than 227 pints per person per year. They're world-famous for Tuborg and Carlsberg, two civic-minded companies that sponsor art and science as well as making superb light beers.

The U.K. is about average amongst the beer drinkers of the world. This may come as a blow to the prestige of our bitter drinkers – but they can take heart. We still consume over 200 pints per man per year. On average, that means that each of us drinks over five pints a week.

Americans and Canadians drink over 151 pints of beer a year – mostly while watching baseball and football games.

The Italians drink less beer than anyone else on the list. They get through a mere 22.5 pints per person per year. But they make up for it when it comes to drinking wine. The average Italian downs an extraordinary 189 pints of *vino* annually. That translates into about 175 bottles, or around half a bottle of wine a day for every man, woman and child in the land.

The French aren't far behind the thirsty Italians. They drink just over 182 pints of wine per person. The Spanish each drink nearly 134 pints a year.

Despite all their budding wine connoisseurs, Americans consume only a

204

ALCOHOL CONSUMPTION

	U.K.	AUSTRALIA	AUSTRIA	BELGIUM	DENMARK	FRANCE	(WEST) GERMANY	IRELAND	ITALY	JAPAN	NETHERLANDS	NORWAY	SPAIN	SWEDEN	SWITZERLAND
	1975	1975	1975	1975	1975	1975	1975	1975	1975	1975	1975	1975	1975	1975	1975
	207.0	250.1	182.7	237.1	227.0	79.7	259.4	230.6	22.5	62.3	138.9	79.4	79.9	103.5	127.8
	9.2	19.7	61.8	30.3	20.2	182.5	51.0	6.2	189.2	0.8	18.0	5.9	133.8	13.5	77.3
	3.5[x]	4.5	5.8	5.2	6.3	9.1	11.9	7.2	7.0	NA	11.9	6.5	9.1	10.3	7.0

[x] Great Britain   NA not available

little more than 11 pints of wine apiece every year – surprisingly low for a wine-producing country. The Canadian rate is just about the same. But our rate is lower. In the U.K., we drink only half the amount of wine the Dutch or Australians drink. But our rate has been going up rapidly during the 70s, possibly because we can now buy wine off the supermarket shelf.

Nobody drinks more hard stuff than the Canadians. (Except maybe the Russians with their vodka – but they're not on the table.) They get through over 12 pints of spirits per person every year.

The second biggest spirit drinkers are the Germans. In addition to all that beer, they imbibe over 11 pints of spirits per person annually. That's a lot, no matter how you sip it.

We drink just over three pints of spirits per person every year. It's not that we don't like the stuff. It's the tax. This is an insidious device imposed by the government that keeps the frugal Scots from drinking all their Scotch – and just about everyone else.

Sources:
+Brewers' Society
0 The World Atlas of Wine, revised edition by Hugh Johnson, Mitchell Beazley Ltd, ©1977
  Spirit consumption figures are based on an average strength of 50% alcohol; conversions from metric measures by Heron House.

Annual per capita consumption (lbs) of meat

Who are the most carnivorous nations in the world? If you expect to find the U.S. at the top of the list, you're wrong. Each year Australians consume a whopping 255lb of meat.

The West Germans are close behind — they get through 220lb. That's a lot of *wiener schnitzel*. In France, the arts of *haute cuisine* transform 198lb of meat per person into delicious dishes like *boeuf bourguignonne*.

Finally we get to the U.S., where only 123lb. is consumed annually. Americans are getting more cholesterol – conscious every year, which may have something to do with meat's comparative lack of popularity.

When faced with a choice between steak and raw fish, the Japanese don't hesitate. They each eat only 10lb of meat a year.

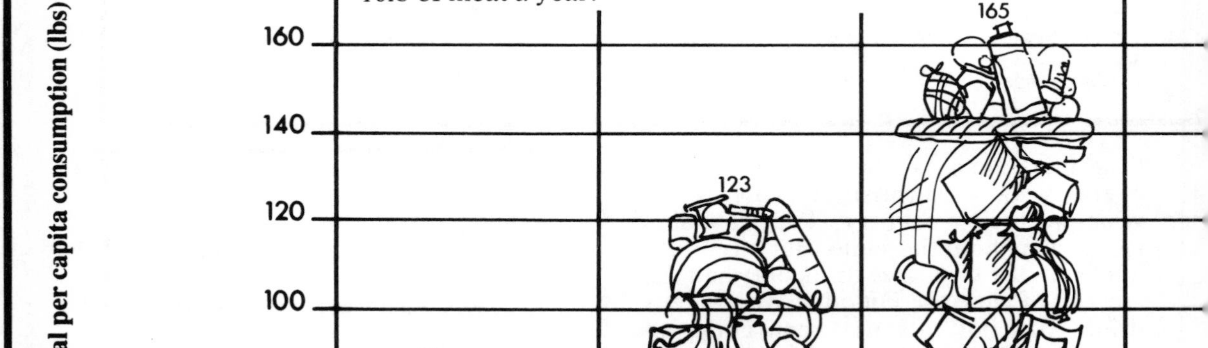

260
240
220
200
180
165
160
140
123
120
100
80
60
40
20
10

JAPAN   U.S.   SWITZERLAND

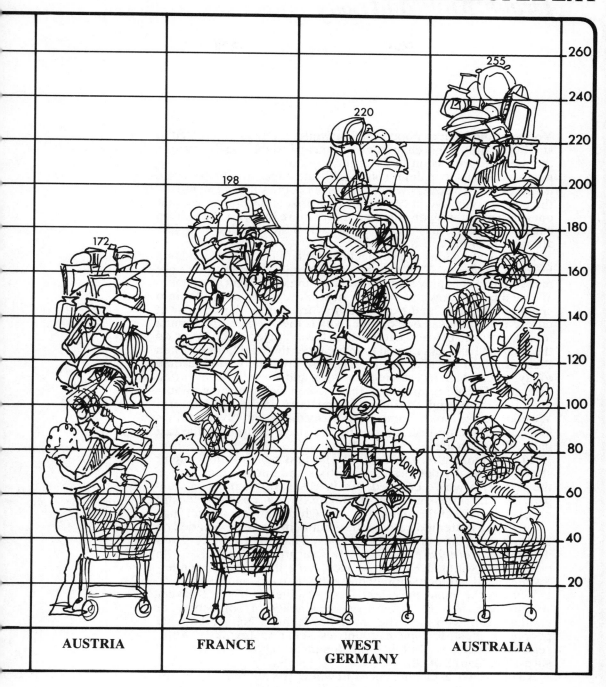

260

240

220

200

180

160

140

120

100

80

60

40

20

255

220

198

172

AUSTRIA

FRANCE

WEST GERMANY

AUSTRALIA

Health: Mind and Body

Young or old, rich or poor, black or white, we're all interested in staying healthy. This chapter gives you an insight into the state of the medical profession in different parts of the world, and how much it costs us to maintain our health and cope with illness when it strikes. It also covers which illnesses people suffer from most – the country you live in can make a surprising difference when it comes to meditating on whether you're more likely to suffer from cancer or heart disease, or from self-inflicted illnesses, like alcoholism or narcotics addiction.

We also take a look at an aspect of health that's both painful and inevitable – death. This may be our inevitable end but there are many factors that play a part in determining what form it will take. We've broken down the many horrible causes of death into specific categories and examined how frequently these occur in different countries. For example, why is the average Frenchman less likely to die of a heart attack than the average Swede? Why do the Spanish have less reason to fear cancer than the Dutch? Then there are the aspects of health that are increasingly troublesome today – mental health, suicide, abortion and sexually-transmitted diseases. These are problems that have grown more common – and often increasingly difficult to cope with – in the 20th century. We take a close look at some of the many factors involved – things like changing moral standards and the stress caused by the fast pace of modern life.

So who looks after our health? The first table in this chapter shows how many doctors (and dentists and nurses) there are in the countries we've covered. As you'll learn from the accompanying text, the number of doctors in a country doesn't necessarily reflect the amount of care which patients are getting.

Perhaps a better indicator is the amount of money expended on health care in the individual countries. That's the subject of our next table. Here there are quite a few surprises. The U.K., with our famed National Health Service, *isn't* the biggest spender. Swedes shell out more money on health than any one else.

What is the most prevalent form of illness? It's not a very serious one. We all suffer from headaches at some time or another. But who gets the most headaches? Nearly half the people in most European countries seem to regularly suffer serious headaches. But it's the Japanese who have the most surprising figures of all.

All of us have accidents once in a while. It's illuminating to find out what kind of accidents we suffer. The next table shows us how many people die from accidents and what type of accidents they are. The most accident-prone people are the Canadians. Around ten per cent of all deaths in Canada are the result of accidents. Unfortunately, we can offer no explanation as to why the Canadians are the people most plagued by accidents. Are they, as a nation, careless – or is it just bad luck?

A lot of our illnesses are self-inflicted. Alcoholism and narcotics addiction come into this category – and these are the subjects of another two tables. And we deal with the sexually transmitted diseases, which have become almost epidemic in some countries. But certainly the worst thing of all we can do to ourselves is to commit suicide. And that's the subject of the next table. How many people kill themselves? The accompanying text to this table gives us some illuminating insights here.

Certainly one of the reasons people commit suicide is as a result of mental illness. Mental health is much in the news these days. Another controversial and much discussed subject is abortion. As you'll see, the figures in our table on this subject are closely linked with the laws of the countries involved. We've tried to obtain (where possible) figures for both legal and illegal abortions, although this is a notoriously difficult field.

We've all got to die – so it's worth finding out what you're liable to die from. The last tables in this chapter show the principal causes of death in the countries we've covered. Further tables show the incidence of some of the principal causes of death – such as heart disease and cancer – and the age at which you're most vulnerable to them.

As the philosophers tell us, our health is more valuable than gold. This chapter will furnish you with at least a few insights on how to look after your most valuable possession.

	U.S.A.	CANADA
	1974	1974
1. Doctors per 10,000 inhabitants	16.5	16.6
people per doctor	610	600
2. Dentists per 10,000 inhabitants	5.1	3.8
people per dentist	1970	2650
3. Nurses (including midwives) per 10,000 inhabitants	63.9	NA
people per nurse	160	NA

In the U.K., the number of doctors per ten thousand population is about average for all the countries covered, at 15.3. But it's even lower across the sea in Ireland, only 11.8. That's 850 people to every doctor, nearly 150 more than we have in the U.K.

To a greater or lesser degree, the doctor for population figures overstate the availability of medical help (as you no doubt noticed last time you tried to get a house call). That's because a varying, but sometimes significant, percentage of doctors don't see patients at all (or don't see very many). They may be busy teaching, or in administration or research. What's significant is that they're not busy seeing you. So all of the figures on this table overstate the true doctor to patient (as opposed to people) ratio.

Among the countries shown, Japan is lowest with just one doctor for every 870 citizens. They're also lowest in terms of nurses. One reason for this poor showing is that a large number of companies have medical staff on the premises for their employees. It may not be *personal* but it is *efficient* (and, presumably, a pretty good check on absenteeism).

Australia, like the U.K. and Ireland, has a low rate when it comes to doctors. How much a doctor can earn – and keep – plays a part in all three cases. Ireland has excellent medical schools but in many ways the nation is still a poor, developing country. So doctors are underpaid and they emigrate, often to the U.K., sometimes to Australia. At the same time Australian and British doctors are heading overseas, often to "greener" pastures in North America.

All the countries shown on the table have far more doctors to tend the population than many other countries in the world. While the U.S. and Canada are about equal, Mexico (not shown on the table) presents a depressingly different story. South of the border, there's just one doctor per 1,500 inhabitants.

If you're at an airport with a screaming toothache, head for Norway not Spain. When it comes to dentistry, all the Scandinavian countries are way above average for the countries shown.

The U.K. only has three dentists per

DOCTORS, DENTISTS, NURSES

U.K.	AUSTRALIA	AUSTRIA	BELGIUM	DENMARK	FRANCE	(WEST) GERMANY	IRELAND	ITALY	JAPAN	NETHERLANDS	NORWAY	SPAIN	SWEDEN	SWITZERLAND
1973	1972	1974	1974	1972	1974	1974	1972	1973	1973	1974	1974	1973	1973	1974
15.3	13.9	20.3	17.6	16.3	14.7	17.4	11.8	19.9	11.5	14.9	16.5	14.8[+]	15.5	16.8
707	720	500	570	620	680	520	850	500	870	670	610	670	650	590
3.0	4.1[+]	2.0	2.2	7.6	4.8	5.1	2.2	NA	3.6	3.0	9.2	1.0[+]	8.6	4.0
3270	2460	4950	4550	1310	2090	1960	4570	NA	2740	3290	1090	9650	1160	2510
49.0	NA	35.3	NA	81.2	55.3	37.0	64.0	NA	32.6	34.5	68.2	NA	69.1	NA
207	NA	280	NA	120	180	290	160	NA	310	290	150	NA	140	NA

[+]Number registered — not all working in the country   NA not available

Source:
World Health Organization

ten thousand people, a low rate. In the U.S. they have just over five – the same as West Germany, which has the highest figure for the Common Market countries.

Austria, so big on doctors, is exceptionally poorly off for dentists, with only one for about five thousand countrymen.

The numbers also show that the figures for dentists vary far more between countries than those for doctors.

We've only been able to obtain accurate nursing figures for 11 countries. These reveal Scandinavian concern with the health of their citizens. America does well on the nursing front, compared to most Western European countries, with one nurse for every 160 men, women and children in the population. That means they're roughly equal with Ireland, but that's where most of the similarity ends. Ireland's nursing figures include mid-wives, and most babies there are born at home. What's more, with far fewer doctors to look after patients, nurses fill the gap in Ireland; while in the States, they supplement, rather than replace, a doctor's care.

In Austria and Germany, the doctor to population ratios may be better than in the U.S., but the lack of nurses might indicate that their doctors do more of the medical work themselves. This is even more true of the Netherlands, where the number of doctors is below the average for the countries listed. When considered with their low nursing figures, the picture here is bleak.

211

	U.S.	CANADA
	1974	1973
Annual expenditure on health (UK £):		
government expenditure, per person	88	112
private expenditure, per person	126	39
total, per person	214	151

If you get sick, how much will it cost you to recover and who's going to foot the bill? This table gives the average cost of health care in different countries and tells how much of that bill the government picks up.

In the U.K., where the National Health Service has provided free health care to the entire population for over 30 years now, the total cost of the programme to the government is only £87 a year per person. An additional £11 is spent by the few of us who wish to be private rather than government patients. One reason for these low costs is that it's more efficient to have a central authority running the entire hospital system. Beds can be used at near maximum capacity, and unnecessary duplication of expensive health care machinery and facilities can be kept to a minimum.

Doctor and hospital bills cost the Americans over twice as much as they cost us. However, the table shows that they have less to complain about than many other people. Total cost of health care in the U.S. is £214 for each inhabitant – a high figure, but not when it's compared to the staggering £344 paid by each West German, or £281 per Frenchman, £266 per Australian or £250 per Swede.

The Swedes are the highest government spenders at £197 a person per year, and the Danes are second with £184. These two Scandinavian countries are strenuous supporters of socialised medicine, so it's not surprising they're the superspenders. The West Germans and Australians come next in government outlay per person. There's another very important point. Although you may not have to pay as much directly in one of these countries when you visit your doctor or are admitted to a hospital, the system is being financed by taxation. So indirectly you are shelling out. For example, in the U.K. about £4,000 million of taxpayers' money goes towards running our National Health Service. West Germany has the most expensive health care system in the world. In addition to government funds, Germans pay out a whopping £193 a year for private health care. The Australian figure is more modest, £108 is spent privately – £18 a year less than Americans pay. With a huge but sparsely populated land, Australians have had to shoulder the cost of building and maintaining modern hospitals in localities where the population is too small to support them efficiently. This accounts for the high government spending down under.

U.K.	AUSTRALIA	AUSTRIA	BELGIUM	DENMARK	FRANCE	(WEST) GERMANY	IRELAND	ITALY	JAPAN	NETHERLANDS	NORWAY	SPAIN	SWEDEN	SWITZERLAND
1975	1975	1974	1974	1974	1974	1974	1975	1975	1975	1972	1973	1974	1974	1974
87	158	74	95	184	131	151	69	71	77	71	106	32	197	113
11	108	38	100	NA	150	193	15	81	11	31	NA	29	53	NA
98	266	112	195	184[+]	281	344	84	152	88	102	106[+]	61	250	113[+]

NA not available [+] *Partial total*

Costs are lowest in Spain, where the total government and private expenditure is £61 per person per year. Ireland comes next (£84), followed by Japan (£88) and the U.K. (£98). The personal expenditure of the average Japanese is almost as low as that of the average Briton. However, in Japan, fewer tax pounds go towards the cost of subsidising medicine.

Today every government in the world – including Red China with its squadrons of semi-trained "barefoot doctors" – realises it has a responsibility to keep its citizens in good physical and mental health. Unlike the U.S., nations like the U.K., Scandinavia, Ireland, Japan and others have opted for a universal health care system in which the State pays everyone's medical bills.

Which system works better – ours or the private insurance health plans? We complain we have to wait months before getting National Health Service treatment for minor ailments. In the U.S. they grumble about the high cost of health bills. It seems that no system is perfect.

Source:
Organization for Economic Co-operation and Development

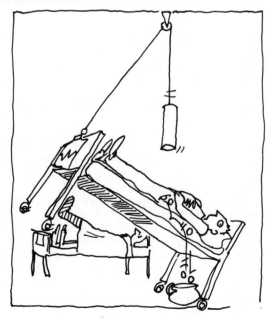

The mind-blowing fact is that one person in three throughout the world has a headache regularly. Overall, the figures are complicated and can't be explained by social customs and habits. For instance, the U.K. has had a National Health Service for 30 years. With free medical treatment, we have grown accustomed to dash off to our doctor every time our throbbing temples need relief. Forty-two per cent of men and women in the U.K. suffer regularly. This despite all those T.V. adverts for patent medicines. We must consume all those pills and time-release capsules that we're encouraged to buy but those over-the-counter remedies don't have enough effect to stop us plaguing our G.P. Only 13% of Americans, on the other hand, visit their doctor because of headaches.

However, in Sweden – where health care is as liberal and efficient as in the U.K. – the headache toll is far lower. The stoic, phlegmatic Swedish temperament must make a difference.

In Spain, throbbing guitars pale beside the throbbing temples of 42% of the Spanish who are afflicted with aching heads. The pain drives them to consume over 7,716 pounds of aspirin every year. Over-the-counter headache remedies may work in America, but in Spain they're obviously less effective.

This tormented nation can't even blame socialised medicine for pampering everybody into hypochondriacal hysterics. There is no national health system. This neglect of basic health needs shows in more pain from minor ailments than in other countries. Fourteen per cent of Spanish women suffer from regular period pains; more Spanish than anyone else endure toothache – 13% of them are in dental pain.

Perhaps the pain threshold varies too. The Spanish and French (also high on the headache table) are voluble and impassioned. Swagger and braggadocio are expected in their men. A man is expected to complain about a pain, seek a remedy, and then – at the top of his lungs and with pride inviolate – beseech God to send relief.

You'd have thought that the peace and serenity of the Austrian and Swiss

	U.S.A.	CANADA
	1977	–
1. Percentage of population (men and women) who suffer regularly from:		
headaches	13^x	NA
toothache	NA	NA
2. Percentage of women in the female population who suffer regularly from:		
period pains	NA	NA

U.K.	AUSTRALIA	AUSTRIA	BELGIUM	DENMARK	FRANCE	(WEST) GERMANY	IRELAND	ITALY	JAPAN[0]	NETHERLANDS	NORWAY	SPAIN	SWEDEN	SWITZERLAND
1977	–	1977	1977	1977	1977	1977	–	1977	1977	1977	1977	1977	1977	1977
42	NA	36	40	34	39	28	NA	39	.47‡	38	31	42	27	36
8	NA	6	8	5	6	5	NA	7	NA	5	6	13	4	6
11	NA	9	10	12	12	7	NA	11	NA	9	9	14	6	9

NA not available [x]*This figure relates to the percentage of the adult population who consult a doctor because of severe headaches* ‡*Refers only to cases of neuralgia*

Source:
Euromonitor
[+]Confidential industry source
[0]*Statistics Year Book,* 1975

mountains would have smoothed the most furrowed brow. Not so. Thirty-six per cent of Austrians and Swiss suffer from headaches. Belgium, France, Italy and the Netherlands are ahead of them but behind the U.K. and Spain.

Upper lips are stiffer in Sweden and West Germany. To feel pain is a sign of weakness, and to talk about it is contemptible. In these two countries, medicine is taken furtively and not many headaches are reported. Period pains aren't common in Swedish and German women. Six and seven per cent respectively admit to suffering with menstrual cramps.

At four per cent, the Swedes also have the lowest occurrence of toothache. The West Germans are only just above them at five per cent.

However, this Northern European fortitude is completely eclipsed by that of the Japanese. Only 0.47% of the population there ever complains of a headache – an astoundingly low figure. Perhaps oriental inscrutability hides suffering.

Some nations suffer more pain than they admit to. Others agonise more loudly about *less* pain. And there may even be countries where the pain of minor ailments goes unacknowledged and unfelt.

The only conclusion is that pain is more a product of expectation and psyche than physique. And mind, once again, triumphs over minor matters.

Canadians are the most accident-prone people on the table. Over seven out of every hundred Canadians die as a result of some kind of accident. Canadians top the table in car accidents, fatal poisonings and industrial mishaps. Almost twice as many Canadian workers die of job-related causes as Americans – and five times as many die on the job as in France.

The French may have relatively safe factories, but they have the second highest overall figure for accidental deaths for every one hundred thousand deaths. For some reason, the French are always falling down. Almost 2.5% of French fatalities result from accidental falls. This would be more understandable if they lived in Switzerland, with all those mountains to negotiate.

West Germany is the next highest country on the list in terms of total accident-related deaths per hundred thousand population. Almost seven per cent of all German deaths are caused by mishaps. There are no outstanding figures here, but they all add up. The only areas where West Germany is exceptionally low are in the categories of poisoning and death by firearms. Only seven out of every hundred thousand Germans die because of a gun accident. Compare that to the American rate of 13 for every ten thousand – far higher than

	U.S.A.	CANADA
	1974	1974
1. Total number of deaths in accidents per 100,000 deaths	5409	7344
2. Number of deaths per 100,000 deaths:		
in motor vehicle accidents	2399	3792
from accidental poisoning	286	417
from accidental falls	845	1098
by accidents caused by firearm missiles	130	73
by accidents mainly of an industrial nature	289	529
in other types of accident[+]	1459	1853
3. Total number of accidents (000s)	104.6	12.2

any other country in the table. With their liberal gun laws, it's not surprising that the U.S. easily heads the list for accidents caused by firearms.

Austria, Switzerland, Australia and the U.S. come next, with around 5.5% of all deaths caused by accidents.

The Austrian figure is largely due to fatal motor accidents and a high rate of accidental falls. The same applies to Alpine Switzerland.

It's a different story in Australia, where the car accident rate – the main cause of accidental death – is second only to Canada's.

The U.S. figure is also largely made up

	U.K.	AUSTRALIA	AUSTRIA	BELGIUM	DENMARK	FRANCE	(WEST) GERMANY	IRELAND	ITALY	JAPAN	NETHERLANDS	NORWAY	SPAIN	SWEDEN	SWITZERLAND
	1974	1974	1975	1974	1975	1974	1974	1974	1974	1975	1975	1974	1974	1975	1975
	2861	5572	5683	5241	3695	6919	6564	4456	4793	4800	4713	5081	4612	4455	5590
	1126	3294	2585	2181	1677	2139	1958	1609	2287	2023	2033	1391	1609	1401	2212
	142	114	86	226	156	120	66	115	76	125	75	253	115	296	97
	957	1033	1979	1646	1197	2494	1626	1397	1652	674	1800	1847	919	1906	1976
	7	53	11	20	2	23	7	31	24	4	5	30	24	10	13
	125	292	337	122	99	99	175	283	193	379	119	218	200	125	291
	505	1116	684	1045	565	2044	671	1020	560	1595	682	1340	1745	717	1002
	19.0	6.4	5.4	6.0	1.8	38.1	47.7	1.5	25.4	33.7	5.3	2.0	13.6	3.9	3.1

of road accidents. Its record in most other departments is surprisingly good. With two exceptions: accidental deaths by firearms, and the "other types of accident" category, which covers everything from falling through the ice while skating, to choking to death on a chicken bone or being strangled by your pet boa-constrictor.

The U.K.'s figures are the lowest on the table. Only 2.8% of U.K. deaths are the result of accidents. The main factor in Britain is the low car accident figure. There are over 200 fewer deaths per hundred thousand each year in the U.K. in this category than in any other country

Source:
Heron House estimates based on World Health Organization figures
+This includes deaths in accidents caused by fires, in drowning and transport (excluding motor vehicle) accidents.

listed. Amazingly, fewer people were killed on U.K. roads in 1976 than in 1930.

Spain is high in the "other types of accident" category; a traditional Spanish sport like bullfighting would probably be included in this category. The risk of death in the bull-ring is much higher than it is on the tennis court.

	U.S.A.	CANADA
	1976	1976
1. Narcotic addicts (000s)	540.0$^+$	10.8$^+$
2. Drug abusers (000s)	92.6^0	NA

We all take drugs: an aspirin or two for that headache, a couple of barbiturates to get us through that sleepless night, or a tranquilliser to ease us through a lonely day's housework. Then there's the aperitif on hand for those pre-dinner pick-me-ups.

For most people these habits never reach serious proportions but it's clear that abuse of drugs – and even addiction – exists. The ones we mentioned earlier are legal drugs. The late 60s – the era of flower power and psychedelia – saw an increase in the use (or abuse) of illegal drugs. Rock Festivals were fragrant with marijuana and hallucinatory drugs such as L.S.D. became widely available. The 60s have gone, but addiction is still with us in the 70s.

Real figures for illegal drug use are hard to obtain. Some countries see drug abuse as more of a health problem than a criminal matter. So many of the statistics won't appear on the crime records.

Also, drug takers – from Naples to New York – are discreet. They prefer to practise their habit in the privacy of their own homes – so who can really estimate how many solid citizens are turning on with something?

This is especially true in the case of marijuana, which has lost a lot of the stigma attached to "drug-taking". Abuse of tranquillisers is also a hidden problem. In countries like the U.S. and the U.K., doctors prescribe literally millions of them each year to help people cope with the stress of modern life. However, it's impossible to know just how many people have become dependent on these drugs.

The hard core of drug offences involves narcotics like heroin and morphine, which are derived from the opium poppy, or cocaine. America tops the list with over half a million known addicts, mainly in the larger cities. Estimates often put this figure much higher, since many addicts have never been arrested. Because most U.S. addicts obtain their supplies on the street, street trafficking is a number one crime problem. Narcotics users also account for a large proportion of violent urban crimes. As heroin addiction increases the user needs larger and larger quantities of the drug. That means he needs more and more money, and often the only way he can get it is by theft, often mugging.

One way the U.S. government tried to tackle the problem was to cut the drug off at its source of supply, by discouraging the cultivation of opium poppies in Turkey and the Far East for example. The U.S. even tried paying farmers in some countries to grow other crops. But the valuable opium crop is vitally important to the economies of some of them. The vast amount of money that can be made in the international drug market means that controlling the production and sale of opium is very difficult.

NARCOTICS ADDICTION

U.K.	AUSTRALIA	AUSTRIA	DENMARK	FRANCE	(WEST) GERMANY	IRELAND	JAPAN	NETHERLANDS	SWEDEN
1977	1976	1976	1976	1976	1976	1976	1976	1976	1976
131.1[x]	3.4[+]	0.50[+]	NA	NA	NA	0.30‡	6.4‡	NA	2.0[+]
327.7[x]	180.0**	NA	5.0[+]	100.0[+]	400.0[+]	0.29‡	9.7‡	10.0[+]	10.0‡

NA not available

In 1977, there were over 131,000 registered heroin addicts in the U.K. And yet, this year, the number of police convictions for illegal heroin use was under 8,000. This is because British addicts get their supplies from clinics. This way of treating the problem makes for a high addiction figure. However, by keeping most addicts off the streets, it minimises street peddling and the violent crimes associated with the need to obtain heroin illegally.

The figures for the use of soft drugs in the U.K. is much higher. A total of 327,669 people get high on illegal substances such as cannabis or amphetamines ("speed") or trip out on the hallucinogen, LSD. Cannabis, the extract of the marijuana weed, is the most popular and campaigns are underway to legalise it in Britain as in many countries.

The use of cannabis has become quite common in many parts of the U.K., especially among the young. Many people, including some in the medical profession, claim that it's not a dangerous drug. Detailed studies have been made of marijuana, but there's no conclusive evidence to prove that it's either harmful or harmless.

The number of people taking LSD or "acid", is much smaller. Most people,

Sources:
[+]Commission on Narcotic Drugs
[0]*Annual Abstract of Congression*
[x]*Man Alive*, BBC, April 4th 1978
‡National statistical offices
**Twenty-ninth International Congress on Alcoholism and Drug Dependence, 1970
These are illustrative figures as the definitions of narcotic addiction and drug abuse vary from country to country.

including users, agree that it's a more serious drug.

When it comes to pot, the Scots seem disinterested. Scottish health authorities consider alcoholism a far greater menace. The French figures for soft-drug use – cannabis and so on – are much lower than in the U.S. and Britain. That just might have something to do with the cheap free-flowing wine there.

In America, marijuana use is probably more common than the figure seems to show. In states like California "smoking dope" is no longer a felony. That's just as well, since the U.S. government finds it virtually impossible to cut off supplies of marijuana, which is grown everywhere from Columbia, Mexico and Oregon to hothouses in Spanish Harlem.

VD

The sexual revolution has brought the singles' bar, gay liberation movements, integrated college dormitories – and a lot of V.D. The U.S. leads the world in gonorrhoea – over 450 cases reported per hundred thousand population. That's nearly one case for every 200 people.

Scandinavia, for so long regarded as the home of sexual freedom, is next worst off. Their figures are in a class of their own compared with the rest of Europe. In Sweden the gonorrhoea rate is 313 per hundred thousand, in Norway 248 per hundred thousand, and in Denmark 212. We can see that the Scandinavian countries lead the field in reported cases of the major sexual diseases in Europe. This may not only be due to their sexual revolution – especially in Sweden – but also because of their well-publicised health facilities. The likelihood of catching one of these diseases may be higher, but at the same time there is not the same stigma attached in their treatment.

In Canada where the sixties also speeded up the sexual revolution there is a very high rate for venereal diseases. Canada ranks ahead of Denmark in the incidence of gonorrhoea.

Japan is the lowest of all, with its reported rate of only four gonorrhoea cases per hundred thousand. The Catholic countries come next lowest – which is no surprise to anybody. Italy and Belgium (with a large Catholic popu-

	U.S.A.⁰	CANADA
	1976	1975
1. Reported cases of gonorrhoea: (000s)	1001	50.7
per 100,000 inhabitants⁺	455	223
2. Reported cases of syphilis: (000s)	76.7	3.9
per 100,000 inhabitants⁺	35	17

lation) have only eight cases per hundred thousand. And in Catholic Ireland they have a mere nine per hundred thousand. These figures are noticeably lower than all the others in the table, with the exception of Japan's.

Doctors used to say that gonorrhoea was "no worse than a bad cold" and were usually able to cure the disease with one shot of penicillin. However, new strains, which are resistant to penicillin, have recently appeared. These penicillin-resistant strains originated in south-east Asia and Africa and have migrated to both the U.S. and the U.K. – though thankfully not to any large extent. Researchers are working to find antibiotics that doctors can use to counter this new threat.

America has the dubious distinction of also leading the world in syphilis. They have about 35 cases annually per hundred thousand people, almost double the fig-

	U.K.×	AUSTRALIA	AUSTRIA	BELGIUM	DENMARK	FRANCE	(WEST) GERMANY	IRELAND*	ITALY	JAPAN	NETHERLANDS	NORWAY	SPAIN	SWEDEN	SWITZERLAND
	1975/6	1976	1974	1976	1976	1972	1974	1976	1975	1976	1976	1973	–	1976	–
	26.7	11.4	9.6	0.8	10.8	16.9	78.9	0.3	4.5	5.1++	7.2	10.1	NA	26.4	NA
	47	78	126	8	212	31	128	9	8	4	52	248	NA	313	NA
	4.0	2.6	1.2	0.3	0.46	4.1‡	5.9	0.06	3.8	3.2++	NA	NA	NA	0.3	NA
	4	18	16	3	9	8	9	2	7	3	NA	NA	NA	4	NA

NA not available
++Based on WHO monthly figures minus one month for which an average figure was added

ure for the next highest country, Australia. Canada is third in syphilis cases with 17 per hundred thousand population.

Sweden – one of the leaders in gonorrhoea – rates extremely low for syphilis with only four cases per hundred thousand inhabitants.

The good news about V.D. is that the rapid rise in cases of venereal disease experienced by a number of countries during the turbulent 1960s has now slowed down. Due to the efforts of governments and private health agencies, more V.D. sufferers are seeking treatment for this condition. They're also encouraged to help trace the partner who gave them the disease and encourage him or her to go in for treatment.

Health authorities in the U.S. and Canada are hopeful that the rise in V.D. will soon be checked and that the figures will go down. However, the situation is

Sources:
World Health Organization
+Heron House estimates
0Center for Disease Control
×On the State of Public Health, HMSO
‡Office of Health Economics
*Dept. of Health

not the same in western Europe. In many European countries, the sexual revolution is just beginning to take hold. With the standard of living going up, people moving around more freely, and a greater acceptance of casual sexual contacts, the V.D. rates are expected to rise in the Common Market countries.

This trend could create problems for women. If they begin to adopt more relaxed moral standards, their V.D. rate will begin to climb. In the U.K., the figure for contracting gonorrhoea is three times greater for men than for women.

	U.S.A.	CANADA
	1970	1970
1. People (estimated % of drinkers) who consume over 6.3 fluid oz absolute alcohol per day	3.8	3.1
2. People over the age of 15 (% of population) who drink over 6.3 fluid oz absolute alcohol per day	2.7	2.5
	1978	1978
3. Age at which alcohol may be bought legally	18-21[+]	18-19[0]

About 1,900 years ago, Paul wrote to Timothy: "Drink no longer water, but use a little wine for thy stomach's sake."

The operative words here are "a little", because a lot of wine can and often does lead to alcoholism.

It's only in comparatively recent times that excessive drinking or a dependence on alcohol has been viewed as anything other than a bad habit or weakness. Today, in many countries, alcoholism is officially classed as a disease, although authorities disagree on what the symptoms of that disease are. Some classify any heavy drinker as an alcoholic, while others say that symptoms such as delirium tremens or cirrhosis of the liver are a surer indication.

The amount of alcohol consumed by an individual is not a very reliable indicator because some people have a low tolerance to drink and can develop signs of alcoholism while ingesting relatively small amounts. Others have a high tolerance and can consume large amounts of booze with relative impunity.

In our table, we have established a standard few could argue with: a consumption of 15 centilitres or 6.3 fluid ounces of absolute alcohol per day. That's 12.6 drinks of 140° proof spirit. Not everyone who consumes this amount will be an alcoholic. But then others with a low tolerance for alcohol will show symptoms of alcoholism while drinking less. Our figure gives an accurate basis for comparison between countries.

France and Italy, two leading wine-producing nations, also lead the world with the number of alcoholics. In France, almost one drinker out of ten consumes alcohol at or above the alcohol level, while in Italy the figure is slightly more than one in 12. Although the U.S. is also a fairly large wine producer, fewer than one out of 25 exceed the limit.

Just because a country produces a lot of alcohol doesn't mean its population will become drunkards. The U.S. and the U.K. – both of which manufacture huge quantities of distilled spirits – are fairly low on the alcoholism tables.

Spain – another wine-producing country, and one that has no legal minimum drinking age – is third highest in the table. The Swiss – with no minimum drinking

ALCOHOLISM

	U.K.	AUSTRALIA	AUSTRIA	BELGIUM	DENMARK	FRANCE	(WEST) GERMANY	IRELAND	ITALY	JAPAN	NETHERLANDS	NORWAY	SPAIN	SWEDEN	SWITZERLAND
	1970	1970	1970	1970	1970	1970	1970	1970	1970	1970	1970	1970	1970	1970	1970
	2.8	4.1	3.9	3.8	2.7	9.5	5.1	2.6	8.2	NA	2.1	1.6	5.9	2.5	5.2
	2.1	3.3	3.7	3.6	2.5	9.0	4.8	1.8	7.4	NA	1.9	1.1	5.3	2.0	4.4
	1978	1978	1978	1978	1978	1978	1978	1978	1978	1978	1978	1978	1978	1978	1978
	18	18	16	18ˣ	18	‡	18	18	16	20	18**	18	‡	18	‡

NA not available   +Varies from state to state   0Varies from province to province
ˣ14, if accompanied by an adult   ‡No legal age   **16 for any alcohol, 18 for spirits

age – are next, followed closely by West Germany. Although West Germany is a country that's renowned for wine, there's a minimum drinking age of 18. On the whole, the country isn't known for its addiction to the bottle.

This table totally destroys the popular image of the drunken Irishman. The Irish, despite their reputation, come lower in the table than anyone except the Norwegians and the Dutch, showing up the important difference between drinking a lot of alcohol and actually being an alcoholic. Studies have shown that the Irish are rather allergic to alcohol and are affected by comparatively small amounts of it. There's also a strong temperance movement in Ireland.

Sweden has a low figure: it would seem that a few cheering glasses of *aquavit* during those long, winter nights don't lead to alcoholism. In any case the Swedish government makes it difficult to buy alcohol, and you can be banned from driving if you're caught behind the wheel with any alcohol in your bloodstream.

Norway – like Ireland – has a strong temperance movement, and whole Norwegian cities often elect to go dry – putting Norway at the bottom of our table.

Australia – with a strict licensing system – is in the middle of the table, despite all those cans of Foster's lager. Overall, the figures indicate that the problem of alcohol really lies in the hands of society. If people can get booze for breakfast, they'll drink it.

Sources:
1 & 2 Addiction Research Foundation, Toronto
3 Government sources

223

We all have our own ideas about why people commit suicide, and there are plenty of different psychological theories to confirm our particular views. Perhaps suicide is the most difficult subject of all to analyse in terms of figures. Each individual case represents a particular individual tragedy, in which innumerable personal factors have played a part.

It may sound like a truism, but people seldom succeed in killing themselves accidentally. For every successful suicide, there are at least 15 attempted suicides. Some which succeed but are never meant to be successful, could be called accidents. A suicide attempt is often a cry for help. Or it may represent a temporary failure of nerve in the face of overwhelming difficulties. For example, a person might expect to be discovered in the attempt and stopped but, tragically, isn't. Or they underestimate the strength of the pills they take, or suffer from the unforeseen but deadly effect of a mixture of drugs. So the motives and intentions of successful suicide must always remain something of a mystery.

There are many myths about national suicide rates, but these aren't really borne out by the statistics. For example contrary to popular belief, Sweden does not have the highest rate on our table. This doubtful distinction is held by its Scandinavian neighbour, Denmark.

The Danish suicide rate is 26 per hundred thousand inhabitants. That's nearly double the average rate for the countries shown (13.5 per hundred thousand).

Austria comes after Denmark with a rate of 24 per hundred thousand. Then there's a comparatively big drop to the third highest countries on the table: Switzerland and neighbouring West Germany (20 per hundred thousand) – looks like the introspective German-speaking people are the ones most prone to suicide, rather than the Scandinavians as a whole. Sweden comes fifth in the table with 19 suicides per hundred thousand people.

It's interesting that the top five countries have affluent, modern societies where wealth is evenly distributed and there's very little poverty. Their high suicide figures indicate that riches and social welfare don't necessarily bring happiness. With the exception of Austria, they have one common revealing factor: they aren't strongly religious countries. However, it's difficult and perhaps dangerous to generalise about superficial resemblances, especially when dealing with such a problematic subject.

The table shows that suicide can hardly be called a popular practice – even Denmark's figures account for only one in every 4,000 people. Nonetheless, this does mean that in the equivalent of every small country town and every urban neighbourhood, there's at least one potential suicide. To say nothing of the other 15 who try and fail. However, suicide is one of the least common causes of death.

There have been societies in which suicide has held a traditional and respected place. The ancient Greeks and Romans both thought it was a perfectly honourable form of death. And in Imperial Japan, suicide was an accepted practice when performed by the traditional method *hara-kiri*: ritual self-disembowelment, more politely known as *seppuku*. Although Japanese society

Suicides per 100,000 inhabitants		Date
U.S.A.	12	1974
CANADA	13	1974
U.K.	8	1974
AUSTRALIA	11	1974
AUSTRIA	24	1975
BELGIUM	15	1974
DENMARK	26	1974
FRANCE	15	1974
(WEST) GERMANY	20	1975
IRELAND	3	1974
ITALY	5	1974
JAPAN	17	1974
NETHERLANDS	10	1975
NORWAY	10	1974
SPAIN	4	1974
SWEDEN	19	1975
SWITZERLAND	20	1975

Source:
World Health Organization

has undergone a great transformation since the old imperial days, their present suicide rate is still high (17 per hundred thousand) – easily the highest outside Western society.

What prevents people from committing suicide? A major factor is, undoubtedly, fear. This probably applies most strongly in religious countries, notably Roman Catholic, where suicide is a mortal sin and brings disgrace to the family of the person who has killed himself. The result is that many suicides are simply hushed up. They're often recorded as accidents to spare the feelings of the bereaved.

This may well be the reason why the lowest figures in our table are all for predominantly Roman Catholic countries. Ireland comes lowest with three suicides per hundred thousand. Perhaps their more easygoing way of life also has something to do with this figure. One of the major contributory causes of suicide is often stress – which is hardly a national characteristic of the Irish. Catholic Spain is next lowest with a figure of four per hundred thousand. The Italians are third lowest with five per hundred thousand.

North America is about average for the countries on the table, with a figure of 13 per hundred thousand in Canada and 12 in the U.S. This blanket figure hides certain facts. For instance, in San Francisco – the city with the highest suicide rate in the U.S. – the figure is a staggering 2,561 per hundred thousand. But in rural sections of the continent – which tend to be heavily religious – it's very much lower than the average. Another surprise is the U.K., with a mere eight per hundred thousand. Maybe our British "stiff upper lip" has something to do with it.

Psychiatry is big business in the United States, where "I'd be lost without my analyst" has become a popular cliché. Many people stay with their analyst for years, even though they're not particularly ill. But in most of the world, psychiatrists are usually referred to only in times of mental illness or nervous breakdown.

So it's no big surprise to find that the U.S. has the highest ratio of psychiatrists to people – nearly one to every nine thousand. Nor is it any great revelation that in America there are more psychiatrists than anywhere else. But the size of the difference in numbers is staggering. With over 23,000 psychiatrists, they have nearly eight times as many as the next highest country, West Germany. The U.K. is low down in this category. In Britain, there are more than 20,000 people for every psychiatrist. Maybe the famed British "reserve" accounts for this figure or possibly the much maligned National Health Service.

The Dutch rank second from the top in the ratio of psychiatrists to people. Holland is a country that's hardly renowned for its deranged citizens. In fact, the familiar image of the solid and dependable Dutch would lead you to expect the very opposite. But this low ratio probably has as much to do with the efficiency of the Dutch health service. The result is that in the Netherlands there

	U.S.A.	CANADA
	1970	1970
1. Psychiatrists:		
number of psychiatrists (000s)	23.2[+]	1.4**
people per psychiatrist[+] (000s)	8.9	15.5
	1975	1974
2. Patients:		
mental hospital beds per 10,000 inhabitants	12.8	21.8
admissions per 10,000 inhabitants	30.4	27.7

are less than 13,000 people to every psychiatrist.

Another surprise is that Norway has the third lowest ratio of people to psychiatrists – just over 13,000. Obviously those long gloomy Scandinavian winter nights might have something to do with it. And the post-war transformation of provincial Norwegian society into a modern industrial state may play a rôle in the need for such services.

Rapid changes in society, with added stresses and uncertainty, apply elsewhere too. Changing values are probably the main factor in the next-ranking country on the list. Japan has just over 14,000 people for every psychiatrist. But here certain national characteristics must also come into play. Many experts link heart disease with stress, and in the table on heart disease, you'll find that Japan's

226

	U.K.	AUSTRALIA	AUSTRIA	BELGIUM	DENMARK	FRANCE	(WEST) GERMANY	IRELAND	ITALY	JAPAN	NETHERLANDS	NORWAY	SPAIN	SWEDEN	SWITZERLAND
	1970		1970			1970	1970	1970		1970	1970	1970	1970	1970	1970
	2.8‡	NA	.32**	NA	NA	2.1**	3.0**	.13	NA	.74	1.0**	.30	1.2	.46	.32‡
	20.1	NA	23.3	NA	NA	24.4	19.8	24.0	NA	14.1	12.8	13.1	29.2	17.7	20.3
	1975	1972	1976	1974	1970	1974	1975	1976	1975	1975	1975	1975	1974	1975	1975
	30.5	20.7ˣ	35.4°	26.9°	21.1	37.4°	18.2	47.9°	28.9°	18.4	19.2	31.2	12.1	40.5	27.6°
	31.9	21.9ˣ	NA	25.1°	63.8	50.8°	31.6	47.2°	24.9°	13.2	16.9	27.6	13.2	99.7	NA

‡ *Includes child psychiatrists*   **Includes neurologists*   *NA not available*

figures are by no means the highest. It's possible that stress affects the Japanese in a more immediately psychological way and that people in most other countries are affected physiologically.

What about those countries where psychiatry began? Austria – the fatherland of Sigmund Freud – has a ratio of one psychiatrist to over 23,000 people. And in Switzerland, birthplace of Carl Jung, there's only one psychiatrist for more than 20,000 people.

Spain heads the happy list of countries who seem to have the least need of psychiatry. Here there's only one psychiatrist to nearly 30,000 people. The French come next, followed by the happy-go-lucky Irish.

Let's look at the more serious cases. One measure – which we've used – is the number of hospital beds per 10,000

Sources:
World Health Organization

+ Heron House estimates
° National statistical offices
ˣ *Hospitals and Nursing Homes*

population which are allocated to mental patients. The U.K. is sixth in this list, which is headed by Ireland.

Ireland is also high in the number of admissions per 10,000 inhabitants, though France and Denmark are higher. But by far the highest in this category are the Swedes. One in every hundred Swedes is admitted to a mental hospital every year – an astonishingly high figure. And Denmark's 64 per 10,000 is also high. Neurosis and introspection appear to flourish in northern climates.

Abortion is a highly controversial subject. Some people regard it as murder, others as a basic human right.

It's defined as the terminating of a pregnancy before the foetus is fully formed. In some countries, Japan for example, abortion is actually one of the main methods of birth control.

	U.S.A.	CANADA
	1976	1976
1. Legal abortions (000s)	114.7[+]	54.0[0]
as % of live births[+]	35	15
2. Estimated illegal abortions (000s)	200[x]	NA

There are a number of recognised medical ways of inducing abortion. And there is also a vast mythology of "fringe" methods – all dangerous. One of the chief recognised medical methods includes surgery: the cervix is dilated and the uterus evacuated by scraping, suction or vacuuming. However, terminating pregnancies over 12 weeks is more complicated and another method is used; a strong saline solution is injected into the womb causing a miscarriage within 48 hours.

Most religions have condemned abortion, but at present the Roman Catholic Church's stand on this matter is the firmest. To them abortion is murder. As a result there has been very strong opposition to the introduction of legalised abortion in Roman Catholic countries.

The first country to introduce legalised abortion was the U.S.S.R. in 1920, early in their social experiment. Since the 1940s, most advanced nations have had pro-abortion pressure groups. With the rise of Women's Lib, the strength of these movements has increased. Feminists demand each woman's "right to choose".

Japan was the next country to legalise abortion. Then it was introduced into some of the Eastern European countries. Scandinavian countries soon followed, so did Switzerland. In these countries, abortion was selective, with certain legal procedures necessary for authorisation.

The U.K. and certain states in the U.S. liberalised their abortion laws in the later 1960s. But in France, where there is strong Roman Catholic influence, abortion wasn't made legal until 1975. And this only after a large group of respected female public figures (including Simone de Beauvoir and Simone Signoret) circulated a statement that they had all had abortions, openly inviting a prosecution which would have made a mockery of the law. Abortion has recently been made legal in Italy.

When abortion is legalised, the number of abortions tends to rise dramatically, but only for the first few years. Even so, the percentage of pregnancies terminated by induced abortion can be extremely high.

Even in countries where abortion has been legalised, some criminal abortions

U.K.	AUSTRALIA	AUSTRIA	BELGIUM	DENMARK	FRANCE	(WEST) GERMANY	IRELAND	ITALY	JAPAN	NETHERLANDS	NORWAY	SPAIN	SWEDEN	SWITZERLAND
1976	1976	1976	1975	1975	1976	1976	1976	1975	1976	1975	1975	1974	1974	1976
128^{0++}	60.0^{0}	NA	NA	26.8^{+}	133.6^{+}	7.8^{+}	1.8^{+0C}	NA	664.1^{+}	16.0^{+}	15.1^{+}	14.1^{C}	32.4^{+}	NA
18	26	NA	NA	37	18	3	3	NA	33	9	27	2	33	NA
$30^{‡***}$	NA	100^{0**}	30^{0}	0.4^{0++}	50^{X}	NA	NA	500^{+}	1300^{X}	NA	NA	NA	0^{+}	50^{X}

$^{++}$England and Wales only; includes 26,900 non-residents ^{00}Abortions obtained in England and Wales   **Pre-1974 figure ‡‡1975 figure   ***1966 figure

do occur. Deaths from criminal abortions are estimated at between 35 and 95 per hundred thousand abortions. Compare this with the rate for legalised abortion – just over one death per hundred thousand pregnancies terminated within the first three months. This figure is slightly misleading – about 40 per thousand die when legal abortions are included.

A high proportion of legal abortions are performed on married women. Most illegal abortions are carried out on young, unmarried girls, women who've had previous abortions, and women who're pregnant when they marry.

The illegal abortion figures on our table show that 1.3 million illegal abortions were carried out in Japan – almost twice the legal figure.

France's legal total is disproportionately high; 1976 was its first year of legalised abortion, and, as mentioned previously, the figures are always high at this stage.

In the U.K., the figure is under 130 thousand. But as a percentage of live births, the rate of abortions is the same as in France.

Sources:
$^{+}$Population Council, New York
^{0}International Planned Parenthood Federation
^{X}National statistical offices and relevant commissions
$^{⊦}$British Pregnancy Advisory Service

Arguments for legal abortion

- Women have a right to control their bodies.
- A foetus is not a human being.
- Rather than have a baby, some women will resort to dangerous "back street" abortions.
- Unintended miscarriages and accidents terminate more pregnancies than abortions do.
- There is less physical risk in an abortion than in a completed pregnancy.

Arguments against legal abortion

- No-one should be able to deny a human being's right to live.
- From conception, a foetus has the genetic information that will make it a unique individual.
- Legalised abortion implies society does not consider it wrong.
- Abortions have to be performed quickly and take priority over other gynaecological cases.

	U.S.A.	CANADA
	1974	1974
Main causes of death (as % of total deaths):		
neoplasms/cancer	18.9	20.4
diseases of the circulatory system	53.2	49.3
diseases of the respiratory system	5.6	6.5
diseases of the digestive system	3.8	3.7
ill-defined conditions	1.6	0.9
accidents, poisonings and violence	8.1	10.1

In all the countries shown on the table, the main causes of death are diseases of the circulatory system. These are diseases which affect the tissues of the heart, arteries and veins. Hardened arteries, for instance, which occur when fatty deposits thicken the artery walls, cause the bloodflow to slow down and put strain on the heart. This can lead to coronary thrombosis, in which one or more coronary arteries become blocked and the patient collapses with acute chest pain – the standard "heart attack". If the occlusion of the heart is more gradual, it may produce angina pectoris (painful attacks which are less serious than thrombosis) or cause the heart to degenerate. Deaths from heart attacks during old age are included in our figures.

Cancer is the next biggest cause of death. The medical term for this is neoplasm, which literally means "new growth". This happens when an expanding mass of useless cell tissue accumulates and causes a tumour –which can be either benign or malignant. The cancerous ones are often, but – thanks to the miracles of modern surgery – not always, fatal. A lot of research is being carried out into the prevention and cure of cancer, which – hopefully – will reduce the number of deaths from it.

Pneumonia, pleurisy, bronchitis, 'flu and asthma are the main killer diseases of the respiratory system.

The most striking fact to emerge from the table is that causes of death vary very little in the countries on our table. People tend to die of the same diseases in the same proportions.

Australia is the highest risk country for diseases of the circulatory system: 54% of its inhabitants die from heart or rheumatic complaints. Sweden comes next. The U.S. is third on the list, and in the U.K., Norway and Ireland, the figures in this category are also high.

France comes bottom of this category: at just under 38%, the French suffer the smallest number of deaths from heart attacks. On the other hand, they have the most deaths from digestive complaints, ulcers, and cirrhosis and other diseases of the liver.

The table shows that the Mediterranean countries – where olive oil, low in cholesterol, is the most popular oil for cooking and salads – have fewer deaths.

	U.K.	AUSTRALIA	AUSTRIA	BELGIUM	DENMARK	FRANCE	(WEST) GERMANY	IRELAND	ITALY	JAPAN	NETHERLANDS	NORWAY	SPAIN	SWEDEN	SWITZERLAND
	1974	1974	1975	1974	1975	1974	1974	1974	1974	1975	1975	1974	1974	1975	1975
	20.8	17.4	20.2	21.1	23.8	21.7	21.4	17.6	20.4	20.3	25.8	19.3	17.3	22.3	24.1
	51.9	54.0	49.8	43.2	50.2	37.9	46.4	50.8	47.7	42.8	44.7	51.1	44.2	53.4	47.4
	13.3	7.6	7.0	6.5	6.9	6.6	5.9	13.1	7.4	7.7	6.4	9.1	11.9	4.7	5.9
	2.5	2.6	5.8	3.5	3.1	6.2	5.6	2.2	6.2	5.7	3.1	2.3	5.5	4.2	4.0
	0.6	0.8	1.6	9.1	2.5	8.0	3.8	1.6	3.2	5.4	4.7	4.8	5.6	0.5	1.2
	3.9	7.7	7.8	6.8	6.5	8.9	6.6	5.1	5.5	8.0	5.9	6.2	5.1	6.9	8.3

In most countries, cancer accounts for about a fifth of all deaths. The cause of the disease is still one of the great medical mysteries. Theory has it that stress may be a contributory factor. This is certainly backed up by the fact that the highest proportion of cancer deaths is in densely populated Holland. And that the Spanish, followed by the Irish, have the lowest figures. Life in Spain and Ireland is relatively easy-going, and in both countries a relatively high proportion of their population lives in rural areas (30.5% and 44.7% respectively). However, Australia is also low in the cancer league – and a high 86% of Australians live in cities.

Spain and Ireland may be low in the cancer figures, but a large number of deaths in both countries are caused by chest complaints. The U.K. is even worse: 13.3% of deaths here are caused by bronchitis, 'flu or some other chest ailment.

Source:
World Health Organization

Medically mysterious deaths occur with greatest frequency in Belgium where 9.1% of all deaths are reported due to "ill-defined conditions". France is close behind with eight per cent deaths undiagnosed. These figures are very high when compared with the 1.6% in the U.S. and the 0.6% in the U.K.

In which countries are we likely to be cut down before completing our expected span? Canada is the leader here with 10.1% of deaths due to accidents, poisonings and violence. The French live dangerously also – 8.9% of their dead succumb in an untimely manner. Switzerland ranks third in violent deaths – just ahead of the U.S. and Japan.

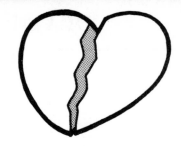

Most of us in the western world lead the good life. But, as with everything, the good life has its price. The main cause of death in affluent urban societies is heart disease. In North America and Europe, far more people die of heart disease and related illnesses than from any other cause.

Heart disease can take many different forms, including inflammatory infections of the membranes, diseases of the valves, enlargement of the heart, degeneration of the tissues and disturbances of the heart's rhythm. All of these forms of disease affect the heart's main function: to keep the blood circulating through the body by pumping blood from the veins through the lungs into the arteries.

The percentage of people who died from heart disease is surprisingly uniform in the countries listed on our table – all of which have a high standard of living compared to the rest of the world. With a range of only 16% the figures show a closer correspondence than almost any other statistics in the book.

Australia has the highest percentage of people who die from heart disease and related diseases – over 54%. At first glance, it doesn't seem to be a typical urban affluent society. But if you look at the table showing the percentages of people who live in rural areas (see table on urban versus rural populations) you'll see that Australians head the list of urban dwellers. More Australians face the stresses of urban life than the rest of us and this is, undoubtedly, a major factor in these high figures.

The U.S. comes second on the table, with 53% of its deaths due to heart disease or related illnesses. Sweden has the same figure as the U.S., and the U.K. comes next, only one point behind these two. Many experts think that these high figures may be due in part to a high cholesterol level in the national diet. In Ireland, one out of every two dies of heart disease – approximately the same as Norway and Denmark.

Who is best off on our table in terms of heart disease? In France, only 38% die

	U.S.A.		CANADA	
	1974		1974	
1. Deaths from heart or related diseases:				
as % of total deaths	53		49	
male as % of total male deaths	51		48	
female as % of total female deaths	56		51	
2. Number of male/female deaths from heart diseases as % of heart disease deaths, by age groups:	M	F	M	F
0 - 24	.4	.3	.3	.3
25 - 44	3	1	4	1
45 - 54	8	4	8	3
55 - 64	18	10	18	8
65 and over	71	85	70	88

HEART DISEASE

	U.K.	AUSTRALIA	AUSTRIA	BELGIUM	DENMARK	FRANCE	(WEST) GERMANY	IRELAND	ITALY	JAPAN	NETHERLANDS	NORWAY	SPAIN	SWEDEN	SWITZERLAND
	1974	1974	1975	1974	1975	1974	1974	1974	1974	1974	1975	1975	1974	1975	1975
	52	54	50	43	50	38	46	51	48	43	45	51	44	53	47
	50	51	45	42	50	35	43	50	43	40	43	50	40	53	44
	54	58	54	45	50	42	49	52	53	45	46	52	48	54	51

	UK M	UK F	AUS M	AUS F	AUT M	AUT F	BEL M	BEL F	DEN M	DEN F	FRA M	FRA F	GER M	GER F	IRE M	IRE F	ITA M	ITA F	JAP M	JAP F	NET M	NET F	NOR M	NOR F	SPA M	SPA F	SWE M	SWE F	SWI M	SWI F
	.2	.1	.2	.2	.2	.1	.3	.2	.1	.1	.4	.3	.2	.1	.3	.1	.3	.2	.7	.5	.2	.2	.1	.1	1.0	.9	.1	.06	.3	.1
	2	.8	3	1	2	.6	2	.9	1	.9	2	.8	2	.8	2	.9	2	1	5	2	2	1	1	.6	3	2	1	.6	2	.9
	8	3	9	4	5	2	6	3	5	2	6	2	5	2	6	3	6	3	7	4	7	2	6	2	6	3	4	2	5	2
	18	8	20	10	11	6	15	7	15	7	12	4	13	6	16	9	14	7	13	9	15	6	16	6	14	8	13	5	13	5
	72	88	68	85	82	91	77	89	79	90	79	93	80	91	76	77	78	89	74	85	76	91	77	91	76	86	82	92	80	92

Source:
World Health Organization

from heart attacks or related diseases. The French have long been celebrated for their hedonistic way of life, with its emphasis on enjoyment – so perhaps this figure isn't so extraordinary. It appears that one way to avoid heart disease is to relax. This is borne out by the fact that Japan, which still has a stable, traditional way of life, comes after France in the list of "least worse off" countries. The Belgians obviously share the Gallic secret of relaxation also.

In all the countries listed, except Denmark, women are slightly more likely to die from heart disease. Australian women top the list. The average difference between men and women is about four per cent – though in Italy it's as high as ten per cent, and in Austria it's nine per cent. Possibly the men in these countries have learned how to take life more easily – and pass the burden on to their womenfolk.

The age group sections of the table verify this possibility. As you'd expect, deaths from heart disease soar in the 65-plus age group for both sexes. But in the 45 to 54 age group, the figures for men are over double the figures for the women in most countries. This trend is magnified in the 55 to 64 age group. These figures obviously reflect the fact that more men are in full-time occupations at these ages and at greater risk from heart disease.

233

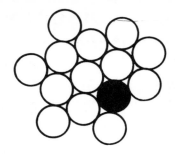

We're all afraid of cancer, but what exactly is it?

Basically, cancer is a malignant growth formed by the patient's own tissues; a collection of cells separate from their neighbouring cells, which multiply at a disorderly and uncontrolled rate. Eventually this change becomes so total that; seen under the microscope, they no longer bear any resemblance to their original structure or to the surrounding tissue. They continue to draw on the body's supply of nourishment but cease to perform their original, useful functions.

This is why cancer is so dangerous. Since the cancer cells are no longer under the control of the host body they continue to reproduce themselves. If their growth is not checked, the cancer cells will invade and destroy the adjacent healthy cells. Also, some of the cancer cells may be transported in the bloodstream to distant parts of the body where the process is repeated.

The extent of. the problem of cancer can't be exaggerated. In the U.S. and many other countries nearly one in five people die from cancer. And even the treatment is at best destructive. (For instance, if diagnosed early, a high proportion of breast cancers can be cured – but the cure often involves a mastectomy.)

There is amazingly little discrepancy in the figures for the different countries on our chart. On average, cancer causes 20% of all deaths. This is a surprising uniformity when you consider the different climates and social conditions in the various countries.

		U.S.A.	CANADA
		1974	1974
1.	**Deaths from cancer (000s)**		
	male	199.2	18.8
	female	166.3	15.3
	as % of total deaths		
	male	18.6	19.5
	female	19.3	21.7
2.	**Male deaths due to various types of cancer[+] (as % of male deaths caused by cancer):**		
	trachea, bronchus, lung	30.9	28.9
	prostate	9.6	9.6
	stomach	4.6	8.1
	oesophagus	2.5	2.3
	intestine (excluding the rectum)	9.4	9.2
3.	**Female deaths due to various types of cancer[+] (as % of female deaths caused by cancer)**		
	breast	19.3	20.4
	stomach	3.6	5.8
	intestine	12.8	14.3
	trachea, bronchus, lungs	10.4	7.7
	uterus	3.4	3.1

	U.K.	AUSTRALIA	AUSTRIA	BELGIUM	DENMARK	FRANCE	(WEST) GERMANY	IRELAND	ITALY	JAPAN	NETHERLANDS	NORWAY	SPAIN	SWEDEN	SWITZERLAND
	1974	1974	1975	1974	1975	1974	1974	1974	1974	1975	1975	1974	1974	1975	1975
	75.0	11.4	9.7	14.0	6.3	68.9	77.5	3.3	61.8	80.4	17.1	4.1	28.3	10.5	7.6
	64.3	8.7	9.6	10.6	5.8	50.9	78.3	2.9	46.4	62.4	12.3	3.5	22.7	9.2	5.9
	22.2	17.8	20.8	22.9	22.7	24.1	21.5	17.2	21.9	21.3	26.9	19.0	18.5	21.7	25.4
	19.5	17.0	19.5	19.2	25.2	19.2	21.3	18.0	18.7	19.2	24.6	19.7	15.9	23.0	22.6
	39.8	28.9	27.4	32.9	27.5	17.6	25.0	26.5	24.5	13.3	35.2	14.2	18.7	16.0	26.5
	6.5	9.2	8.4	8.7	8.7	8.7	8.9	9.4	6.6	1.6	9.0	14.5	8.7	16.8	11.2
	10.5	8.2	16.1	10.6	9.5	8.2	14.8	12.9	15.5	37.8	10.7	12.9	17.5	11.2	10.5
	2.8	2.4	1.9	1.8	1.9	7.1	1.9	3.3	2.6	4.8	1.5	1.5	3.6	2.1	4.2
	7.0	9.2	7.2	6.9	7.9	7.0	7.2	9.1	6.2	3.5	6.5	6.1	4.8	8.2	7.3
	19.6	18.8	13.6	17.2	17.6	15.7	14.4	17.9	16.6	5.2	20.2	15.9	12.1	16.1	21.5
	8.9	7.0	13.9	10.4	6.8	8.6	13.0	11.3	14.5	31.2	9.0	10.6	16.9	8.2	9.9
	11.4	14.3	10.5	12.2	10.7	11.8	10.3	12.9	9.1	4.9	11.0	8.5	8.5	9.1	8.6
	11.7	6.5	5.7	4.6	7.9	3.4	3.9	10.8	5.0	6.5	3.5	3.9	4.6	5.4	3.9
	2.9	2.3	6.5	5.0	3.3	6.3	3.8	2.8	9.0	7.3	3.0	2.8	8.0	3.6	4.3

Sources:
World Health Organization
+Main causes of death by cancer

Another general trend is that males are slightly more prone to this disease than females – on average about 2%. The exceptions here are in the U.S., Canada, Denmark, Ireland, Norway and Sweden, where the women are more prone. The Netherlands suffers most from cancer, the cause of 26% of all deaths there. Switzerland is second, where the rate is 24%. Also high are Denmark and Sweden.

Spain has the lowest prevalence of cancer deaths. Here the rate is only just over 17%, a figure closely followed by Australia and Ireland.

Over a quarter of a million women die from breast cancer throughout the world each year. In western Europe and the U.S., it's the major cause of death amongst women, a malady that strikes all ages. It is the leading form of death in women between the ages of 35 and 54, and is second only to cardiovascular disease in the higher age groups.

However, this pattern doesn't hold true worldwide. The breast cancer death rate is lower in eastern and southern Europe, and drops dramatically in Africa, Latin America and Asia.

Experts disagree over the exact causes of the disease, but statistics show that married women are less likely to have breast cancer than unmarried women; and women who have had children are less prone to the disease than those who haven't. Women who have had children in their teens or early 20s are even less likely to suffer from this form of cancer.

What other factors are there? The stresses of contemporary life, emotional pressures, weight problems, environmental and occupational hazards have all been cited as possibly contributing to the disease.

Several trends emerge from our table. On the whole it seems that in the Western countries, those that are densely populated have the worst figures. Overall, the U.K. (population density: 611 per square mile) is the highest, with 39 deaths per hundred thousand, and the Netherlands

Deaths from breast cancer per 100,000 women[+]	U.S.A.	CANADA
All ages	27.6	25.4
35-44	20.5	21.6
45-54	52.2	56.5
55-64	77.8	83.4
65-74	95.2	99.4
75 and over	132.2	153.8

(1,065 per square mile) and Belgium (851 per square mile) are also high – both have more than 34 deaths per hundred thousand. (On the other hand, why should relatively sparsely populated Denmark be so high?)

At the other end of the scale, Australia and Canada, with their wide open spaces, have relatively "good" figures. But why should Italy (492 per square mile) be lower than either, with only 23 per hundred thousand? It is questions like these that still keep the experts guessing.

Japan (804 per square mile) is obviously a special case. It has far and away the lowest figures in the table – four per hundred thousand deaths overall. We can only speculate as to the reasons for this. Diet, geographical situation, and cultural factors may all play their part. Does eating fish and rice help?

There are other factors. Japan is still a highly formalised society, which has only recently become industrialised. It has also been slow to adopt many Western

U.K.	AUSTRALIA	AUSTRIA	BELGIUM	DENMARK	FRANCE	(WEST) GERMANY	IRELAND	ITALY	JAPAN	NETHERLANDS	NORWAY	SPAIN	SWEDEN	SWITZERLAND
38.9	23.6	30.3	34.2	36.8	27.6	30.2	28.9	22.9	4.2	34.3	28.1	—	30.3	33.9
22.0	15.0	13.9	19.3	19.8	13.2	15.9	24.1	16.8	5.4	22.8	15.5	NA	12.2	16.3
57.8	43.5	40.4	51.8	55.0	37.8	41.3	55.5	39.4	11.1	60.7	38.5	NA	41.3	46.4
83.4	64.7	62.0	68.4	83.5	58.0	61.9	69.5	54.9	12.8	87.1	65.9	NA	66.9	76.9
106.8	85.4	81.5	92.3	112.1	77.4	82.6	94.7	71.7	12.3	113.9	83.5	NA	84.6	115.9
161.8	147.3	122.3	156.5	184.6	133.3	135.1	129.9	101.3	17.7	208.2	122.8	NA	134.4	190.2

NA not available

Source:
World Health Organization
+Data for years 1965-69

social habits. In several spheres women are still second-class citizens relegated to the tranquil backwaters of the social stream.

But take a closer look at those Japanese figures. They are proportionally very much higher in the lower age groups. For instance, Japan's overall figure is about a tenth of the U.K. figure – but the 35 to 44 age group is nearly a quarter of the U.K. figure. Trends like this can be almost as important as the figures themselves. And from this trend it looks as though Japan's days of comparative immunity are sadly coming to a close.

Vast sums are now being allocated to cancer research in all the countries on the table. Indeed, several experts believe that we are on the verge of a major breakthrough in the field. Also, preventative measures are now widely used – screening is available to women in most areas of every country listed in the tables.

Today, breast cancer can be halted if it is treated early on.

EARLY SYMPTOMS

Breast cancer is a disease that all women fear. However, if it is discovered early enough, it can be cured, so it's important for women to check their breasts for early symptoms. These are a small, painless lump or thickness which can be felt with the flat of the hand; swelling or inflammation; the indrawing of one nipple; roughness and thickening of the skin of the breast; change in the shape of one breast; blood-stained discharge from the nipple; eczema of the nipple.

These symptoms do not necessarily indicate cancer, but, in any case, they should be diagnosed and treated.

To investigate a lump or swelling, the doctor may take a small piece to examine under a microscope. Should it prove to be cancerous, he may advise treatment by X-rays, surgery, or possibly both.

The number of babies who arrive stillborn or die within a few days after birth worries doctors around the world. Figures vary from country to country and within each country from year to year.

It's hard for even an expert to make much sense of these statistics. The U.K., for instance, is twelfth on the table in infant mortality: in nearly 23 out of every thousand births, babies are either stillborn or die within six days. Another 17 die in their first year of life. This puts us again near the bottom of the pack.

You'd expect a country like ours, with extensive state-subsidised health care to have a much better record than that – and for the countries that spend less on health to have higher infant mortality rates. This is only partially true. Sweden, Denmark and Norway are more successful than we are in holding down infant mortality. The U.S., too, has a better record than ours. At the other end of the scale, Italy far exceeds us in deaths, but France, not usually renowned for excellence of health care, has fewer (a total of 32.2 per thousand compared with our 40.5).

The period of greatest danger for babies is the first week of life. The majority of infants who don't survive their first year perish within six days of birth. In the four countries with the best record on infant mortality – Sweden, Norway, the Netherlands and Denmark – the average is 14.6 deaths per thousand within six days of birth. In the countries with the worst records (Italy, Ireland, West Germany, Austria, the U.K., Australia and Belgium) the average death rate is almost nine more per thousand – 23.6.

After the first week, death rates are lower, and more uniform from country to country. Sweden is the lowest with 8.3 deaths per thousand and Austria the highest, 20.8.

	U.S.A.	CANADA
	1975	1975
1. Live births per 1,000 inhabitants	14.7	15.7
2. Infants who died within 6 days of birth per 1,000 total births[x]	21.9	17.6
3. Infants who died in their first year per 1,000 live births	16.1	15.0
	1974	1974
4. Principal causes of infant mortality (as % of male/ female infant deaths):	M/F	M/F
anoxic and hypoxic conditions[‡]	26/21	20[++0]
congenital anomalies[**]	15/17	23[++0]
pneumonia[xx]	4/5	5[++0]
other[00]	55/57	52[++0]

INFANT MORTALITY

	U.K.	AUSTRALIA	AUSTRIA	BELGIUM	DENMARK	FRANCE	(WEST) GERMANY	IRELAND	ITALY	JAPAN	NETHERLANDS	NORWAY	SPAIN	SWEDEN	SWITZERLAND
	1975	1975	1975	1975	1975	1975	1975	1975	1975	1975	1975	1975	1975	1975	1975
	14.1	17.2	12.3	12.3	14.3	14.1	9.7	21.6	14.8	17.2	13.0	14.0	18.2	12.6	12.4
	22.8	22.1	23.0	22	14.5	18.6	23.0	23.2	29.2	16.7	15.4	15.4	NA	13.2	14.1
	17.7	16.1	20.8	16.2	10.7	13.6	19.7	18.4	20.7	10.8	10.6	10.5	13.8	8.3	10.7
	1974	1974	1975	1974	1974	1974	1974	1974	1974	1974	1975	1974	1974	1975	1974
	M/F	M/F	M/F	M/F	M/F	M/F	M/F	M/F	M/F	M/F	M/F	M/F	M/F	M/F	M/F
	22/18	12/10	15/14	8/6	26/20	20/19	21/20	16/15	12/10	23/21	23/19	10/8	NA	20/18	26/24
	21/28	20/26	7/7	23/26	24/31	21/24	18/20	20/26	19/21	14/15	24/29	24/34	19[++]	28/35	26/27
	8/7	4/4	4/4	3/3	2/1	2/2	2/2	17/13	8/10	8/9	2/1	4/4	13[++]	3/2	3/5
	49/47	64/60	74/75	66/65	48/48	57/55	59/58	47/46	61/63	55/55	51/51	62/54	10[++]	49/44	45/44

NA not available   [++] Male and female

Sources:
1, 2 & 3 Department of Health and Social Security, London
4 World Health Organization
[X] Demographic Year Book 1975
[O] Canada Year Book 1976-1977
‡ Absence or deficiency of oxygen. 1974 figures
** Includes spina bifida, congenital anomalies of the heart, other congenital anomalies of the circulatory system, cleft palate and lip, and all other congenital anomalies.
[XX] Viral and other pneumonia
[OO] Includes infective and parasitic diseases, enterilise diarrhoeal diseases, whooping cough, meningococcal infections, measles, avitominoses and other nutritional deficiencies, diseases of the nervous system, meningitis, acute respiratory infections, influenza, bronchitis, emphysema and asthma, intestinal obstruction and hernia, birth injury and difficult labour, condition of placenta and cord, haemolytic disease of the newborn, other causes of perinatal mortality, symptoms and other ill-defined conditions, accidents.

Percentage of Population

The French really hit the alcoholic jackpot when it comes to hitting the bottle. Nearly one in ten of France's entire population is medically at risk from alcoholism. (The French not only have the greatest wines – they also have the greatest hangovers.) The Italians aren't much better off. In Italy one out of twelve adults has a drinking problem. It can't be a coincidence that the top five countries on the chart all have large wine industries. Compared with these, the USA, home of Alcoholics Anonymous, fares comparatively well. In Britain, only one in 35 drinkers consume enough to have a problem.

U.S.A.	AUSTRALIA	WEST GERMANY
3.84	4.11	5.07

ALCOHOLISM

9.53

8.21

5.94

5.20

SWITZERLAND   SPAIN   ITALY   FRANCE

Sex

Love and marriage go together like a horse and carriage. Or so the song says. But a lot of things have changed since those lines were written, including the way we run our sex lives. Things really seem to swing in some countries. After reading this chapter, you'll know if you're missing out on the action or leading the sexual revolution.

The first set of tables will give you an idea of just how widely people's views can differ on controversial issues like the double standard, the so-called "permissive society" and the results of changing moral standards. The "Swinging Sixties" have come and gone but the arguments they raised are still raging as fiercely as ever. You'll see how factors like age and nationality make a big difference in the responses to questions like "Are attitudes to sex today too permissive?". And can you guess where 70% of the men think their sex lives are very enjoyable, while under half the ladies are really enjoying their encounters with the opposite sex?

After looking at what people think about sex, there's the equally intriguing question of what they *do* about it. The tables on the frequency of sexual activity among different age groups in four countries may provide some shocks. For instance, there's a country where the 45 to 49-year-olds outdo all the young marrieds and swinging singles.

Pre-marital sex is a particularly explosive topic. What do the British think of the "new morality"? Does it mean a breakdown in our country's moral standards? This section points out just how attitudes vary from one country to another.

So much for opinions. But if you want to know what your teenage daughter really gets up to when she goes out on a date, have a look at the section on pre-marital sexual experience. Has the permissive society become a reality for girls in their mid-teens? The tables even break the figures down according to race, so you can see how different cultural patterns affect young people's sexual behaviour. Then there's an eye-opening comparison of U.S. pre-marital activities with British and Dutch.

After looking at what the kids are doing, the next section provides some clues about *when* they start doing it. The tables here cover sexual activities from the tender age of under 12 right up to the over 30s. Again, the comparisons range from different age groups and sexes to different countries.

There's another angle to the whole question of pre-marital sex – that's whether increased sexual freedom has led to promiscuity. Are teenagers really "sleeping around"? Is such behaviour universal? Or do different countries and cultures produce different sexual standards? These tables deal mainly with young people, but the Swedish figures include 30 to 60-year-olds. Which provides a good yardstick for measuring just how much change there's been in sexual behaviour in the last generation.

In the days when love and marriage actually *did* go together, it was quite likely that the first sexual partner would be the one you walked down the aisle with. Have things changed since then? The tables on first sexual partners suggest they have – at least in the two countries surveyed. Who's replaced spouses? You've guessed it – the steady date.

If you're a young single still playing the dating game, you'll be especially interested in the next section on the average age at marriage. The table is broken down into sections dealing with rural and urban dwellers' marriage trends. You can also check the average age gap between marrying couples in the various different countries. So check the figures here carefully if you have your heart set on a wedding ring – and start making your holiday plans accordingly.

As the section on average age at marriage shows, wedding bells are still pealing for a lot of young couples. The tables here offer some invaluable facts for the marriage-minded.

So much for marriage – what about the other side of the coin? The table on marriage and divorce leads to some inescapable conclusions about the effect of changing moral standards on permanent relationships.

Needless to say, one of the prime causes of broken marriages is infidelity. The next two sections, on people's attitudes towards extra-marital sex and their extra-marital experiences, seem to indicate that a lot of people just aren't practising what they preach. For instance, while nearly nine out of ten British men *thought* that a husband ought to remain faithful, only six out of ten had actually managed to do so.

In any discussion of sex, there's one area that's especially controversial – homosexuality. The tables here give some revealing insights into people's feelings about homosexuality – and where it might be safest to "come out of the closet".

Other than politics, there's one topic that's absolutely guaranteed to start an argument. That's sex. Do you believe in the "double standard"? How do you feel about pre-marital sex? Has the permissive society gone too far? Are sex crimes increasing because of lower moral standards? Whatever your opinions on these questions, someone's sure to feel exactly the opposite.

So have a look at the tables in this section. You'll get a good idea of who thinks the way you do – and who doesn't. You'll also see how your responses to the above questions and others about sex may well depend on whether you're male or female, and what country you come from.

Single women in the U.K. were asked if most people's attitudes to sex are now much too permissive. The lowest number on the table (eight per cent) was for those who strongly agreed. That's only about one in 12 feeling things have gone too far in the sexual revolution. More than twice as many felt exactly the opposite – 17% strongly disagreed. Of course, youth's views tend to be more extreme than those of society at large. The young women questioned were all 18 to 26 years old, an age group certainly in a position to enjoy the advantages of greater sexual freedom. So it's not so surprising that a majority of them, (38%) disagreed, either strongly or

	1970	1970
WEST GERMANY	Men	Women
Men/women+ (%) who agree:		
Sexual freedom leads to demoralization of young people	28	40
Sex has made people freer. People aren't frightened to talk about things that must be discussed	57	51
With sexual freedom going so far, people will not only harm their bodies but lose their spiritual values	33	45
Sexual freedom enables young people to get to know each other better before they marry	41	34
The open attitude to sexual matters must lead to an increase in sex crimes	28	34
Sexual freedom threatens marriage and the family	28	32

Source:
Allensbacher Berichten, 1970, Institut für Demoskopie Allensbach
+Aged 16 and over
Percentages do not add up to 100 as more than one answer was accepted.

slightly, that there's too much permissiveness, while only 27% agreed. Quite a few – over a third – stayed completely neutral on this issue. Why so many? We can only speculate. But maybe they're just too young to remember anything but today's atmosphere of increased sexual freedom. The "Swinging Sixties" seem to have left a deep mark.

It's interesting that the responses in this all-female survey were a complete contrast to the attitudes held by the women of West Germany. What accounts for the difference – is it the generation gap? Or could there be a cultural gap? Maybe the British aren't as

ATTITUDES TO SEX

Women[+] (%) who strongly agree that most people's attitudes to sex are now much too permissive	U.K. 1976
strongly agree	8
slightly agree	19
neither agree nor disagree	34
slightly disagree	21
strongly disagree	17
no answer	1

Source:
Honey magazine, May and June 1977. Survey conducted by the Schlachman Research Organization Ltd. London.
[+]Single women 18-26

straitlaced and reserved as legend has it.

In West Germany, one thing seems clear. Women have a much more traditional, conservative approach to the issue of permissiveness and its consequences. For example, when asked if greater sexual freedom would not only harm people's bodies but also lead to the loss of spiritual values, 45% of the women said yes. That's 12% more women than men feeling that increased sexual freedom is a threat to the deeper values of society.

Herren and *damen* again disagreed about whether young people are demoralised by sexual freedom. Forty per cent of the women thought so, but 12% fewer men were of the same opinion. Forty-one per cent of the men thought that sexual contact between young people before marriage was a good way for them to get to know one another. The women, no doubt feeling maternal concern for their nubile young daughters, weren't so quick to agree. Only 34% of them thought that people need to know each other *that* well.

There was only one issue where over half of both sexes agreed things had changed for the better. That was the increased freedom people now feel to discuss formerly taboo subjects that should be brought into the open. Just over half (51%) of the females agreed that sex has liberated people in this area. But once again the males went further – 57% agreed with this trend. So we can assume that some of the conversation in those *bier kellers* is spicier these days.

What about the question of sexual freedom as a threat to stable marriages and the family? Understandably, more women than men were worried about this possibility. It used to be bad enough worrying about that attractive new secretary. But now temptation for the bored husband abounds on all sides. Nearly a third of the men were also worried about the state of the family in these permissive times.

As if that wasn't enough to worry about, there's the serious social problem of sex crimes. (Take a look at the tables on this subject and you'll see why the West Germans have good reason to worry.) Just under a third of the men (28%), and over a third of the women (34%), were convinced that modern attitudes to sex lead to an increase in sexual offences. This is an issue that's been hotly debated in the media for a long time, but a lot of Germans have apparently made up their minds. Sexual deprivation is perhaps felt more keenly in a society where many people are openly permissive. Resentment and jealousy are common feelings in sex offenders.

The Germans as a group showed a pretty high degree of concern over the

	AUSTRALIA	
Men/women (%) who rate their sex life as:	1974 Men	1974 Women
very enjoyable	70	45
mostly pleasant	18	30
occasionally pleasant	4	14
neither pleasant nor unpleasant	2	3
mostly and/or very unpleasant	0	2
not applicable	6	6

Source:
Cleo magazine, August and September 1974. Survey conducted by Roy Morgan Research Center Ltd

amount of sexual permissiveness in their society. Even though men generally felt less worried than the women, over a third of them were worried about each of the issues raised.

In the U.S. there have been big changes in sexual attitudes in the past few decades. The table opposite shows the responses of American men and women to some very controversial questions on the subject of sex. On the first issue, that of the "double standard", both sexes are in close agreement. Three-quarters of men and women felt that for many years men have been free to live by one moral code, while women were expected to follow another. In other words, the fellows are free to sow their wild oats, but "nice girls don't".

So how do Americans feel about that situation? The responses to the next question showed that 77% of the males thought single women should in fact enjoy the same freedom as single men. That's nearly eight out of ten. Maybe their men aren't so chauvinistic after all. Either that or they think the more sexually liberated single girls around, the better off they are. The surprising thing here is that only 74% of the women agreed with the men that they should have equal freedom. Maybe it's just the traditional female fear of being branded a "loose woman" that makes some of them hesitate to claim the same sexual freedoms. Either that, or Women's Lib has done a better job of winning over men than it has of convincing women that equality in one sphere of life should be equality in all spheres. Many women would rather define their own freedom than simply accept the *same* freedom as men.

On the ever-touchy subject of pre-marital sex, the ladies once again appeared to be more conservative than the men. Over half the women (53%) thought it was immoral to have sexual relations before marriage. But considerably fewer men felt the same way. In fact, 46% of them thought it was fine to indulge before that long walk down the aisle.

Second, a look at what's going on Down Under. Those high-living, fun-loving Aussies were asked to rate their sex lives. The most striking fact about the results is that a whopping 70% of Australian men find their sex lives very enjoyable. That's seven out of every ten Australian males who are content, at least in this rather vital area of life. Another 18% – just under two out of ten – rated their sex lives as mostly pleasant. That makes a grand total of 88% who don't find much to complain about. No wonder Australia attracts so many immigrants! Four per cent found sex

	1974	1974	1974	1974	1974	1974
		Men			Women	
	Agree	Disagree	Not Sure	Agree	Disagree	Not Sure
Reactions of men/women (%) to:						
One moral standard for men and another for women have existed for many years	74	18	8	75	16	10
Single women shouldn't enjoy the same kind of freedom as single men	14	77	9	15	74	11
Pre-marital sex is immoral	44	46	10	53	36	11

U.S.

Source:
The Virginia Slims American Women's Opinion Poll, The Roper Organization Inc, 1974

pleasant only occasionally, and a minority (two per cent) had a distinctly "blah" approach – they said their sex lives were neither pleasant nor unpleasant. Now for the really amazing fact – absolutely *no* Australian men rated their sex lives as unsatisfactory. At this point you may find yourself wondering if their capacity for sexual fulfilment is only equalled by a talent for exaggeration.

So let's turn to the other half of this issue – the women's point of view. It looks as if the men are having a lot more fun than their partners. Under half the ladies (45%) find their sex lives very enjoyable. It seems Australia is faced with a dissatisfaction gap. And compared to that, the missile gap and the generation gap are nothing.

Still, 30% of the women found things "mostly pleasant" in the area of sexual relations. That's a total of 75% of women who are either very or reasonably satisfied with their sex lives.

On the question of whether things were at least occasionally pleasant, more than three times as many women as men made this rather cool assessment. No doubt about it, Australian women are either hard to please or just plain honest. There are even two per cent who dare to admit that things are just plain awful in their sex lives.

It's hard to escape the feeling that maybe – just *maybe* – that Australian tendency to hyperbole has crept into some of the men's claims in this table.

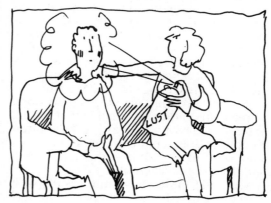

One of the biggest – and most publicised – upheavals in everyday life has been the so-called "sexual revolution." Almost every newspaper and magazine frequently carries articles on some aspect of the "permissive society". Nonetheless, sexual intercourse is still a pretty touchy subject. People's prejudices and misconceptions often get in the way of an objective assessment of what's really going on. Some would have you believing that all standards of decency and morality have been abandoned in a headlong rush towards decadence. Others maintain that things haven't really changed that much since the earlier part of the century.

Certainly, there are still a lot of powerful fears and taboos that prevent people from freely indulging in sexual activity. Despite the Pill, there's still the fear of unwanted pregnancy. Guilt feelings, moral training and social disapproval are other big factors in many people's minds. Then there's the fear of venereal disease. That's something that's had a resurgence as certain strains have developed resistance to modern drugs.

In this section, we've tried to shed some light on the whole subject of sexual intercourse by looking at just how often people in four countries, of both sexes and differing marital status and ages, indulge in sexual activity.

Our first table shows the results of a survey carried out among over 800

AUSTRALIA	1974	1974	1974
	By sex		
	M	W	Under 20
People (%) who, on average during the last six months, have had sexual intercourse:			
never	12	15	31
a few times	11	8	21
once or twice a month	11	13	8
once or twice a week	38	34	15
3 or 4 times a week	19	21	13
over 5 times a week	6	7	8
daily or more often	3	1	0

Source:
Cleo, August and September 1974. Survey conducted by Roy Morgan Research Center Pty Ltd.
Totals do not add up to 100 because of a proportion of "no answers".

Australians, both married and single.

If you look at the overall figures for men and women, you'll see that only one man in every eight hadn't had any sexual experiences in the six months preceding the survey. Slightly more women had abstained. The great majority of both sexes – 38% of men, 34% of women – said they had intercourse once or twice a week. The next most common frequency was three or four times a week – here the women were slightly ahead, with about one in five indulging that often. The lowest figures are in response to the question "did you have intercourse daily or more often?". Still, three per cent of the men boasted they'd had a very active six months.

Next we looked at how these responses break down according to age groups. In the under-20s, a mere 31% hadn't had sexual intercourse at all. That means about seven in every ten of the Australian young people surveyed *had* had some

						AUSTRALIA
1974	1974	1974	1974	1974	1974	1974
			By age			
20-24	25-29	30-34	35-39	40-44	45-49	50 & over
13	4	3	1	5	6	28
8	5	4	5	3	7	15
3	15	13	13	9	21	16
29	42	38	41	52	59	31
27	28	29	34	22	3	6
16	6	7	4	6	0	1
4	0	5	2	3	4	0

sexual encounters. But their sexual encounters tended to be sporadic. One-fifth of them had only had intercourse a few times in the whole six-month period. On the other hand, 15% managed to have sex once or twice a week, and 13% three or four times per week on average. That's far less than the overall average, but still a high figure for this age group.

The highest figure for sexual activity by a particular age group is 59%. That's how many 45 to 49-year-olds have sex once or twice a week. One in 25 of those in the same age bracket claimed to indulge daily or more often. That's the second highest average on the table. Maybe life really does begin at 40!

Things seem to cool off rapidly after the half-century mark. Over a quarter of the over-50s hadn't had any sexual encounters in the six-month period.

If we look at the combined figures for regular weekly sex, both once or twice a week and three or four times, it's the late 30s group that's most active – 75% said they had sex that frequently. Next came the late 20s (70%), then the early 30s (67%). Strangely, the early 20s group, with all those young marrieds and swinging singles, ranks comparatively low on

our tables. Only about one in three had intercourse as often as once or twice a week, and about one in eight hadn't had any sexual encounters at all.

So in Australia, one's sex life seems to get increasingly active as the years go on, at least till that daunting half-century.

Turning to the United States, young unmarried women were asked how often they had intercourse in the past four weeks. The same question was asked in 1971 and in 1976.

The 1971 study revealed that most unmarried teenage girls had intercourse infrequently. As measured by the frequency of intercourse during the month preceding the study, approximately two in five abstained. Teenagers in 1976 appear even more abstemious. About half of the respondents hadn't had sex during the preceding month, and less than three in ten hadn't had intercourse as many as three times in the month.

In the 15 to 17-year-old group, another ten per cent hadn't had sexual relations in 1976 compared to 1971. The same was true of six per cent of the 18 to 19s.

But the trend reverses when we come to the question of intercourse six times in a four-week period. The youngest group gave the same response – (8.6%) in both years. However, the 18 and 19-year-olds who were having relations that frequently in 1971 increased by only six per cent in 1976.

Now let's have a look at Sweden, one of the pioneers in the field of sexual liberation. Two groups, one of 18 to 30-year-olds and another of 30 to 60-year-olds, were asked how often they'd had intercourse in the past month.

In both groups, over 70% of both men and women had intercourse either two to four times or five to ten times. The men claimed greater frequency in both age groups, but the women weren't far behind. Interestingly, over 20% of the 18 to 30-year-old women had relations up to 20 times during the four week period, as compared to a mere 12% of the men.

The highest frequency claim by any age group was the 30 to 60-year-old males. Fifty-three per cent said they had intercourse two to four times that month.

Around two per cent of the younger age group said they had intercourse over 20 times a month – that's almost once a day.

Finally, what about the U.K., where our famed British reserve supposedly keeps emotions in check? The table shows the responses of nearly 400 people, married, unmarried and divorced. The vast majority of married people – 203 out of 257 – had sexual intercourse at least once a week over the preceding year.

	1969	1969
	Married	Separated/divorced
1. People[+] who have never had sexual intercourse	0	0
2. People[+] who, during the past year, have had sexual intercourse:		
0 times	0	1
1-4 times	0	2
5-24 times	13	2
25-50 times	30	1
over 50 times	203	3
don't know	11	1
total	257	10

Source:
Michael Schofield: *The Sexual Behaviour of Young Adults* (Allen Lane, 1973) p. 170. Copyright © Michael Schofield, 1973. Reprinted by permission of Penguin Books Ltd.
[+]Aged 25. The figures are absolute numbers not percentages.

Another 30 said they had intercourse 25 to 50 times.

A third of the divorced people questioned claimed to have intercourse over 50 times a year.

Moving on to the unmarried category, we find that about half of those with a steady boy or girl-friend had intercourse no less than 25 times in the year. About a quarter of them claimed a rate of over once a week. That's a lower percentage of unmarried couples having regular sexual relations than married couples, but it does indicate that it's become a commonplace in Britain for a lot of "steadys" to sleep together regularly.

Singles without a regular partner had sex less frequently. About two out of five had never had intercourse at all. A quarter had intercourse five to 24 times.

U.K.

1969	1969
With a steady boy/girl-friend	No steady boy/girl-friend
7	19
3	6
6	3
8	13
13	5
12	4
3	2
52	52

SWEDEN

	1973	1973	1973	1973
	30-60 years old		18-30 years old	
	M	W	M	W
Men/women (%) who, during the last month, have had sexual intercourse:				
once	12.9	13.4	8.1	5.8
2-4 times	53.4	49.6	36.0	36.0
5-10 times	28.6	30.4	41.6	36.0
11-20 times	4.3	6.1	12.0	20.2
over 20 times	0.8	0.4	2.3	2.0

Source:
Riksförbundet für Sexuell Upplysning, Stockholm

U.S.

	1976	1976	1971	1971
	15-17	18-19	15-17	18-19
Unmarried women (%) who, during the last four weeks have had sexual intercourse:				
0 times	51.0	43.5	41.5	37.7
once or twice	30.1	19.6	34.5	25.7
3-5 times	10.3	13.4	15.4	19.5
6 times or more	8.6	23.5	8.6	17.1

Source:
Reproduced, with permission, from *Family Planning Perspectives,* Vol 9, number 2, from "Sexual and Contraceptive Experience of Young Unmarried Women in the USA, 1976-77" by Melvin Zelnik, Ph.D and John F Kantner, Ph.D.

BUCK CLUB

Our attitudes to sex have undergone a profound change during the last decade. The result has been nothing less than a sexual revolution. Or has it? There's been a lot of talk and publicity on this topic – but what are the facts? What really are our attitudes to sex? Perhaps the central issue here is our attitude to pre-marital sex. That's what this table is about. From the look of the figures there hasn't been as much of a revolution in our attitudes as many would have you believe. Over in the U.S., an overwhelming majority of men and women think that the new morality doesn't make for better, more successful marriages. The women are slightly more conservative on this point than the men, but in both cases a majority of over three to one incline to the opinion that the new morality won't help marriage. There's also an almost two to one majority of both sexes in that country who think that the new morality doesn't help people to make a better choice of marriage partners.

Yet does the new morality enable couples to have more honest relationships? It certainly does in the view of the experts in the field. But they look as if they're out on a limb here. Most Americans don't think that the new morality enables people to have more honest relationships, according to our figures. Though opinions are more divided here – around 50% against and 36% for, amongst both sexes. As with the previous ques-

	1974 Agree	1974 Disagree
	M	M
Men/women+ (%) who think the new morality will:		
make for better, more successful marriages	21	68
enable people to make a better choice of marriage partner	33	55
enable couples to have more honest relationships with each other	39	48
encourage more people to live together without marrying	73	19
encourage more people to stay single	53	35
weaken the institution of marriage	61	31
cause the breakdown of the country's morals	49	41

tions, it's the women who are marginally more conservative in their views. In general, you'll see that this is the case in almost all the answers.

American men and women are in accord on the issue of whether the new morality encourages more people to live together without marrying. Seventy-three percent of both men and women agree that it does, with a mere 19% who disagree. It's interesting to note that this is the question with the most agreement, as well as having the fewest "don't know" answers (a mere eight percent of both sexes). So Americans seem to be deeply convinced on this matter – with greater unity of feeling than when they elect the President or watch the World Series. It's not surprising then to find that the next highest majority is of people who think that the new morality weakens the institution of marriage – just over 62% for both sexes.

On the other hand, Americans are not convinced that the new morality encour-

ATTITUDES: PRE-MARITAL SEX

			U.S.
1974	1974	1974	1974
Don't know	Agree	Disagree	Don't know
M	W	W	W
10	17	72	11
13	28	58	14
13	34⁰	52	14
8	73	19	8
11	54	33	13
8	64	28	9
10	56	33	11

Source:
The Virginia Slims American Women's Opinion Poll, The Roper Organization Inc, 1974
+Aged 18-50

CANADA	
	1977
People (%) who think teenage sex without love is:[+]	
absolutely wrong	46
wrong	19
quite wrong	11
quite acceptable	12
perfectly acceptable	2
no answer	10

Source:
Weekend Magazine, Toronto, December 3rd, 1977
[+]The original question asked: "How do you rate teenage sex without love on a scale of 1 to 7 ranging from "absolutely wrong" (level 1) to "perfectly acceptable" (level 7)?" The other levels were undefined and we have interpreted them so that levels 2-3 = "wrong"; level 4 = "quite wrong"; levels 5-6 = "quite acceptable".

ages people to stay single. Only around 53% of both sexes think this is the case – with more than a third of the people questioned disagreeing.

So what is the general American view on the new morality? Do people think that it'll cause a breakdown of the country's morals? The opinions of men and women are more divided on this topic than on any other – though even here the difference is only seven per cent. In both cases more people thought that the new morality would lead to a breakdown of the country's morals – 56% of the women questioned thought so, and 49% of the men. What does this mean? These answers would seem to reflect widespread American misgivings about present attitudes to pre-marital sex.

It's the same story when you come to Canada. Here people were asked what they thought about teenage sex without love. From the look of these figures the sexual revolution is a long way off.

Almost half the people questioned (46%), were of the opinion that teenage sex without love is absolutely wrong. A further one in five (19%) considered it wrong, and on top of that one in ten (11%) thought it quite wrong. This means that over three-quarters of the people questioned (76%) were of the opinion that teenage sex without love is, to a greater or lesser degree, wrong. A pretty bleak picture as far as permissive attitudes are concerned. Just one person in eight questioned (12%) considered teenage sex without love to be quite acceptable, and a mere two per cent considered it to be perfectly acceptable. These opinions seem to be out of line with

	FRANCE	
	1970	**1970**
	M	**W**
Young men/women[+] (%) who think it is alright for two single people to have sexual relations if: they're engaged	80	74
they've known each other a long time	75	65
they've known each other a short time	48	25

Source:
Rapport sur le Comportement Sexuel des Francais by Dr. Pierre Simon, published by René Julliard, Pierre Charron, 1972. Survey conducted by l'Institut Francais d'Opinion Publique.
[+]Aged 20-29

the facts (as you'll see if you look at some of the other tables in this chapter).

What do people think about pre-marital sex in other countries? What do all those romantic French think about love and sex, for instance? From the look of these figures, they're heavily in favour of sex – at least amongst the 20 to 29 age group. Eighty per cent of the French men questioned thought that it was all right for two people to have sexual relations if they were engaged. Slightly fewer women (74%) agreed. This looks like being the category of least disagreement amongst the sexes in the French figures. Seventy-five per cent of Frenchmen between the ages of 20 and 29 think that it's all right for two single people to have sex if they've known each other for a long time – that's ten per cent more than the women. When it comes to opinions on whether it's all right for two people to have sexual relations if they've known each other for a short time, the opinions of the sexes are even more divided. Forty-eight per cent of the men think it's all right, but only

25% of the women – a difference of almost two to one. It looks as if there may be a few lovers' quarrels on this score.

Now we come to the attitudes to pre-marital sex in the U.K. The figures here are from a survey of young women between the ages of 18 and 26 who were asked why they didn't have pre-marital sex. In many cases they gave more than one answer, but it's interesting to see that the most popular answer (given by 46%) was that they hadn't met the right man. The second most popular answer (37% gave it) was that they thought they should wait until they got married. Twenty years ago this would almost certainly have been top of the list, followed by the next most popular reply – that they didn't because they were afraid of getting pregnant (26%). This answer will obviously decrease with the increasing availability of contraceptives. Almost a quarter of those questioned (24%) said they didn't have pre-marital sex because they were afraid of letting their parents down, and slightly less (23%) said they didn't as a

NETHERLANDS		
	1968	1968
Men/women[+] (%) who agree a girl may have sexual intercourse with a boy if she's in love with him	M	W
agree	50	20
disagree	46	80

Source:
"Sexualiteit in Nederland", published in the women's magazine *Margriet,* 1968.
[+]Aged 21-34

U.K.	
	1976
Young women[+] (%) who do not have pre-marital sex because:	
they're not interested in men	7
they're embarrassed at being thought inexperienced	9
they think pre-marital sex is wrong	23
they're scared of being thought promiscuous	7
they've not met the right man	46
they think they should wait until marriage	37
of religious reasons	9
they feel they're letting their parents down	24
they're frightened of getting pregnant	26
they think it would reduce the chances of a happy marriage	3
they're nervous at the thought of it	16
they're scared of catching VD	7
no answer	9

Source:
Honey magazine, May and June 1977. Survey conducted by the Schlachman Research Organization Ltd, London.
[+]Young women 18-26
Percentages do not add up to 100 because more than one answer was accepted.

matter of principle – because they thought pre-marital sex was wrong.

It's also interesting to note the least popular replies to this question. Only three per cent said they refrained from pre-marital sex because they thought it would reduce the chances of a happy marriage. This figure would certainly have been higher 20 years ago. So would many of the other less popular replies. Only seven per cent replied that they refrained because they weren't interested in men, or because they were scared of being thought promiscuous, or because they were scared of catching V.D.

Now we come to those figures for the Netherlands on how many people agree a girl may have sexual intercourse with a boy if she's in love with him. Here the differences between the sexes are very acute indeed – around 30%. Even so, the men are almost evenly divided (50% for, and 46% against) – but the women are four to one against. However, these Dutch figures are for 1968, and if anything they simply reflect how much our attitudes have changed since then. Though whether they'll go on changing at the same rate over the next decade is anyone's guess.

Gallup International has conducted a ten-nation survey on young people between the ages of 18 to 24 to determine their attitudes towards pre-marital sex.

As the table shows, love and marriage don't go together like a horse and carriage, but love and sex certainly do. With the sole exception of India, well over 50% of the people questioned approve of pre-marital sex.

The disapproval rate varies wildly from country to country. On the "will she or won't she" question, the answer in the U.K. is that 14 out of a hundred girls *won't*, and in America, the figure is just 23 out of a hundred. In Sweden – a nation long renowned for its liberal attitudes – 94 out of 100 *will*. In general, the more traditional societies have the highest disapproval ratings. In India, three out of four people disapprove of pre-marital sex under any conditions. Two predominantly Catholic countries follow: Brazil and the Philippines. Yet in France – another Catholic country – only ten per cent disapprove of sex before marriage. So French priests must hear some rather colourful stories during confession.

Japan is a society that's steeped in tradition, and it seems that chastity is regarded as an important part of traditional values.

The Yugoslavs come out on top in believing that sex is all right if the couple are in love. This appears to be true for all Communist countries – where there are a large number of single parent households.

Sweden is the country that approves of sex without love most wholeheartedly: 38% of Swedes believe that sex for the sake of sex is O.K. That's over twice the U.K.'s rate. Pre-marital sex in Scandinavia is condoned by 94 out of a hundred people, compared with 83 in the U.K. It's the supposedly staid Swiss who are in second place with 91%.

Overall, Europe is way ahead of other continents when it comes to believing that sex without love or marriage is perfectly acceptable.

The table also shows that in virtually all countries, young people have decided points of view on these questions. Only in Germany do over five out of a hundred have no answer on these important personal matters of choice.

Finally the entire chart clearly shows that the sexual revolution has arrived. Had the survey been conducted in 1953 – or even 1963 the answers would probably have been very different.

	U.S.A.	BRAZIL
	1973	1973
Young[+] people's (%) attitudes to pre-marital sex:		
should be avoided under any circumstances	23	40
all right if the parties concerned are in love	57	48
all right even if the parties concerned are not in love	19	12
no answer/no opinion	1	0

U.K.	INDIA	PHILIPPINES	YUGOSLAVIA	FRANCE	(WEST) GERMANY	JAPAN	SWEDEN	SWITZERLAND
1973	1973	1973	1973	1973	1973	1973	1973	1973
14	73	37	18	10	6	27	4	8
68	23	55	75	65	65	68	56	68
15	4	7	7	22	23	4	38	23
3	0	1	0	3	6	1	2	1

Source:
Gallup International affiliated institutes in the countries concerned

The question asked: "What one statement best describes your feelings about pre-marital sexual relations?".

+Male and female, 18-24 years

Sexual activity is no longer a taboo subject, but it's still a touchy one. That's especially true when it comes to the issue of sexual experience before marriage. So how many people have sex before they get married? That's what this table is all about. And there are a number of surprises.

THE NETHERLANDS		
	1968 Men	1968 Women
People (%) who have had pre-marital sexual relationships, aged:		
under 21	25	20
21 to 25	21	17

Source:
Sexualiteit in Nederland, published in the women's magazine, *Margriet,* 1968.

If you're wondering what those teenagers get up to when they go out on a date, here's the answer. This may be the permissive age, but it still comes as something of a shock to find that nearly one in five of all American 15-year-olds have had sexual experience. By the time they reach 16, the figure is one in four.

When the figures are broken down, black girls compared with white girls, they show that at 15, 38.4%, or nearly two in five of the black girls have had sex, compared to under 14% of the white girls. By 16, over half the black girls in the U.S. have had sexual experience, while only around one in five of white girls have.

Not surprisingly, the figures go up as the girls get older. But the biggest leap comes between the age of 16 and 17. Only 25% of 16-year-old girls have had sex, but it goes up to 40% when they're 17. Amongst the black girls, well over two-thirds have had sex by this age.

These are the figures for 1976. By then, the sexual revolution, begun in the 60s, had resulted in big changes in attitudes and actions. Now look at the 1971 figures. Back then, American teenagers had noticeably tamer sex lives. Only 13.8% of all 15-year-olds had had sexual experience, and almost ten per cent fewer

16-year-olds, than five years later. The 17-year-olds are most affected by changing standards: over 14% more 17-year-olds had sexual experience in 1976 than in 1971. That's true both generally and for white girls alone. However, amongst the black girls in that age group, nearly 10% more were having pre-marital relationships in 1976 than in 1971.

The age group least affected by society's increasingly relaxed attitudes to sex was the white 15-year-olds. In 1971 just under 11% had sexual experience, and in 1976, 13.8% had experienced intercourse — a difference of only 2.9%.

On average between the ages of 15 and 19 about twice as many black girls have had sex as white girls in any given age group. By the time all American girls are 19 well over half of them have had sexual experience. Amongst the black girls, only one in seven *hasn't* had sexual experience by this age!

How does this compare with girls in the U.K.? Sixty-one per cent of unmarried women between the ages of 18 and 20 in the U.K. say they have had pre-marital sexual experience. Just 36% claim they haven't. That leaves only three per cent who gave no answer. (Does this mean ''Mind your own business'', or simply

PRE-MARITAL EXPERIENCE

	U.S.					
	1976 All	1976 White	1976 Black	1971 All	1971 White	1971 Black
Unmarried women (%) who had sexual experience, aged:						
15	18.0	13.8	38.4	13.8	10.9	30.5
16	25.4	22.6	52.6	21.2	16.9	46.2
17	40.9	36.1	68.4	26.6	21.8	58.8
18	45.2	43.6	74.1	36.8	32.3	62.7
19	55.2	48.7	83.6	46.8	39.4	76.2

Source:
Reproduced, with permission, from *Family Planning Perspectives*, Vol 9, number 2, from "Sexual and Contraceptive Experience of Young Unmarried Women in the USA, 1976-71" by Melvin Zelnik, Ph.D. and John F. Kantner, Ph.D.

"Don't know"?) So in 1976 over three out of five British women between 18 and 20 had experienced intercourse before marriage. These figures suggest that it's now commonplace for young women in the U.K. and the U.S. to have slept with someone before taking marriage vows.

Overall the U.K. figures seem to be on a par with those for the U.S. – especially if you remember that the figures increase with the older age groups.

Neither set of figures bears much relation to the only Dutch figures we've been able to get, which are for 1968. The sexual revolution obviously hadn't happened in Holland by then. Only 25% of Dutch men under 21 said they'd had pre-marital sexual relationships, and only 20% of the girls. If you look at the American figures, you'll see that they increased by between five and ten per cent between 1971 and 1976 – and the Dutch figures have probably followed suit. That would still mean the Dutch figures for both sexes are much lower than the other two countries. As it is, more American 15-year-olds had sexual experience in 1976 than 21 to 25-year-old Dutch women in 1968.

	U.K.
	1976
Unmarried women[+] (%) who have had pre-marital sexual experience	61
have not had pre-marital sexual experience	36
no answer	3

Source:
Honey magazine, May and June 1977. Survey conducted by the Schlachman Research Organization Ltd, London.
[+]Aged 18-20

	U.S.	
	1976	1971
Average age at first sexual intercourse of unmarried women aged:		
15	14.7	14.7
16	15.5	15.9
17	16.4	16.4
18	16.8	17.2
19	17.1	18.0

Source:
Reproduced with permission from *Family Planning Perspectives,* Vol 9, no 2, from "Sexual and Contraceptive Experience of Young Unmarried Women in the USA, 1976-77" by Melvin Zelnik Ph.D. and John F Kantner Ph.D.

Today's children are constantly exposed to advertising, television programmes, magazine and newspaper stories that all emphasise sex one way or another. Many schools have begun sex education programmes covering everything from conception to venereal disease. So when do young people start making their own discoveries about the birds and the bees?

The first table here deals with unmarried American women from 15 to 19 years of age. Almost one in seven said their first experience of sexual intercourse was at the tender age of 15. If that strikes you as pretty young, remember that teenagers today generally have a lot of freedom, at least over in the U.S. It's often been said that America is a youth-oriented society. Since they have a whole sub-culture of their own, it's not surprising that American adolescents are beginning their sexual activities at an early age. Also, sex is big business. It's used to sell everything from cars to cigarettes. Could heavy exposure to that sort of advertising have the effect of increasing awareness of sex, leading to earlier experimentation?

The highest figure on this table is the one for 19-year-olds in 1971. That year 18% of the girls surveyed of that age had sex. By 1976 the figure had dropped slightly to 17%.

The table on Canada will give you some insight about its young people's experience. Unmarried students of both sexes were asked what age they were when they first had sexual intercourse.

	SWEDEN	
	1967	1967
	M	W
Men/women[+] (%) who first had sexual intercourse at the age of:		
under 16	38.0	25.1
17	19.8	18.6
18	17.7	26.2
19	10.8	12.3
20-30	13.7	17.8

Source:
Riksförbundet för Sexuell Upplysning, Stockholm
[+] Aged 18-30 years

Apparently things there are steamier than the weather reports might have led you to believe. In fact, 2.1% of the males aged 22 to 24 had begun their sex lives at the startlingly low age of

AGE AT FIRST EXPERIENCE

	CANADA			
	1974	1974	1974	1974
	19-21	19-21	22-24	22-24
	M	W	M	W
Unmarried male/female students (%) who first had sexual intercourse at the age of:				
under 12	2.4	–	2.1	1.3
12-13	2.4	1.7	1.1	0.0
14-15	8.4	4.5	6.3	1.3
16-17	38.6	29.4	20.0	15.8
18-19	39.2	48.0	28.4	39.5
20-21	9.0	16.4	26.3	34.2
22-23	–	–	14.7	6.6
24-25	–	–	1.1	1.3

Source:
Sexual Experience, Birth Control Usage and Sources of Sex Education among Unmarried University Students by Dr M Barrett and Dr M Fitz-Earle, 1974

under 12. About half as many girls had been as adventurous. Even more of the younger males questioned (19 to 21) had engaged in sexual relations before the age of 12 - 2.4%. Around the same numbers had experienced intercourse between the ages of 12 and 13, making a total of nearly five per cent of 19 to 21-year-old males and over three per cent of 22 to 24-year-olds who had sexual relations by the beginning of their teens.

Eighteen to 19 was definitely the "big year" for most of the Canadians questioned. Nearly half of the 19 to 21-year-old women had their first experience then. The boys who hadn't made their move by then — also took (or seized) the opportunity at that age. So by the age of 20, 57.9% of both male and female members of the older age group had experienced sexual intercourse.

Sweden is a country with a reputation for very liberal views on sex. How has this affected young people there? Well, when it comes to the age of the first sexual encounter, Swedish girls seem even more precocious than their North American counterparts. One-quarter of them first had intercourse before the age of 16, as compared to 14% of the American girls.

But the Swedish boys look like real Casanovas – a whopping 38%, almost two out of five, claimed their first experience had occured before their 16th birthday.

Obviously there's lots of sexual activity during the teenage years in Sweden. But what about the slow starters? Over 13% of the men questioned hadn't had any amorous adventures until between the ages of 20 and 30. By that time they must have been faced with a lot of competition from those busy 15-year-olds. Considerably more women – nearly 18% – had abstained from sex till over 20 years of age. These over-20s were definitely in the minority.

In fact, 70% of the females and over three-quarters of the men had had intercourse before the age of 19.

So much for North America, and swinging Sweden. Now let's look at the Netherlands. Things are a lot quieter for those Dutch teenagers. It looks as if they spend all their time doing homework. A mere nine per cent of the boys and four per cent of the girls had had intercourse at the age of 15 to 16. Strangely the number

Single women (%) who first had sexual intercourse at the age of:	U.K.		
	1976	1976	1976
	18-20	21-23	24-26
under 16	26	22	3
-16	30	20	13
17-19	42	43	47
20-22	1	15	27
23-26	–	–	10
no answer	1	–	–

Source:
Honey magazine, May and June 1977. Survey conducted by the Schlachman Research Organization Ltd, London.

of young boys having sex actually dropped as their ages went up. So at 18, only four per cent of the youths in Holland took the plunge for the first time. What can their minds be on? Maybe they've forgotten that all work and no play can make Jan a dull boy.

So by the time they'd reached 20, only one in five Dutch youths had had intercourse. The girls were even shyer. Only 18% had had relations with a man. The Netherlands certainly isn't a land of Lolitas.

In fact it seems to be a real outpost of traditional morality when it comes to adolescent sexual activity. By the age of 21, eight out of ten Dutch youths, and slightly more girls, hadn't yet had sexual relations. In contrast, think of the Swedish figures, where 86.3% of the males and 82.2% of the women had experienced intercourse by the time they were 20. Can it be that those Dutch teenagers just don't know what they're missing? Or maybe parents in that part of the world have discovered a way to keep a tight rein on their children's love lives? If they have, it's a secret that could probably be marketed for a vast profit in a lot of other countries.

In the U.K., young unmarried women of differing ages were asked at what age they'd first had intercourse. It seems the times are definitely changing. Girls who entered their early teens in the 70s began sleeping with someone at an earlier age than their older sisters did. In fact, well over half of the 18 to 20-year-olds had experienced intercourse by 16, compared to only 16% of the 24 to 26-year-olds.

Now a look at the country that's raised "*l'amour*" to the status of an art. French youth has a lot to live up to in the area of loving – so how are they doing? They certainly don't believe in starting too young. Only one French lad in a hundred had begun his amorous adventures before the age of 14. Girls of a comparable age hadn't got around to any sexual activity at all – at least none they were admitting to. Things picked up a year later, though. Then, seven per cent and two per cent respectively had sexual encounters. In fact, the boys became increasingly bold right on through their 18th year. That's when the highest percentage (17%) of any male age bracket made their first conquests. By that point, nearly half the French youths questioned had experienced intercourse.

But the *mesdemoiselles* were really playing hard to get. Either that or Gallic parents keep a closer eye on their daughters than on their sons. That wouldn't be too surprising, in such a heavily Roman Catholic country. Whatever the reason, a mere 20% of those

NETHERLANDS		
	1968	1968
	M	W
Young men/women (%) who first had sexual intercourse at the age of:		
15-16	9	4
17	3	5
18	4	6
19-20	4	3

Source:
"Sexualiteit in Nederland", published in the women's magazine, *Margriet,* 1968.

FRANCE		
	1970	1970
	M	W
Men/women[+] (%) who first had sexual intercourse at the age of:		
under 14	1	—
14-15	7	2
16	10	2
17	11	5
18	17	11
19	8	10
20	13	13
21-24	12	33
25-29	7	11
over 30	2	3
no answer	4	3

Source:
Rapport sur le Comportement Sexuel des Francais by Dr Pierre Simon. Published by René Julliard, Pierre Charron, 1972. Survey conducted by l'Institute Francais d'Opinion Publique.
[+]Aged 20-65

jeunes filles had succumbed to the advances of their more sexually active male counterparts by their 19th birthdays. Of course, young men are traditionally more sexually adventurous than young women. There's the fear of pregnancy which keeps many teenage girls from pre-marital sex. Also society tends to wink at the sexual activities of male adolescents as they "sow their wild oats". However, it still frowns on young girls who follow the same course.

It's after they're 19 that the action really starts for the ladies in France. Over a third of them were initiated into the national pastime between the ages of 21 and 24. Better late than never, as the saying goes. This would suggest that many French women save themselves for marriage. Another 11% put off the big moment until they're between 25 and 29, while three per cent are definitely in the "older woman" category, at over 30. Only in their 20th year do an equal number (13%) of young men and women have intercourse for the first time. Before that most of the fellows are busy gaining the experience they need to live up to their reputation as great lovers.

	1976	1976
	15-19 years old	15-17 years old
Young women (%) who have had sexual intercourse with:		
one partner	50.1	54.0
2-3 partners	31.4	31.5
4-5 partners	8.7	8.4
6 partners or more	9.8	6.1

Source:
Reproduced, with permission, from *Family Planning Perspectives,* Vol 9, number 2, from "Sexual and Contraceptive Experience of Young Unmarried Women in the USA, 1976-71" by Melvin Zelnik Ph.D. and John F Kantner Ph.D.

Shock stories of teenage, and even pre-teenage, immorality are the staple diet of sensation-seeking newspapers all over the world. These tables give us a chance to look at the facts. Is it true that young people are becoming more and more ready to go to bed with each other? Has the increased availability of contraception encouraged more youngsters to have sex? And what about the older generation? Are they as pure and moral as they'd like the youngsters to believe? Look at the tables. It's all there for young people in Europe and U.S. – also a revealing glance at the number of partners that Swedes of all ages have had.

The figures dealing with the sex lives of young American women from 15 to 19 appear, at first glance, to show a surprising drop in the number who have had sexual experience – only 50% in 1976 as opposed to 61% in 1971. Yet take another look. In 1976, it's true, there were fewer girls having sex with one partner than there were five years earlier, but it's no cause for celebration among those who are yearning for a return to stricter sexual morality. What the table shows is a great increase in the number of sexual partners the girls are taking. It isn't that over 11% fewer girls are indulging. On the contrary, they are bestowing their favours on a wider range of partners. There was an increase of six per cent in the figures of those who are now having sex with two or three partners. One per cent more were sleeping with four or five partners. Most staggering of all, though, is the four per cent increase among those whose sex lives involved six or more partners. It seems that the rumours and speculation about the snowballing of promiscuous behaviour are true.

A closer look at the figures reveals even more alarming news for parents of young girls. The biggest increases in sexual activity were in the age range 15 to 17. A massive 14% more of these youngsters were having brief sexual encounters with a variety of partners. One-night stands are on the increase. And it is the young ones who are leading the way. Their older sisters aged 18 to 19 showed only a four per cent increase between 1971 and 1976 in the numbers who had two or three partners. There was even a drop of almost one per cent in the numbers of those who had sex with four or five different men. This decrease means that the table shows that almost as many of the younger girls were sleeping with that number of men. The one area where the older girls showed a greater increase in promiscuity was at the top end of the scale. There, the number of girls who, with their six or more different

		U.S.	
1976	1971	1971	1971
18-19 years old	15-19 years old	15-17 years old	18-19 years old
45.3	61.5	66.5	56.1
31.3	25.1	22.7	27.7
9.1	7.8	5.9	9.9
14.3	5.6	4.9	6.3

	U.K.
	1969
People[+](%) who have had sexual experience with:	
one partner	74
one other partner on one occasion	1
one other partner on several occasions	2
several partners, each on one occasion	2
several partners many times	12
none	9

Source:
Michael Schofield: *The Sexual Behaviour of Young Adults* (Allen Lane, 1973), p170. Copyright © Michael Schofield, 1973. Reprinted by permission of Penguin Books Ltd.
[+]Aged 25 years

partners could really be accused of "sleeping around", increased by a stunning eight per cent.

The figures in the next table show the sexual habits of 25-year-olds in the U.K. Only nine per cent of all this group claimed that they had never had a sexual experience. Of all the rest, nearly three-quarters had remained faithful to one partner. This high rate of fidelity may be a reflection on a high standard of sexual morality which we like to think still exists here. On the other hand, two factors have to be taken into account. Firstly, this table was compiled in 1969 when the full effect of the sexual revolution was just beginning to be felt. Secondly, the age band in question was less fully exposed to the changing moral standards than their younger brothers and sisters were. Even taking these points into consideration, the figures still show a tendency towards a freedom in sex that would, in all probability, have shocked earlier generations. Over seven per cent of the group had indulged in sexual intercourse with more than one partner. Of these, well over half had known "several partners". There's no way of telling from this vague phrase what numbers are involved. It could be that the responses referred to a modest two or three. On the other hand, the phrase also covers those whose "conquests" range well into double figures.

"Toujours l'amour" is one traditional aspect of French life that the young people of France seem determined to keep alive and kicking. Despite pressures from that bastion of morality, the Catholic Church, French men and women have, for centuries, been generally regarded by the world as the premier exponents of the arts of love. From Abelard and Heloise right up to Brigitte Bardot, the French have been associated with racing pulses, tender caresses and the language of love.

This fabled aspect of French life is, as these 1977 figures reveal, in safe hands with the next generation. Well over half of the 15-year-old boys who were questioned had already had their first taste of sex, and well over a quarter of 15-year-old girls were no longer virgins. As could be expected, it's the boys whose sexual activity is most marked and a surprisingly high percentage of them (47%) had experienced sex several times. Only 11% of the girls, on the other hand,

	FRANCE	
	1977	1977
	Boys	Girls
1. Boys/girls[+] (%) who have had sexual experience	54	28
2. Boys/girls[+] (%) who have had sexual experience:		
once	33	18
several times	47	11
with a girl/boy their own age	16	26
with an older woman/man	21	10

Source:
Reproduced by permission of L'Express, September 5th-11th, 1977.
[+]Aged 15 years

	SWEDEN			
	1967	1967	1967	1967
	M	W	M	W
	18-30 years old		30-60 years old	
Men/women (%) who have had sexual intercourse with:				
1-5 partners	56.3	87.5	52.3	92.3
6-12 partners	18.2	9.1	23.2	6.1
13-19 partners	7.4	1.0	7.3	0.6
20-50 partners	18.0	2.4	17.1	1.0

Source:
Riksförbundet für Sexuell Applysning, Stockholm

had repeated their sexual experimentation. The girls, too, were the ones who had largely chosen their partners from the same age group. An interesting point to note, though, is that 21% of the boys had been initiated into the mysteries of sex by an older partner. This was true for only ten per cent of the girls. Perhaps the consideration and encouragement that the boys found in the arms of these more mature and experienced women is one of the reasons why nearly half of these 15-year-old boys had been eager to go back for more of the same. Whether this is the case or not, the overall figures certainly seem to indicate that, along with their other traditional loves, wine and food, the youth of France will keep up the country's reputation as a nation of lovers. Under the bridges of Paris, and everywhere else in the country, love is still in the air.

The final table reveals some interesting statistics about Sweden, the nearest rival for the title, "Land of Love". One of the first things that becomes obvious is that the Swedes must be meticulous record-keepers. Some of them are even able to remember as many as 50 partners with whom they have had sexual intercourse.

Over half of the young men and well over three-quarters of the young women had known only one to five people sexually. Presumably, the majority of these involved their married partner and perhaps one or two affairs. The number of young men who had experienced sex with over five different partners is, however, striking. Over 43% of men aged between 18 and 30 had had sexual intercourse with six or more partners. A phenomenal 18% of Casanovas obviously had their eye on a world record and managed to attract from 20 to 50 people into their beds. The Swedish women, however, seem to be much less energetic and ambitious because only just over 12% of them had experienced sex with more than six partners. An interesting sidelight is that for both men and women the figures in the column listing 13 to 19 partners are lower than any of the others. It's probably a case of all or nothing with no dawdling around in the middle.

There's certainly a generation difference. Even though the 30- to 60-year-olds have had more time to play the field the figures show that they've had fewer sexual encounters. This is particularly true of the women in this age range. Ninety-two per cent of them had had sex with five or fewer partners. Less than two per cent of women over 30 had advanced to as many as 13 or more partners.

These Swedish figures are intriguing because while the other tables give details of the sex lives of young people, this one takes into account men and women up to the age of 60. The other tables *do* show that sex among young people is on the increase. They *do* show that some of the old taboos that kept sexual urges in check are nowadays being ignored. Yet the Swedish figures reveal that the older generation were not averse to hopping on and off love's roundabout when it suited them. For example, over 17% of all Swedish males over the age of 30 claimed to have bedded between 20 and 50 partners. This is only marginally fewer than the 18% of younger men who claim

to be in this Olympic category. Any sexual revolution that is taking place today is only the continuance of the attitudes of the past. One stereotype this table shatters is that of the Swedish *femme fatale* who's so sexually liberated she sleeps with scores of men. Those Nordic women, famed for their cool blonde beauty, simply don't distribute their favours indiscriminately. In fact, more American girls of 15 to 19 had slept with over six partners than the 18 to 20-year-old Swedes. Swedish women of all ages seem to be anything but promiscuous, so who are all those male sexual athletes performing with? Forty-three per cent of Swedish men say they've had intercourse with six or more people, but only 12 per cent of the women make the same claim. There must be long waiting lines forming outside the doors of some Swedish ladies.

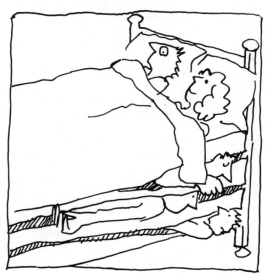

Once upon a time, it was commonly assumed that the first sexual partner should and would be one's husband or wife. But times have changed, especially when it comes to the sexual activity of single women.

Our first table shows the results of a survey conducted among unmarried women in the U.K. Three different age groups were asked who their first sexual partner was. The majority in every group, from the single woman aged 18 to those aged 26, said that their first sexual partner had been a steady boyfriend. In fact, nearly seven out of ten agreed on this. Of course, we haven't got a firm definition here of "steady boyfriend". Is it someone you've been dating for a year, or a month? Nevertheless, that high figure does seem to imply that the majority of single women require a reasonably secure relationship before indulging in the first sexual relationship. It may be the modern day equivalent of "saving yourself for marriage".

It was the women aged between 18 and 20 who were most likely to have first slept with a steady boyfriend. Perhaps this is because women in this age group are in that awkward stage – a bit too young to think seriously about marriage, but old enough to feel ready for the greater personal and sexual freedom of adults. The romantic attachment to a "steady" may seem to provide the security a young girl wants before committing herself to having sexual relations.

The next highest proportion of women

	U.K.		
	1976	1976	1976
	18-20	21-23	24-26
Single women (%) whose first experience of sexual intercourse was with:			
an acquaintance	5	11	7
a friend	15	13	17
a steady boyfriend	70	65	67
their fiancé	9	11	10
no answer	1	–	–

Source:
Honey magazine, May and June 1977. Survey conducted by the Schlachman Research Organization Ltd, London.

in every age group, said "a friend" was their first sexual partner. That term would seem to indicate a more casual, rather than romantic, relationship. But it is a meaningful connection implying some degree of trust and understanding between the partners.

This is borne out by the number who listed their fiancés as first partners. Fiancés rated third on this table, with about one in ten of the women in each age group saying that their first experience occurred with their intended. (We haven't any figures on whether these women went on to marry those particular males, and subsequently divorced them, but assume that they did intend to marry at the time of first sleeping with them.)

Finally, we come to the more casual sexual encounters. The highest figure here was for the women aged 21 to 23. Eleven per cent of them – more than double the number of 18 to 20s – admitted that they first experienced sexual intercourse with an acquaintance. This group may represent young women who had left their parents' home and thus had more opportunity to have casual sexual rela-

FIRST PARTNERS

AUSTRALIA	
	1974
People (%) whose first experience of sexual intercourse was with:	
their spouse after marriage	23
their fiancé	16
a steady date	26
someone they know	23
a prostitute/relative/stranger/ casual acquaintance	12

Source:
Cleo magazine, August and September 1974. Survey conducted by Roy Morgan Research Center Pty Ltd.

tions. Or it may just be that when questioned, women in a slightly older group feel more free to admit to having initially had intercourse with someone who wasn't a regular boyfriend or fiancé.

Our next table deals with the initial sexual encounters of Australians, and records both men's and women's experiences. Once again, many said it was with a steady date. However it is a smaller proportion than the all-female responses in the first table. (About one in four as opposed to seven in ten.) The next most common experience was either with someone they knew, or with a fiancé. It's an interesting fact that the wedding night, as the time of sexual initiation, has been replaced by a less clear-cut sexual encounter with a boy- or girl-friend. Sixteen per cent of the Australians questioned listed their fiancés as the first partner, putting them in fourth place on this table. That's a higher percentage than for any of the three female age-groups in the U.K. So perhaps Australians do set more store by traditional relationships than the British. Either that or they get formally engaged more often.

Least common as a first sexual partner was a prostitute, relative, stranger or a casual acquaintance. Still, 12% of the Australians had been sexually initiated with such a partner – that's about one in eight. It's a higher percentage than in any of the age-groups on the table for British women. One likely reason is that the Australian figures include men. Traditionally, men in all societies have been more free to have casual sexual relations than women, without fear of scandal or social disapproval. In fact, the notorious "double standard" has always implied that men could be experienced lovers on their wedding nights while the blushing bride would have had no previous sexual experience. Undoubtedly, the lingering belief that "nice girls don't" still influences at least some men and women. But the times certainly seem to be changing in this respect. Certainly, single women are now much more free to experiment in sexual activity. As this section shows, a lot of them are doing just that. But even if changing sexual patterns are altering the traditional choice of the first sexual partner, good old-fashioned love and affection seem to be holding their own in 20th century Britain and Australia.

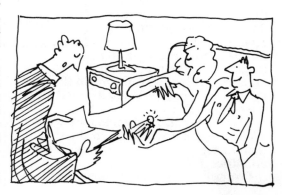

Who said marriage is going out of style? For the great majority of us, love and marriage still go together like a horse and carriage. What's more we're hitching up earlier – especially in the U.S. American men can hardly wait to get to the altar. They're the youngest bridegrooms in the world, marrying at an average age of 23.4 in the cities and even earlier (23.3) in the country. Among women, too, Americans are the youngest newly-weds, averaging 20.8 years of age in the cities and 20.2 in rural areas. It's interesting that in the past 15 years the age gap between British brides and grooms has been narrowing. Perhaps the growing emphasis on equal rights for women has encouraged girls – and boys too – to expect age equality in marriage rather than having a pseudo daddy-daughter relationship. Spanish males hang around in bachelorhood much longer than their peers in other countries. In rural Spain the average groom is nearly 30 (29.4 years) and in the cities 27.2.

The pattern is different among women. Women get married later in the country than in the towns in only three of the 12 countries surveyed.

Overall, American women head the procession to the altar. Canadian, British, West German and Australian brides follow – on average, in their twenty-first year. Young American wives also head the rush for a divorce, and remarry early.

If the figures for first marriages are compared with the number of years men and women expect to live unmarried

	U.S.A.	CANADA
	1960	1961
1. Expected years lived in the unmarried state:		
Men aged 15-65 [0]	11.4	14.0
in urban areas	11.4	13.4
in rural areas	11.4	16.0
Women aged 15-65 [0]	8.7	10.9
in urban areas	9.1	11.4
in rural areas	7.3	9.0
	1960	1961
2. Average age at first marriage:		
Men [0]	23.4	25.1
in urban areas	23.4	24.7
in rural areas	23.3	25.4
	1960	1961
Women [0]	20.7	21.7
in urban areas	20.8	21.8
in rural areas	20.2	21.5

MARRIAGE

	U.K.	AUSTRALIA	AUSTRIA	BELGIUM	DENMARK	FRANCE	(WEST) GERMANY	IRELAND	ITALY	JAPAN	NETHERLANDS	NORWAY	SPAIN	SWEDEN	SWITZERLAND
	1966	–	1971	–	1960	1968	1971	–	–	1970	–	1970	1970	1970	1970
	13.7	NA	13.6	NA	13.7	14.6	10.4	NA	NA	13.0	NA	14.0	15.4	16.2	14.9
	13.6	NA	13.2	NA	13.4	13.8	10.2	NA	NA	13.2	NA	14.0	14.4	15.1	14.3
	14.0	NA	14.0	NA	15.0	17.3	11.4	NA	NA	12.5	NA	15.5	17.8	21.9	15.6
	11.6	NA	11.8	NA	12.3	11.8	9.6	NA	NA	10.8	NA	11.2	14.3	12.4	13.1
	11.8	NA	12.0	NA	12.7	12.0	9.7	NA	NA	11.1	NA	12.2	14.3	12.3	13.7
	11.1	NA	11.5	NA	10.2	11.3	9.2	NA	NA	10.1	NA	10.4	14.4	12.9	12.3
	1966	1975	1971	1975	1960	1968	1971	1971	–	1970	1975	1970	1970	1970	1970
	25.5	24.5+	26.1	26.1+	25.9	26.8	26.0	27.5+	NA	24.5	26.6+	25.2	28.3	27.5	26.1
	25.6	NA	26.0	NA	25.6	25.9	24.5	NA	NA	27.8	NA	25.4	27.2	26.4	26.4
	25.3	NA	26.2	NA	26.2	27.7	25.4	NA	NA	27.1	NA	24.9	29.4	28.5	25.7
	1966	1974	1971	1975	1960	1968	1971	1971	–	1970	1975	1970	1970	1970	1970
	21.8	21.9+	22.2	23.5+	22.5	23.2	21.9	25.0+	NA	25.2	23.9+	22.4	24.0	23.8	22.9
	21.9	NA	22.4	NA	22.7	23.2	21.9	NA	NA	25.4	NA	22.8	23.8	23.8	23.2
	21.6	NA	22.0	NA	21.8	23.0	22.0	NA	NA	24.6	NA	22.1	24.6	24.1	22.5

NA not available

there are some intriguing implications. It's clear that in countries where both sexes tend to marry late, men and women can expect to spend the longest periods single. But they may also expect to spend up to five years longer in the single state than can be accounted for by that teen to 20 waiting time.

Sources:
United Nations
+National statistical offices
0Heron House estimated averages

271

One of the biggest changes in recent years has been attitudes towards divorce. "Till death us do part" once meant what it said. Now in many countries divorce is almost as prevalent as marriage.

In the past divorce cost a British king the throne, members of Parliament their seats and, many think, Adlai Stevenson the U.S. presidency.

Contrast this with the recent relatively passive acceptance of Princess Margaret's divorce from Lord Snowdon.

The institution of marriage is strongly challenged by the increasingly common tendency of modern couples to set up house together, open joint bank accounts in their own names, invest in property together – and generally carry on in old-fashioned married style, but without benefit of clergy. Even with relaxed laws, people seem to fear the legal obligations and entanglements of dissolving their marriages more than they crave the security of legal wedlock.

But for those who persist in marrying, what shape is holy matrimony in these days? In the U.S., patterns are changing. The table shows that while marriages are flourishing (10.5 people in every thousand got married in 1974), divorce is big business also – with 4.6 people per thousand untying the knot. Nearly half as many people got divorced as got married. There's more marrying and unmarrying – marital activity, if you will – in the U.S. than in any other country.

Many American marriages are repeat trips, with brides and grooms who have already been down the aisle – another example of the popularity of recycling in the U.S.

Divorce laws in the U.S. may also contribute to the high number of divorces. As many alimony-poor ex-husbands paying maintenance can testify, women very often profit handsomely from divorce – though not as often as just a few years ago.

In some states like California, joint property laws require that, in divorce cases, whatever capital or possessions that have accumulated during the years of marriage must be split equally. But in many instances, the ex-wife – particularly when there are children, or when she is considered the wronged party – walks out of the courtroom with all the assets from the past... and a guarantee of a big percentage of any profits in the future.

Canada has 15% fewer marriages than there are next door in the land of the free and the brave. Nearly nine Canadians per thousand get married. Divorces are a mere two per thousand – that's less than half the U.S. break-up rate.

In Sweden there's one divorce for every 1.7 marriages – leaving 0.7 contentedly married Swedes. Hopefully, contentment is more common in

	U.S.A.	CANADA
	1974	1974
1. Marriages per 1,000 inhabitants	10.5	8.8
2. Divorces per 1,000 inhabitants	4.6	2.0
3. Ratio of marriages to divorces	2.3:1	4.4:1

U.K.	AUSTRALIA	AUSTRIA	BELGIUM	DENMARK	FRANCE	(WEST) GERMANY	IRELAND	ITALY	JAPAN	NETHERLANDS	NORWAY	SPAIN	SWEDEN	SWITZERLAND
1974	1974	1974	1974	1974	1974	1974	1974	1974	1974	1974	1974	1974	1974	1974
7.5	8.3	6.5	7.5	6.6	7.6	6.2	7.3	7.3	9.2	8.1	6.9	7.7	5.5	6.0
2.14	1.2+	1.4	1.06	2.6	0.9	1.5+	0	0.3	1.04	1.4	1.3	0	3.3	1.6
3.5:1	6.9:1	4.6:1	7.1:1	2.5:1	8.4:1	4.1:1	0	24.3:1	8.8:1	5.8:1	5.3:1	0	1.7:1	3.7:1

+1973 figures   0Not applicable as divorce isn't recognized

common-law relationships.

For better or for worse, the marriage rate in Sweden is increasing. It's risen in three years from 4.9 to 5.5 marriages per thousand people. Swinging Sweden along with Denmark, porn centre of the western world, are two of the few countries to show an increased marriage rate.

Italy and Ireland both have many marriages per divorces. In Italy, the Roman Catholic Church, which seldom permits or even recognises divorce, is a big factor in the figures.

Most of us are in hot pursuit of some elusive Holy Grail of happiness – a state of bliss that Hollywood and American romanticism has led us to *expect* out of life. *Getting* a mate, getting *rid* of a mate, not having a mate – we try it every which way, trying to be happy, happy...happy.

Sources:
1 & 2 United Nations
3 Heron House estimates based on United Nations figures

How far do people practise what they preach when it comes to sexual fidelity? That's what this first table allows us to check because we can compare it with a table in the previous section.

A high proportion (89%) of men in the U.K. thought that, once married, a man should remain faithful to his partner. According to the earlier statistics, though, only 62% of them actually claimed that they'd done this. True, the seven per cent who admitted to extra-marital affairs is matched here by the seven per cent who don't think that fidelity is essential. What about the 31%, however, who gave no answer in the previous table? Guilty consciences? U.K. women, also, show a difference between their principles and their actions. Here 94% of them state that a wife should be faithful yet only 82% of them swore blind in the last section that they've remained true to this belief.

There must have been some guilty Dutch men in 1968, because while 11% admitted to having sex outside marriage, only three per cent in this table are sure that there's nothing wrong with this. The two per cent of Dutch women who said that they'd had sex with someone other than their hus-

	U.K.					
	Yes		No		Don't know	
	1969		1969		1969	
Men/women+ (%) who think:	M	W	M	W	M	W
A husband should be faithful for the rest of his married life	89	90	7	7	4	3
A wife should be faithful for the rest of her married life	93	94	5	4	2	2

Source:
Sex and Marriage in England Today
by Geoffrey Gorer, published in 1971
by Thomas Nelson & Sons Ltd, London.
+Aged 16-45

	SWITZERLAND	
	1977	1977
Men/women+ (%) who agree totally that fidelity has nothing to do with the true nature of marriage	Men	Women
agree totally	54.1	57.2
more or less agree	19.0	12.6
disagree strongly	16.0	15.6
disagree more or less	7.5	9.3
don't know	3.2	5.0

Source:
Survey conducted by Isopublic, Zürich, on behalf of Weltwoche.
+Aged 18-24

ATTITUDES: EXTRA-MARITAL SEX

People (%) who think a married person having sexual relations with someone other than the marriage partner is:	US[+] 1977 Men	CANADA[0] 1977 Women	NETHERLANDS[x] 1968 Men	1968 Women
always wrong	72	60	52	65
almost always wrong	13	13	44	32
sometimes wrong	10	14	NA	NA
not wrong at all	3	4	3	2
don't know	1	8	NA	NA

NA not available

Sources:
[+]The National Opinion Research Center of the University of Chicago
[0]The Canadian Gallup Poll Ltd
[x]*"Sexualiteit in Nederland"*, published in the women's magazine *Margriet*, 1968.
The original question asked in the Netherlands was: "Do you find extra-marital sexual relationships not permissible, sometimes understandable, sometimes objectionable, or do you have no objections in principle?" We have interpreted the answer "not permissible" as "always wrong", "sometimes understandable" as "almost always wrong", "sometimes objectionable" as "sometimes wrong" and "no objections in principle" as not wrong at all".

bands obviously rested easy in their beds, extra-marital or otherwise, because that figure corresponds with the two per cent on this table who thought that there was nothing wrong with such behaviour.

It looks as if Europeans are less rigid in their marital morality than the North Americans. What Mom told them still held good in 1977 for 95% of men questioned in the U.S. and 87% of Canadian women. They believe that, to a greater or lesser degree, it's wrong to have extra-marital sex. Perhaps their rigid views on this subject account for much of their high divorce rate. In their society there isn't much acceptance of *extra*-marital sex if the marital state is to remain. Compare these figures with the 73% of Swiss men and over 69% of women who feel that fidelity is not an essential prerequisite to the married state. There must be something in all that mountain air because only 16% of Swiss men and 15.6% of women are absolutely sure that a true marriage requires total sexual faithfulness. Even way back in 1968 European attitudes were less hardened than American views in 1977. Admittedly, only a small percentage of Dutch men and women thought that there was nothing wrong at all in a married person having affairs, but far fewer agreed with the North Americans that it was always wrong. The Dutch obviously thought that there were sometimes extenuating circumstances.

Is marriage sacred any more? Do people consider it necessary to be faithful to their marriage partner? Are they actually faithful to their marriage partner? These are among the oldest questions in the world, and we're still asking them. That's what this table is all about.

What percentage of people in the U.K. have made love to someone other than their marriage partner since marrying? Sixty-two per cent of males claim they haven't – and that's their story, and they're sticking to it. Seven per cent of British males admit to infidelity.

British wives are even less promiscuous – 82% of them claim to have been completely faithful since marriage, while only three per cent admit that since the wedding bells they made love to someone other than their husband.

It makes you wonder about the 31% of British men and 15% of women who gave no answer. Are they keeping quiet about their indiscretions or worried about being behind the times?

This trend towards greater male promiscuity is echoed in the Dutch figures. Though on the whole the Dutch are (or claim to be) a great deal less promiscuous than their British opposite numbers.

Seventy-eight per cent of married Dutchmen claim never to have had extra-marital sexual relations. But only 11% admit that they *have* – that's four per cent more than the British, who on the

	U.K.	
	1969	1969
	Men	Women
People (%) who've made love to someone other than their husband/wife since marriage:	7	3
people who haven't	62	82

Source:
Sex and Marriage in England Today by Geoffrey Gorer, published in 1971 by Thomas Nelson & Sons Ltd, London. Percentages do not add up to 100 because there is a proportion of "no answers".

whole are admittedly more promiscuous in these matters. This would seem to indicate that perhaps Dutch husbands are a little more honest. But only one per cent of Dutch married men admit to having regular extra-marital sexual relations.

With whom do those married Dutchmen have their extra-marital relations? From the look of the figures, certainly not with friends' wives. Eighty-six per cent of all Dutch married women claim to have been completely faithful to their husbands. Only two per cent admit to having had extra-marital sexual relations. None of them said they had extra-marital sexual relations regularly – an amazing figure.

Now we come to the figures on extra-marital experience among Australian women. Compared with the women of Holland and the U.K. those Aussie girls seem to have few qualms

EXTRA-MARITAL EXPERIENCE

	AUSTRALIA
	1973
Women (%) with extra-marital experience, aged:	
under 26	25
26-30	37
31-40	45
41-50	46
over 50	17

Source:
Reproduced from *The Sex Survey of Australian Women* by Prof. Robert E. Bell, published in 1974 by Sun Books, Australia.

	THE NETHERLANDS	
	1968	**1968**
	Men	Women
People (%) who have had extra-marital sexual relations:		
never	78	86
sometimes	10	2
often	1	0

Source:
"Sexualiteit in Nederland", published in the women's magazine *Margriet,* 1968.
Percentages do not add up to 100 because of a proportion of "no answers".

about extra-marital sex. No 80% fidelity here – except amongst the over-50s.

The most surprising thing about these Australian figures is how they go up as the age of the women increases. Twenty-five per cent of Australian married women under 26 claim to have had extra-marital experience. In the group aged 26 to 30 this rises to 37%. After that things really start to happen. According to these figures nearly half (46%) the Australian married women between the ages of 30 and 50 have extra-marital sex. What does this mean? Either those middle-aged Aussie women have a very active sex life, or they have very active imaginations. And what do their husbands think about all this going on? The figures for married Australian males just aren't available.

A few decades ago, homosexuality was spoken of (if at all) in whispers. Even hinting that someone was sexually interested in members of his or her own sex has long been considered fighting talk. Today homosexuality is still one of the most taboo aspects of sexual behaviour. But the "sexual revolution" of the recent past has brought some changes in that area. Homosexuals are still frequently the subject of hostility and derision. However the more liberal atmosphere of the 60s and 70s has brought more frank discussion of the whole subject. Many homosexuals have joined militant organisations like "Gay Lib" and started their own newspapers and magazines. They put forward the homosexual point of view and lobby to change discriminatory laws. Many homosexuals, especially in large urban areas, are demanding the right to openly admit their sexual preference without fear of job discrimination or reprisals. In some U.S. cities, candidates in local elections have run as "gay" candidates. In others, homosexuals have demanded that candidates for office declare their views on the rights of homosexuals in society.

In some places there's been a backlash from "straight" (heterosexual) members of the community. They're demanding a return to legal penalties for practising homosexuality and they're demanding that homosexuals be discriminated against in certain ways. For example, by banning them from teaching children. In fact, some states in the U.S. still have legislation on the books prohibiting homosexuality. In the U.K., homosexual behaviour between two consenting males

	CANADA
	1977
People (%) who think homosexual behavior in adults is caused by:	
childhood upbringing and other environmental factors	33
mental illness	14
a hereditary/natural trait	10
sexual preference	5

Source:
Weekend Magazine, Toronto, December 3rd 1977
Percentages do not add up to 100 because we have excluded other optional answers which were given in the original survey.

over the age of 21 wasn't legal until after the famous Wolfenden Report in the 60s. There were no laws prohibiting lesbianism because Queen Victoria refused to believe women were capable of such behaviour.

Oscar Wilde was one famous victim of legal prosecution for his sexual preferences. "The Ballad of Reading Gaol" was the poetic result. Today, in the U.S., some police departments use decoys -- officers pose as soliciting homosexuals and then arrest anyone who responds to their advances. During World War II, the Nazis persecuted homosexuals by sending them to concentration camps along with other "undesirable" minorities.

Not all societies have frowned on homosexuality. The ancient Greeks considered it the highest form of love. Women weren't considered men's equals, and therefore only the love of one man for another was thought worthy of being celebrated in poetry and song.

In some American Indian tribes a boy who preferred feminine behaviour was simply treated as a female. Eventually he could dress and act as a woman without any stigma being attached to his

	U.S.[+]	U.K.[0]
	1977	1977
People (%) who think homosexuality is:		
caused by upbringing/ environment	56	28
inborn	12	31
both	14	21
neither	3	3

Sources:
[+]The Gallup Organization Inc, USA
[0]Social Surveys (Gallup Poll) Ltd, UK
Percentages do not add up to 100 because of a proportion of "don't knows".

	NETHERLANDS	
	1968	1968
	M	F
People (%) who think homosexuality is:		
an innate affliction	28	24
a sickness	23	22
unnatural	21	25
dirty	5	6
unacceptable sexual behaviour	15	10

Source:
"Sexualiteit in Nederland" published in the women's magazine *Margriet*, 1968.

behaviour.

Despite all the publicity it receives today, homosexuality is still a murky topic for many people. Is the preference for members of your own sex a disease? Is it a crime? Is it the result of upbringing and environment? Or is it just a preferred way of behaviour that's nobody else's business?

As you'll see from a look at these tables, feelings on the subject vary widely from one country to another. The majority of Canadians (33%) think that family environment can cause homosexual behaviour. That reflects the increasingly popular view that childhood and upbringing have a lot to do with how people behave later in life. About one in seven Canadians think homosexuality is such a deviation from normal, heterosexual standards that it's actually a form of mental illness. Both those views are shared by some psychiatrists and behavioural psychologists, who have tried to treat homosexuality as an illness, with varying degrees of success. One in ten Canadians thinks homosexual behaviour is "in the blood", an inherited characteristic, like blue eyes. Pre-

sumably they don't think there's much that can be done to change such behaviour.

Turn to the tables on the U.K. and on the U.S. and you'll see that there's a big difference of opinion about the issue. In the U.S. well over half those questioned think upbringing and environment are the determining factors in shaping sexual preferences. Maybe that's not so surprising in the land where "going to see a shrink" is a fairly common practice. Modern psychology has given them the idea that children exposed to certain family patterns — for example, a dominant mother and ineffectual father — become homosexuals. Only 12% think homosexuality is an inborn trait. Their British counterparts disagree. Only 28% think early family surroundings are significant, while the majority feel homosexuality is an inborn trait. In fact, over 2½ times as many Britons as Americans think being homosexual is "just the way you are" from birth. Perhaps we British would be less enthusiastic about having so many single sex schools if we shared the American point of view.

	U.S.[+]	CANADA[0]
	1977	1977
Men/women (%) who think that sexual relations between two adults of the same sex are:		
always wrong	68	60
almost always wrong	5	7
wrong only sometimes	7	9
not wrong at all	14	13
don't know	5	12

Sources:
[+]The National Opinion Research Center of the University of Chicago
[0]The Canadian Gallup Poll Ltd
[x]*"Sexualiteit in Nederland"*, published in the women's magazine *Margriet*, 1968.
The original question asked in the Netherlands was: "Do you think one should let the homosexual live his life as he wants to?" We have interpreted the answer "yes" as "not wrong at all" and "no" as "always wrong".

So much for opinions on the causes of homosexuality. How do people feel about actual homosexual behaviour? Those ancient Greeks may have thought it was the highest form of love, but they'd run into some disagreement in present-day Holland. A quarter of Dutch women feel homosexuality is contrary to nature, and over one in five of the men agreed. An even greater number of Dutch men (28%) think it's an affliction the homosexual is born with. Almost the same number of men and women consider sexual activity with your own sex a sickness.

All these views are generally negative, but not as stern as the one held by ten per cent of women in the Netherlands. They find homosexual behaviour completely unacceptable in society. Around one in 20 people in the Netherlands thinks such actions are "dirty".

So it's surprising to see that only about half as many people in the Netherlands as in North America think sexual relations between two adults of the same sex is always wrong. Sixty-eight per cent of those polled in the U.S. felt that way, while 60% of their northern neighbours agreed. If you include the number who think homosexuality is almost always wrong, 73% of Americans and 67% of Canadians are strongly disapproving. Not a fact to gladden the hearts of "Gay Lib" members. It looks as if the so-called permissive society doesn't extend its liberal views as far in the field of homosexual behaviour as in other areas.

Only around one in seven North Americans thinks homosexual behaviour is absolutely O.K. But in the Netherlands, there's much more tolerance.

Homosexuals complain that once out of the closet (that is, after they've admitted their sexual preference) they're the victims of discrimination.

The table on the question of employing homosexuals as elementary school teachers certainly gives weight to their argument. In the U.S., an overwhelming majority of 65% think known homosexuals should be barred from holding this type of job. That's about the same number as those who think sexual relations between members of the same sex are always wrong. Since the majority of people in the U.S. think that homosexual

	NETHERLANDS[x]		People (%) who agree, homosexuals shouldn't be hired as elementary school teachers	U.S.[+]	U.K.[+]	CANADA[o]
	1968	1968		1977	1977	1977
	M	W				
	32	30	agree	65	68	35
	NA	NA	disagree	27	22	14
	NA	NA	don't know	8	10	8
	58	55				
	10	15				
	NA not available					

Source:
[+]The Gallup Organization Inc
[o]Weekend Magazine, Toronto, December 3rd 1977
The original question asked in Canada was: "Should self-confessed homosexuals be allowed to hold jobs such as school teaching that involve dealing with children?". People questioned were shown a scale of seven steps ranging from "definitely not" (which in our chart corresponds to "people who agree") and "completely acceptable" ("people who disagree"). Only these two answers have been retained from the Canadian survey.

behaviour is a result of the early environment, it's not surprising.

People in the U.K. feel even more strongly on the subject. Almost seven out of ten of us are opposed to letting self-confessed homosexuals work with young children. Only around a quarter of those questioned in both countries think sexual preference in a person's private life shouldn't be a factor in employing him or her as a teacher.

So in the U.K. and the U.S., an overwhelming majority of people are in favour of discriminating against homosexuals.

Canadians take a more moderate view of the issue. Only 35% take a definite stand against employing homosexuals. That's about half as many as in the U.K. On the other hand, significantly fewer (14%) think that this form of discrimination shouldn't exist. Presumably, many Canadians think other factors have to be taken into account before giving judgment.

There are some interesting conclusions to be drawn from this section. One is that in most countries most people still disapprove of homosexuals. Even when the "straight" majority doesn't think homosexual activity is immoral, it looks upon it as a form of deviation or illness. Although the Dutch are much more tolerant than North Americans, even in the Netherlands a third of the population thinks homosexuality is always wrong. When it comes to job opportunities, the message to homosexuals seems to be "keep quiet about your sex life". That's exactly the opposite message to the one put forward by militant homosexual organisations. They're busy urging homosexuals everywhere to "come out of the closet" and demand equal treatment with the rest of society.

It looks like this particular battle of the sexes — or should we say sex — will be fought a while longer.

Ratio of marriages to divorces

25 —

If you're a newly-wed, head for sunny Italy. The bonds of matrimony are more firmly knotted there than anywhere else. There's only one divorce for every 26 marriages. But of course, Italy's the home of the Roman Catholic Church, which frowns on divorce.

Japan has the next-most devoted married couples; there's only one divorce to almost nine marriages. Divorce is really a Western invention, and Japan's traditions are still fairly strong. France runs a close third, probably because of its large Catholic population. The Frenchman's habit of keeping a mistress may have something to do with the low divorce rate.

There are from 5.8 to 6.5 marriages to every divorce in the Netherlands and Australia. Sweden's rate of divorce seems to confirm its reputation as a really liberal country. The ratio there is a staggering 1.7 to one. In other words, almost one marriage breaks up for every two couples that walk down the aisle.

20 —

15 —

10 —

5 —

5.8:1

6.9:1

1.7:1

| SWEDEN | NETHERLANDS | AUSTRALIA |

MARRIAGE AND DIVORCE

7.1:1	8.4:1	8.8:1	24.3:1
BELGIUM	FRANCE	JAPAN	ITALY

Citizenship

Like so many other concepts, our present-day notions of citizenship owe a lot to the ancient Greeks. They were really the first to emphasise the importance of each individual's participation in government. And, of course, the most common form that participation takes is voting. In this chapter, we find that exercising that precious right to vote varies a great deal from nation to nation.

You'll see that in some countries people take voting very seriously indeed. Can you guess which ones? Have a look at our tables on registered voters and see how we in the U.K. rate as staunch defenders of democracy, when it comes to a big turnout at the polling booths compared with, say, the Italians or West Germans. This is especially interesting when you remember that the question of democratic freedom was one of the reasons why World War II was fought.

Although men have waged many a bloody battle over the right to vote, there's the women's side of the story. Even in bastions of democracy, the ladies often had to fight long and hard to gain an equal say in the running of the country. We have some eye-opening facts about just *how* recently they've acquired that right in some nations. For instance, did you know that in Switzerland the fairer sex didn't get the franchise till 1971? Supporters of Women's Lib will get some useful information – or should we say ammunition? – in this section.

After all those hard-fought battles for equal rights with the men, the ladies in some countries seem to have lost interest. Just like a woman, some might say. But to be fair to the females, interest in voting – or lack of it – varies from country to country. First, we take a look at how various nations place in the voting league. Then, we offer some theories on the results. Of course, you're free to draw your own conclusions, which may not agree with ours. But if *thinking* about voting can interest a few more people in actually *doing* it next election day, it was all worth it.

Some nations have decided not to leave voting up to the whim of the individual conscience. In the U.K., you're free to decide whether or not to make the trek to the polling booth come election day. However, there are countries where the government takes a sterner view. So we find countries where voting is compulsory – even in bad weather.

Weather is just one of the factors that affects how many people exercise their democratic right to vote. We've suggested a few more, and given you enough information to draw some conclusions of your own.

The choices made by those who do bother to get out and vote affect all of us – and that includes the foreigners among the native population. In this chapter, we take a look at what's become a very hot topic indeed – immigration.

Foreign workers tend to flock to countries where the economy is booming and there are lots of job opportunities. So our figures in this section are also a sort of economic barometer showing how and where the tides of immigration are flowing today. You'll see for yourself where foreigners actually head for in the greatest numbers.

In some countries, citizenship involves another, more demanding form of participation than voting – service in the armed forces. The draft is a thing of the past in the U.K. Along with Ireland, the U.S. and Canada we now have voluntary armed forces. In Switzerland, by contrast, every adult male is automatically a member of the militia.

Another requirement of citizenship – and probably the most disliked – is taxation. Armies cost money, so we took a look at the money various governments spend on the military. You can probably guess that the U.S. leads the field here – after all, it coined terms like "arms race" and "missile gap", but what about other countries? Could you state with confidence which European nation spends the most on its defence? Or which one seems least concerned with military might? And what about a country like Japan, with a long history as a militant nation and a prosperous economy? This section may well offer some surprising answers to these questions and others you may have.

The final section gives the number of political prisoners in various countries – a subject which receives a great deal of news coverage these days.

We hope this whole chapter will shed new light on some of those sensational headlines you're constantly confronted with, and help you to pick your way through the thorny political arguments of the day.

When election time rolls round, most of us take our right to vote for granted. It's easy to forget that democracy didn't just happen but was achieved only after long, hard struggles. In some countries, the idea of equal voting rights for men and women alike has taken a long time to become reality.

The table shows considerable agreement among nations about the minimum voting age. In all the countries shown, you have to wait until you're at least 18 before you can mark your first ballot paper. In Austria and Sweden, you'll have to wait an extra year before officially coming of age, and in Denmark, Japan, Norway and Switzerland it's another two years. In the U.K., in 1969, the voting age was lowered from 21 to 18.

Although women now have the vote in all our countries, in some places, men have hogged the polling booths much longer than in others. In Switzerland, women didn't get the vote until 1971. This is particularly surprising for a country which considers itself the birthplace of European democracy. Switzerland has a Federal government and also a State Council, made up of members from each of the Swiss cantons – a system which has been very effective in running the government in accordance with the wishes of its citizens. Perhaps the very success of the Swiss democratic system made women slow to insist on their right to participate.

In a number of other countries, women also had to wait until after World War II before they could cast their vote. Belgian women were denied suffrage until 1949. Now they *must* vote: Belgium is one of the few countries that makes voting compulsory. Japanese women weren't permitted to vote until 1946. This delay isn't so surprising; Japan didn't adopt a democratic system along Western lines until the country's post-war redevelopment. France and Italy kept their polls closed to women until 1945. But in Australia men and women have had equal voting rights since 1903.

Now that we've all got voting rights, do we use them? The table shows, for each country, the percentage of registered voters who actually turn out on polling day. It's one thing to have a democratic

	U.S.A.	CANADA
	1978	1978
1. Legal voting age[0]	18	18
2. Year when women could vote for the first time	1920	1920
3. Countries in which voting is compulsory[x]	V	V
	1976	1974
4. People who voted[‡] (000s)	81,556[+]	9672
as % of registered voters	59.2[+]	71.0

	U.K.	AUSTRALIA	AUSTRIA	BELGIUM	DENMARK	FRANCE	(WEST) GERMANY	IRELAND	ITALY	JAPAN	NETHERLANDS	NORWAY	SPAIN	SWEDEN	SWITZERLAND
	1978	1978	1978	1978	1978	1978	1978	1978	1978	1978	1978	1978	1978	1978	1978
	18	18	19**	18	20	18	18	18	18	20	18	20	18	19	20
	1928	1903	1918	1949	1915	1945	1918	1918	1945	1946	1917	1913	1931	1921	1971
	V	C	C	C	V	V	V	V	V	V	C	V	V	V	V
	1974	1977	1975	1974	1977	1978	1976	1977	1976	1976	1977	1973	—	1976	1975
	29189	8128	4663	5712	3106	29142	34064	1603	37600	57236	8314	2156	NA	5457	1956
	72.5	95.0	92.9	90.3	88.7	82.0	90.4	75.6	92.2	73.4	87.5	80.2	NA	91.8	52.4

**On 1st January of the election year   C compulsory   V voluntary
NA not available

system but quite another for everyone to take part. Look at Switzerland, for example. Judging from the figures, women still don't participate in any great numbers. Switzerland has the lowest voter turn-out of any of the nations on the table, with only 52% of registered voters actually casting their ballots. Yet Switzerland's parliamentary system is relatively accessible to everyone. Could that possibly explain the figures? In Switzerland, any group of 30,000 citizens can demand a referendum, which has the power to reverse Federal legislation. Perhaps this safeguard encourages the Swiss to take it easy at election time.

In Britain, less than three-quarters of us turned out to vote in 1974 – one of our key election years. Most other countries have higher election turn-outs. Perhaps we're getting too complacent.

Sources:
Government documents
[+]Elections Research Center, Washington DC
[x]Where voting is compulsory, a fine is incurred for not voting
[‡]Actual *valid* votes at last elections

Many factors influence voter turn-out. For example, the size of the vote – and its results – are often affected by the weather on voting day. Perhaps Australia's good weather partially explains why just about everyone there votes. The figure, at 95%, is higher than anywhere else. Still, the fact is that their voting is compulsory. The figures tend to be high – over 90% – for the other compulsory countries. Except in the Netherlands, where, law or no law, 12.5% of the population stubbornly refuse to vote.

Standards of living vary enormously from country to country. The poorer neighbours of affluent nations often do more than just peer wistfully over the borders – they actually pack their bags and head for that greener grass on the other side of the fence. When they get there, they're employed in the less desirable jobs that the citizens of the host countries don't want. In return they earn wages that would be unheard of in their native countries. By sending most of their earnings back home, they can support their families in a style which would otherwise be impossible.

This table shows which countries attract the greatest numbers of foreign workers, how much of the total work force is made up of immigrant labour, and some of the countries the workers come from.

West Germany tops the table as the Mecca for foreign workers. The healthiest economy in Europe acts as a natural magnet, attracting over two million immigrant workers. The Turkish are the most eager to cash in on some of that affluence, followed by the Yugoslavians. They rub shoulders with quite a few Italians, Greeks and other nationalities. These immigrant workers help to fill the gaps in the home labour force.

However, foreign workers account for only 8.2% of the total German labour force. This is almost the same figure as in France (8.6%), which is the second most attractive country to migrant workers. The Portuguese flock to France in the greatest numbers, closely followed by the Algerians – understandably, considering that Algeria was formerly a French colony and French is its second language. There are 420,000 Algerians now working in France. Italian, Moroccan and Spanish immigrants have also settled in jobs there.

Tiny Luxembourg – sandwiched between France and West Germany – employs only 47,000 foreign workers, yet they account for a massive 31.3% of the total labour force. That means that almost

	U.K.	LUXEMBOURG
	1975	1975
1. Foreign workers (000s)	775	47
as % of total labor force	3.0	31.3
2. Origins of foreign workers (000s)		
Algeria	1	*
Greece	2	*
Italy	56	11
Morocco	1	*
Portugal	4	12
Spain	15	2
Turkey	2	*
Yugoslavia	4	1
others	690	21

AUSTRIA	BELGIUM	DENMARK+	FRANCE	(WEST) GERMANY	ITALY+	NETHERLANDS	SPAIN+	SWEDEN	SWITZERLAND
1975	1975	1975	1975	1975	1971	1975	1975	1975	1975
185	278	43	1900	2191	36	216	64	204	553
5.7	6.9	1.8	8.6	8.2	0.2	4.2	0.5	5.6	17.4
*	3	NA	420	2	NA	*	NA	*	*
*	8	NA	5	212	NA	2	NA	8	*
2	85	NA	210	318	NA	10	NA	2	281
*	60	NA	165	18	NA	28	NA	1	*
*	3	NA	430	70	NA	5	NA	1	4
*	30	NA	250	132	NA	18	NA	2	72
26	10	NA	35	582	NA	38	NA	4	16
136	3	NA	60	436	NA	10	NA	23	24
21	76	NA	325	421	NA	105	NA	163	156

NA not available   *Less than 500

Sources:
Thomas P. Kane, "Europe's Guest Workers", Intercom, January 1978 (Population Reference Bureau, Washington DC)
+National statistical offices
Figures rounded to nearest 000

one worker in three is an immigrant – the highest percentage on the table.

The U.K. employs the lowest percentage of foreign workers – three per cent of the total work force. U.K. passport holders from Commonwealth countries aren't included in this figure. By far the highest number of immigrant workers – 690,000 – comes from "other" countries and Irish workers account for nearly 70% of that figure. Traditionally the Irish have always made the short journey across the Irish Sea to seek employment as labourers. However, if Britain seems a land of plenty to them, it doesn't to many others. Only the Italians have chosen to seek their fortunes here in any noticeable numbers – 56,000 of them.

MIGRANT WORKERS PERMITS

Which country has the highest number of resident aliens? It's not surprising to find it's the U.S. After all, apart from the Indians – who are the only real native Americans – they were all once immigrants. There are nearly five million aliens residing in the U.S. today – more than two per cent of the population.

In the U.K., only three per cent – or 1.5 million people – are registered foreigners. Some of them are workers from Ireland, some are from our old colonies in India and the Caribbean.

However, this percentage is well below average. In Switzerland over 12% of the population are registered foreigners, that's nearly one person in eight. Switzerland is a small country with a high standard of living – and consequently a big manpower problem when it comes to the dirtier and lower-paid jobs. Its geographical position also makes it an obvious target for people who come in search of work. Italians, Yugoslavs, Greeks and Turks all try Switzerland first when looking for jobs in the affluent areas of Europe. The construction work of the Simplon Tunnel kept thousands of immigrant workers busy for years. So it's not surprising to find Switzerland topping the list in percentage terms.

France comes next with nearly eight per cent – over four million – as registered foreigners. A lot of these are Algerians or south Europeans working in agriculture, manufacturing, or the service industries. West Germany – which also relies heavily on immigrant workers – is also near the four million mark.

Who has the fewest foreigners? In Italy, Spain and Japan the figure is well under one per cent of the population. In Italy, they account for only 0.2% – the lowest figure on our table. On the other hand, Italians tend to emigrate more than

		U.S.A.	CANADA
		1975	1975
1.	Naturalised aliens (000s)	131.6	137.5
2.	Naturalised Italians (000s)	8.8	19.8
	as % of all naturalised citizens	6.7	14.4
		1977	–
3.	Number of registered foreigners (000s)	4964[+]	NA
	as % of population	2.3[+]	NA
		1978	1978
4.	Initial cost of a passport (£)	6.29	5.80
	validity (years)	5	5

U.K.	AUSTRALIA	AUSTRIA	BELGIUM	DENMARK	FRANCE	(WEST) GERMANY	IRELAND	ITALY	JAPAN	NETHERLANDS	NORWAY	SPAIN	SWEDEN	SWITZERLAND
1965/75	1945/75	1976	—	1976/77	1970/74	—	—	—	—	1973/75	1975	—	1975	1974/77
37.5	962.7	7.5	NA	5.3	179.8	NA	NA	NA	NA	14.9	1.1	NA	16.7	32.8
1.3	172.0	0.4	NA	NA	48.9	NA	NA	NA	NA	0.5	NA	NA	0.2	11.9
3.7	17.8	6.2	NA	NA	27.2	NA	NA	NA	NA	3.8	NA	NA	1.5	36.4
1976	1971	1971	1970	1977	1975	1975	—	1971	1975	1975	1975	1976	1975	1976
1684[0]	722[x]	177	696	91[‡]	4128	4089	NA	122	752	345	68	159	410	801**
3.0[0]	5.6[x]	2.5	7.7	1.8[‡]	7.8	6.6	NA	0.2	0.7	2.5	1.7	0.4	4.9	12.2**
1978	1978	1978	1978	1978	1978	1978	1978	1978	1978	1978	1978	1978	1978	1978
10.00	13.33	3.18	1.64[++]	10.46	10.59	2.64	3.62	3.47	5.91[00]	9.82	2.27	2.09	36.55	7.11[xx]
10	5	5	1–5	10	5	5	10	5	5	5	10	5	10	1–5

NA not available [+] *Aliens who reported under the Alien Address Programme on January 1st 1977* [0] *Estimate includes Commonwealth citizens and EEC members* [x] *Excludes Commonwealth citizens* [‡] *Excludes Scandinavian citizens* **Excludes foreign dignatories, their families and others* [++] *One year only* [00] *For a single trip only; for more than one trip, the price is £11.82* [xx] *A further £1.57 is payable annually*

most other nationalities. Figures are also given for naturalised Italians in some countries. They show how many of them strike out to make their lives abroad. Australia and Canada attract a steady stream.

The table also shows the cost of passports in various countries. It's interesting to see the wide range – from Australia, where a passport costs £13.33, right down to Belgium, where holidays abroad are no problem; you can get your credentials for just £1.64.

Sources:
1 National statistical offices
2 Relevant government departments
3 & 4 Embassies

		U.S.A.	CANADA
		1977	1977
1.	Size of armed forces (000s)	2088	80.0
	people per member of the armed forces[+]	104	295
2.	Total defence expenditure (£ – millions)	64402	1908
	per person (£)	298	82
	per member of the armed forces (£ 000s)[+]	30.8	23.8
3.	Military service compulsory or voluntary	V	V
	length (months) of compulsory military service	–	–

The U.S. has over two million men and women in uniform. That's the largest combined armed forces on our table – more than the combined forces of all the E.E.C. countries added together. But the U.S. doesn't have the largest combined forces in the world. That dubious honour goes to China (not on the table). They've got nearly four million servicemen and women. In Russia (also not on the table) there are more than 3.5 million people in the military service.

With a military budget of over £64 thousand million, the U.S. places an exceptionally high burden on its tax-paying citizens. Public military expenditure in the U.S. was £298 per person in 1977 – that's over twice the amount each of us forks out in the U.K.

The U.S. combined services are not only larger than the others on our table, and more expensive to equip, they also make up a higher proportion of the population. One in 104 people in the U.S. is in military service. That's just over double the average on the table.

By comparison, the U.K.'s military figures are small. Our armed forces number just under 340,000 – that's one serviceman (or woman) for every 165 people. Our defence expenditure, though, is still a hefty £6,391 million each year.

France comes second on the table, with a combined services figure of just over half a million (almost a quarter that of the U.S.). The average Frenchman pays over 50% less for military expenses than the U.S. citizen. West Germany has just under half a million people in uniform. They spend nearly £1,000 million more than the French on their combined services but only four pounds more per person.

Both France and Germany have compulsory military service – like all other European countries except Ireland and the U.K. The duration of compulsory

U.K.	AUSTRALIA	AUSTRIA	BELGIUM	DENMARK	FRANCE	(WEST) GERMANY	IRELAND	ITALY	JAPAN	NETHERLANDS	NORWAY	SPAIN	SWEDEN	SWITZERLAND
1977	1977	1977	1977	1977	1977	1977	1977	1977	1977	1977	1977	1977	1977	1977
339.2	69.7	37.3	85.7	34.7	502.1	489.0	14.7	330.0	238.0	109.7	39.0	309.0	68.6	18.5
165	205	201	115	147	106	125	217	172	480	126	105	119	121	335
6391	1572[0]	304	1411	629	7831	9462	83	2517	3471	1913	681	1228	1615	729
114	114[0]	39	144	124	146	150	26	44	28	137	168	34	195	116
18.8	22.5	8.1	16.4	18.1	15.5	19.3	5.6	7.6	14.6	17.4	17.4	2.9	23.5	39.4
V	V	C	C	C	C	C	V	C	V	C	C	C	C	C
—	—	8[00]	8-10[x]	9	12	15	—	12-18[‡]	—	14-17[++]	12-15[**]	18	7½-15	4[‡‡]

C compulsory   V voluntary   [0]1976 figures

Sources:
International Institute for Strategic Studies
[+]Heron House estimates
[x]Eight if served in Germany [‡]18 only for naval service [**]15 if navy or airforce
[++] 17 if navy or airforce [00]Six months followed by 60 days reservist training for 12 years
[‡‡]17 weeks recruit training followed by reservist refresher training of three weeks for eight out
of 12 years for Auszug (20-32 years of age) two weeks for three years for Landwehr (33-42
years of age) one week for two years for Landsturm (43-50 years)

service varies between countries. In peacetime, it's usually a nominal one-year period, as it is in France. In West Germany it's 15 months.

Amnesty International is an international organisation, independent of any governments, whose primary concern is with certain specific human rights. Its objectives are precisely defined: "Considering that every person has the right freely to hold and to express his convictions and the obligation to extend a like freedom to others, the objects of Amnesty International shall be to secure throughout the world the observance of the *Universal Declaration of Human Rights*, by...working towards the release of and providing assistance to persons who...are imprisoned, detained, restricted or otherwise subjected to physical coercion or restricted by reason of political, religious or other conscientiously held beliefs or by reason of their ethnic origin, sex, colour or language, provided that they have not used or advocated violence (hereinafter referred to as "Prisoners of Conscience")..."

Amnesty International also opposes "the detention of any Prisoners of Conscience or any political prisoners without trial within a reasonable time, or any trial procedures relating to such prisoners that do not conform to recognised norms to ensure a fair hearing."

It also opposes the "imposition and infliction of death penalties and torture or other cruel, inhuman or degrading treatment or punishment of prisoners or other detained or restricted persons whether or not they have used or advocated violence..."

Amnesty International has National Sections in 35 countries and individual members in 74. Its International Secretariat is in London.

This table (compiled from the *Amnesty International Report 1977*) shows the number of prisoners "adopted" or under investigation by Amnesty International groups in some of the countries mentioned in their report. This "adoption" system is used for Prisoners of Conscience who haven't used or advocated political violence. It means that Amnesty is concentrating particular attention on individual prisoners whose cases have been brought to their attention. Before "adopting" an individual, Amnesty investigates the case in depth. The number of adopted prisoners doesn't necessarily reflect the total number of political prisoners in a given country. In many countries where civil liberties are most restricted, little information is available – hence political prisoners in these countries haven't been adopted by Amnesty International and don't appear on the table. In these countries Amnesty uses other techniques, such as publicity campaigns and fact-finding missions.

One country with a large number of prisoners adopted or investigated by Amnesty International is the U.S.S.R. Even before Solzhenitsyn's devastating *Gulag Archipelago*, the U.S.S.R. had long been notorious for its treatment of political detainees. Increasing international interest has recently been focused on the plight of political detainees in the U.S.S.R. since the Helsinki Agreement. This agreement was intended to guarantee basic human rights to all citizens of the countries that signed it.

Indonesia is another country where civil rights are not always guaranteed. Since Indonesia achieved independence from the Dutch in 1945, several bloody

294

Prisoners adopted by Amnesty International		
U.S.A.	14[+]	1977
ISRAEL	18[+]	1977
YUGOSLAVIA	100[+0]	1977
MALI	28	1977
CZECHOSLOVAKIA	29	1977
BRAZIL	213	1977
CHILE	145[+]	1977
FRANCE	43[+]	1977
POLAND	40[+]	1977
INDONESIA	294	1977
ITALY	1	1976
THE PHILIPPINES	90	1977
SOUTH AFRICA	130[+0]	1977
USSR	300	1977
SPAIN	11	1977
SWEDEN	2	1976
SWITZERLAND	2	1977

[+]Adoption and investigation cases
[0]At least this number

Source:
Amnesty International Report, 1977

political crises have occurred, one of which involved the death of hundreds of thousands throughout the country. There were 294 Indonesian political prisoners adopted by Amnesty International.

Other countries known for severe treatment of those with "unacceptable" views are Brazil and Chile. There were 213 adopted prisoners in Brazil. Chile, whose head of state is General Pinochet, had 145 prisoners under adoption or investigation by Amnesty International groups – a large figure when you compare the size of Chile to that of the other countries on this table.

Next comes South Africa – whose apartheid policy has been officially condemned by many nations. One hundred and thirty prisoners, most of them black, were adopted by Amnesty International.

Two other countries mentioned in this table also had high figures. In Yugoslavia, there were about 100 prisoners, some adopted, more under investigation – amongst them advocates of minority nationalism (the Croatian and Serbian nationalist movements are active) – as well as opponents of the current regime. There were 90 prisoners under adoption or investigation in the Philippines.

Amnesty International also adopts Prisoners of Conscience in Western countries. Fourteen prisoners in the U.S., two in Sweden and in Switzerland, and 43 in France were mentioned in the *Amnesty International Report 1977*.

As a non-political organisation, Amnesty International does not concern itself with political systems of any kind, but simply with Prisoners of Conscience, political prisoners and with the abolition of the death penalty.

Total defence expenditure (£-millions)

70

60

There hasn't been a major global conflict for over 30 years. Nevertheless, some countries spend vast amounts on defence. The U.S.S.R. is the most military-minded nation in the world. It poured a staggering £72 thousand million into its 1976 defence budget. That's sabre-rattling on a truly grand scale.

50

It's not surprising to find that the U.S. is Russia's closest rival. America channelled £64 thousand million into keeping the balance of power at its present level. The defence budgets of these two military superpowers dwarf all the other countries in the world.

40

West Germany, nervously poised between East and West, earmarks £9 thousand million for defence, while France isn't far behind with over £7.8 thousand million. In the U.K., despite our economic problems, we spend over £6 thousand million, nearly as much as our more prosperous European neighbours. Even tiny Ireland puts £83 million into defence.

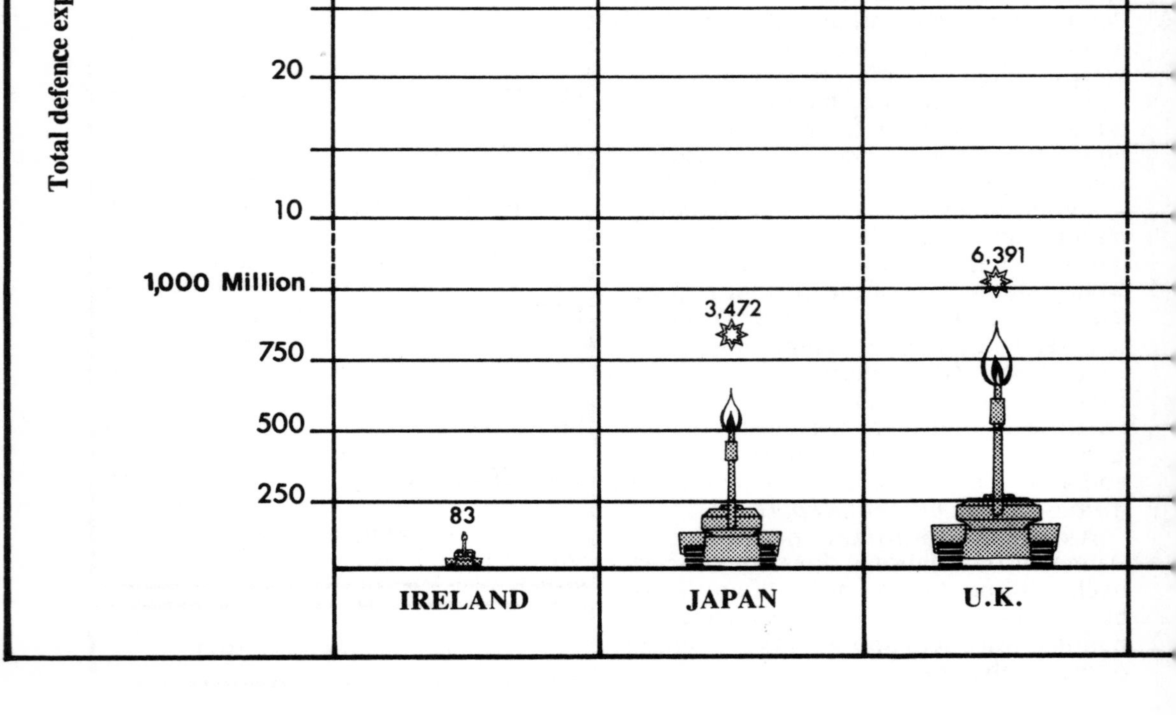

30

20

10

1,000 Million

750

500

250

6,391

3,472

83

IRELAND

JAPAN

U.K.

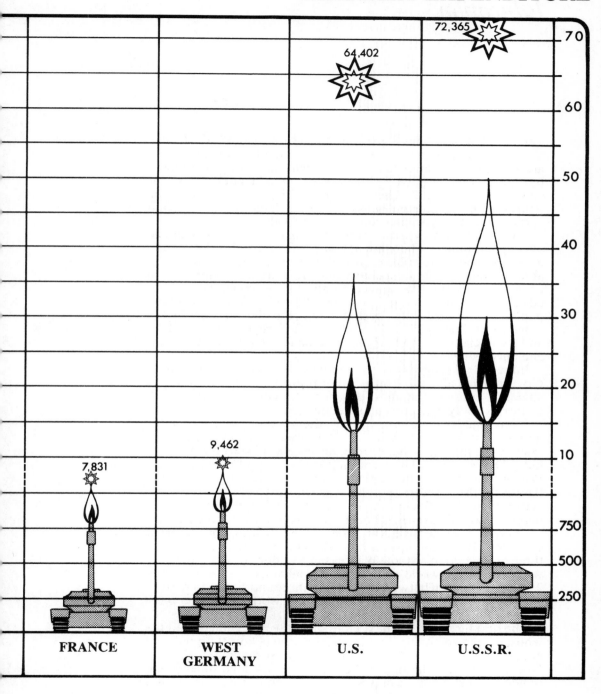

		72,365	
	64,402		70

FRANCE 7,831

WEST GERMANY 9,462

U.S. 64,402

U.S.S.R. 72,365

Education

Are your kids learning all you want them to in school? We all have our own ideas about education. Are the schools getting better or are academic standards falling fast? What should the government be doing about improving the quality of education? Are our classes over-crowded? These are some of the questions which the figures in this chapter will help you resolve. You can see just what sort of education you can expect for your kids.

Education has now become one of the major priorities of almost every government. There are obvious reasons for this. Education gives you a better start in life. A well educated generation is better able to fulfil its role in society. But there are also hard-headed economic reasons for governments to be concerned about education. A country needs an intelligent, informed working population in order to compete successfully in this highly complex technological age. Well qualified engineers, scientists and technicians are not born, they're made – in schools, colleges, and universities.

Educational programmes cost money. And that's the problem. Governments also have many other priorities – and the longer a student's course of studies, the more it's going to cost. Which countries really care most about education?

Many modern experts are of the opinion that the most important years of education are the earliest ones. If a child can read by the time he's four or five, he has a head start on the other kids. That's why early education is the subject of our first table. We've split this question up into four categories. These give the figures for the percentage of children who go to school at an early age of three, four, five or six. Then there are figures on the overall percentage of children in school between the ages of three and six.

There's considerable variation in the figures for different countries. In Belgium, nearly all (96%) of children aged between three and six go to school. Whereas in Sweden, which is in many ways one of the world's most socially advanced countries, only 25% of all children between the ages of three and six are in school. The U.K. and the U.S. are closer to the world-wide average – their figure's around 60%.

The first figures in the next table show the percentage of adults who can read and write. This is an indication of how thorough and efficient an educational system is at its most fundamental level. Needless to say, the figures here are all very high indeed. Do remember that while we may have a 97% literacy rate in the U.K., that illiterate three per cent accounts for well over 1.5 million people. In Canada, surprisingly, only 93% of the population can read and write.

The normal school age enrolment figures are given after those for literacy. Here we give the full time enrolment rates for children aged seven to 14, and the figures for 15-year-olds, 16-year-olds, 17-year-olds and 18-year-olds. Together with this we give figures which show the minimum years of schooling a child is required to have in each of the countries covered; compare these figures with those for literacy.

What's surprising about these figures is the fact, that in all the countries listed, there still appears to be a tiny percentage of children who, in disregard of the laws, simply don't attend school at all.

The next table covers the percentage of men and women enrolled full-time at the third level of education. (The third level is a U.N.E.S.C.O. term, and is meant to include university students – usually people over 18.)

Here we also include figures to show the number of women enrolled full-time. This gives you an idea of how the battle of the sexes is going on the education front. From the look of things, the women are gaining ground rapidly.

Almost everyone can get to some school somewhere, but what happens when you're there? How much do you actually learn? Obviously, there are many factors that will make a difference, but one of the best indications is the pupil-teacher ratio. The fewer students a teacher has to cope with, the more time he has to spend on each individual and the more likely you are to learn. The figures enable you to compare the size of the classrooms in the countries covered.

The figures in this table are broken down into three categories – elementary schools, secondary schools, and colleges and universities. Besides being an indicator of the standard of education, these figures also show where the governments' priorities are. After all, those teachers have to get paid. And if there are cuts to be made in the educational budget, it's interesting to see where the various governments have decided to make them.

	U.S.A.	CANADA
	1970	1970
Children enrolled in school as a % of children aged:		
3	13	5
4	28	20
5	80	85
6	98	98
3 to 6	57	54

"Get an education and you'll get a good start in life." Most people agree on this one. But how early should the start be? Nursery education is a top issue these days. Do children get more stimulation from being in nursery school? Or do they need to be at home with their mothers? And how soon should a child learn to read and write? The table details what sort of importance different countries place on nursery schooling. There are some startling differences. Belgium is top of the table where three-year-olds are concerned. By this tender age 90% of little Belgians are playing with finger paints and learning their ABC's. But across the border, in the Netherlands, there are no three-year-olds in school. The same applies in Ireland. In the U.K. and Canada only one in 20 has started his or her education. Twelve per cent of Spanish three-year-olds are at school.

The French figure is comparatively high. In France, 61% of three-year-olds are enrolled in school. Figures aren't available for all the other countries on the table. However, the overall enrolment figures for children aged three to six give a fair indication of the number of three-year-olds at school. It certainly looks as though the Scandinavians don't go for early education for young children.

The picture changes rapidly if you look at facilities for four-year-olds. In Holland, for example, 84% of them are in school, which puts the Netherlands in the third highest place after Belgium (95%) and France (87%). And 60% of Irish four-year-olds have started a nursery school education. For one of the poorest countries in the Western world, there's a lot of money going into educating under-fives. The Japanese figure is high too – 61%. Considering the large population, that's a lot of nursery classes.

The U.K., Canada and the U.S. aren't putting much of their resources into nursery education. Only six per cent of British tinies and 13% of Americans are in nurseries. And a mere 35% of British four-year-olds are in school.

The French, with their intensive education system have virtually all their children in school by the age of five. In the U.K. and Belgium that's also the compulsory starting age. In Spain, however, only 49% of five-year-olds are sitting at their school desks.

Most countries on the table claim around 100% enrolment when it comes to six-year-olds' education.

EARLY EDUCATION

U.K.	AUSTRALIA	AUSTRIA	BELGIUM	DENMARK	FRANCE	(WEST) GERMANY	IRELAND[+]	ITALY	JAPAN	NETHERLANDS	NORWAY	SPAIN	SWEDEN	SWITZERLAND
1971	1971	1970	1970	1970	1970	1970	1970	1970	1970	1970	1970	1970	1970	1970
6	NA	NA	90	NA	61	NA	0	NA	17	0	NA	12	NA	NA
35	NA	NA	95	NA	87	NA	60	NA	61	84	NA	43	NA	NA
98	NA	NA	99	NA	100	NA	90	NA	80	96	NA	49	NA	NA
99	100	97	99	NA	100	NA	100	NA	100[+]	99	NA	98	NA	NA
60	55[+]	45[+]	96	9[+]	88	45	63	62	65[+]	70	4[+]	56	25	39[+]

Sources:
Organization for Economic Co-operation and Development
[+]OECD estimates

	U.S.A.	CANADA
	1970	1970
1. Adults (%) who can read and write	99	93
	1974	1974
2. Age limits of compulsory education	7-16	6/7-15/16
	1970	1970
3. Children enrolled in full time education as % of age groups:		
7–14	99.0	98.2
15	97.7	98.0
16	93.5	89.1
17	86.2	77.2
18	53.8	45.8

As you'd expect, almost everybody in the countries on our table can read and write. The small percentage that can't includes the educationally subnormal and the mentally deficient.

All the countries listed have a literacy rate of 97% or more, with the exceptions of Spain, Italy and – surprisingly – Canada. In Canada, the rate is just 93%, the same as in Italy. This means that in these three countries, about one adult in 14 has to rely on family or friends in everyday dealings with the printed word.

The 97% figure for the U.K. seems good. In fact, it means that over a million of us are illiterate. No wonder there's been a nation-wide literacy campaign recently. Similarly, that 99% figure for the U.S. means that over two million Americans are unable to read the newspaper or write their own letters.

In most countries listed, children must spend nine years of their lives in school – usually from the age of six to 15, or from seven to 16. The exception to the starting age is the U.K., where children have to trot through the school gates at the tender age of five. The only exceptions to the school leaving age of 15 or 16 are Italy and Belgium. In Italy, children are allowed to leave school at the early age of 13, and in Belgium they need only stay until 14. However, 75% of Belgian 15-year-olds are still at school compared with only 42% of Italians. Italy's early school leaving age could contribute to the low literacy rate there.

Enrolment rate figures for the seven-14 age group are high but not 100%. This is because many of the national figures on the table don't include special schools like religious schools, or schools for the blind or deaf. There's also the truant problem.

The lowest figure is for Italy. Nearly 13% of Italian children between the ages

U.K.	AUSTRALIA	AUSTRIA	BELGIUM	DENMARK	FRANCE	(WEST) GERMANY	IRELAND	ITALY	JAPAN	NETHERLANDS	NORWAY	SPAIN	SWEDEN	SWITZERLAND
1970	1970	1970	1970	1970	1970	1970	1970	1970	1970	1970	1970	1970	1970	1970
97	NA	99	99	99	99	99	98	93	99	99	99	94	99	99
1974	1974	1974	1974	1974	1974	1974	1974	1974	1974	1974	1974	1974	1974	1974
5-16	6-15/16	6-15	6-14	7-16	6-16	6-15	6-15	6-13	6-15	6-16	6-16	7-16	6-15	6-15
1970	1971	1969	1966	1970	1970	1969	1967	1966	1970	1970	1970	1970	1972	1970
98.5	100.0	99.8	98.0	98.1	98.6	98.0	98.9	87.3	99.9[+]	99.0	99.0	90.2	98.8	95.4
73.0	81.5	54.8	75.1	85.2	80.5	54.9	82.4	42.1	83.8[+]	79.7	94.2	35.0	96.7	94.6
41.5	54.2	32.6	61.3	66.8	62.6	30.8	64.3	33.6	79.0[+]	60.6	74.6	29.6	74.0	61.5
26.2	37.2	23.6	47.0	31.8	45.1	20.4	46.5	27.4	74.8[+]	41.5	59.8	22.8	60.8	52.7
17.6	23.6	16.4	33.2	23.2	29.1	15.7	31.8	20.2	29.9[+]	28.4	46.5	19.0	40.8	27.4

[+]OECD estimates

of seven and 14 are missing from the school roll. Some 13-year-olds, of course will already be spending their first wages.

The figures in the higher age groups, largely reflect the national school leaving age. But in Japan, although school pupils can leave at 15, three-quarters of them study until well after that.

In the 18-year-old age group, there are some very big surprises. Who would have expected that the lowest figure would be the country of student youth – West Germany – with an enrolment rate of less than 16%? In Austria it's just over 16%. The U.K. is about equal at 17.6%. One

Sources:
1 World Bank Tables
2 United Nations Educational, Scientific, and Cultural Organization
3 Organization for Economic Co-operation and Development

partial explanation is that these countries have extensive apprenticeship schemes and polytechnic courses for this age group. The U.S. is the country with the most striking high school attendance: 54% of boys and girls are still in class at the age of 18.

	U.S.A.	CANADA
	1974	1974
Full-time enrolment in third level[+] educational establishments:		
men and women (as % of population 18-22)	53.6	34.7
women (as % of female population 18-22)	48.7	30.6

The U.S. has by far the highest number of students enrolled in colleges and universities. About 50% of their youth are taking some form of higher learning. Only Canada, with 34.7% comes close to their figure.

In comparison the U.K. figures are very meagre indeed. Just over 16% of our young people enrol for further education but only 11.4% of the young women. Although it is changing slowly it is still true to say that there is rather an elitist system at work in the U.K. There is still a well-trodden path from public school to those places in seats of higher learning. Other institutions, such as polytechnics, may be challenging the traditional hold that universities have; but we have a long way to go before we can match those countries on the list who obviously offer opportunities to those who may not be the brightest students.

Denmark comes after the U.S. and Canada, with more than a quarter of college-age men and women enrolled. The Danes ascribe a great deal of importance to higher education.

The Italians also favour admitting as many students as possible and have the fifth highest university enrolment rates. Almost one in four Italians in the 18 to 22 age group (one in five women) are taking university courses. The University of Rome, for instance, has so many students that classes have to be given in staggered shifts. If all the students showed up at once, there'd be no place to put them. Italy has high unemployment (see

"employment" table), and at least this high enrolment rate keeps the young people off the unemployment rolls.

France scores fairly low in the percentage of enrolments category – well under 20%. Higher education in France as in the U.K., is inclined to be selective and geared to the brightest students. In France, graduation from a first-rate institution virtually guarantees access to the best jobs. The same is true in the U.K., though we aren't quite as ferociously intellectual as the French.

The Irish come at the bottom of the table, just above the Swiss. In Ireland just over one man or woman in seven attends a college or university. Ireland is a largely rural country with a large working class. Higher education is mostly reserved for the middle classes.

Take a guess which country has the greatest equality, educationally, between the sexes. As expected the North Americans come out on top. The U.S. leads the field by a long way with a healthy 48.7% of young women going on to study in higher education. This near one in two

THIRD-LEVEL EDUCATION

	U.K.	AUSTRALIA	AUSTRIA	BELGIUM	DENMARK	FRANCE	(WEST) GERMANY	IRELAND	ITALY	JAPAN	NETHERLANDS	NORWAY	SPAIN	SWEDEN	SWITZERLAND
	1973	1974	1974	1974	1974	1974	1975	1974	1974	1975	1974	1974	1975	1974	1974
	16.2	22.0	16.8	21.6	28.0	18.0	20.3	15.5	23.9	24.7	23.5	21.3	NA	21.8	13.2
	11.4	17.8	12.5	17.6	25.3	17.3	14.1	11.4	19.0	16.1	14.9	15.8	NA	20.4	7.8

NA not available

ratio makes all other countries look like centres of male chauvinism by comparison. Only Canada, with nearly one in every three going on to third level education, comes anywhere near. The U.K. comes right near the bottom of the list with 11.4%.

It seems as if the old attitude that priority should be given to the nurturing of male brains still applies all over Europe. The best figures in Europe are in Scandinavian countries where, at last, there are signs that they are beginning to realise that women, too, have the capacity and need to deal with advanced studies. In Denmark, just over a quarter of young women enrol in third level educational establishments. A short boat trip away, the Swedes are following; there, one in five girls are educated to university level.

In the rest of the countries shown it looks as if women's liberation movements still have a great deal to do before young women are seen as more than just potential mothers and housewives.

Source:
United Nations Educational, Scientific, and Cultural Organization
+Includes college, university, teacher-training colleges

The effectiveness of any educational institution depends to a large extent on the size of its classes. No matter how good a teacher or professor is, he can't give much individual attention in a class of 50 or 100.

This table shows how many students a teacher or university professor has in an average class. Obviously the figures vary according to the subject being taught. Basic compulsory subjects such as mathematics and English are taught in larger classes than a university course in astro physics. However, these figures can tell us where there is classroom overcrowding.

Where do they have the best pupil/teacher ratio in primary schools? The Scandinavians score highly. The best ratio is in Denmark, where there are only 16 pupils to every teacher. They're almost as well off in Norway, with a ratio of 17 to one. The U.K. is ninth on the table – tied with Japan and Canada. We all have the rather poor figure of 25 to one. The French are third, with 18 students to each teacher.

The Irish are worst off as far as primary education is concerned. They have 31 pupils to one teacher *on average*.

The Spanish are the next most badly off with a ratio of 29 pupils to every teacher – closely followed by the Dutch with 28 to one at the primary level.

There's a definite trend at secondary level. In all countries except Spain and Denmark, the pupil/teacher ratio is lower. Each teacher has fewer pupils and can spend more time with individuals.

Secondary pupils are best off in Belgium. The ratio there is a low seven pupils to every teacher at the secondary level. Norway is right behind with 9 to one, and Sweden is next with 10 to one. Spain has an appalling secondary school situation with a ratio of 39 pupils to one teacher.

Surprisingly, the Americans have the next worst pupil/teacher ratio after the Spanish. There are 19 American pupils at secondary level to every one teacher. The U.K. has a better ratio, 14 to one, and has, we hope, more effective schools.

	U.S.A.	CANADA
	1975	1970
Number of students to teaching staff in:		
first level schools[o]	20	25
second level schools[x]	19	16
	1975	1974
universities and equivalent institutions	16	16
non-university teacher training institutions	14	7[++]
other non-university institutions	22	14
all third level institutions	17	15

STUDENT : TEACHER RATIOS

	GREAT BRITAIN‡	AUSTRALIA	AUSTRIA	BELGIUM+	DENMARK	FRANCE	(WEST) GERMANY	IRELAND	ITALY	JAPAN	NETHERLANDS	NORWAY	SPAIN+	SWEDEN	SWITZERLAND
	1974	1976	1974	1975	1974	1976	1974	1975	1975	1975	1975	1974	1974	1975	
	25	22	21	19	16	18	23	31	19	25	28	17	29	20	NA
	14	14	15**	7++	16	15	14	14	11++	17	16	9[00]	39	10	NA
	1973	1975	1975	NA	1974	1971	1975	1975	1975	1975	1970	1975	1974	1974	1974
	7	14	9	NA	NA	18	8	11	23	12	9	11	NA	NA	9
	10	NA	NA	NA	NA	NA	NA	9	NA	NA	NA	9	NA	NA	NA
	NA	NA	NA	NA	NA	NA	NA	8	NA	10	NA	9	NA	NA	NA
	NA	14	NA	NA	9	NA	8	10	23	12	NA	10	14	12	NA

‡Excludes Northern Ireland   **1973 figure   ++1974 figure   00 1970 figure   NA not available

Our priority is obviously to reduce the high ratio in our primary schools.

At university level, the figures reveal, we have the best student/teacher ratio of seven to one. It seems as if our excellent tradition of graduate and post-graduate work will be continued both in world-famous institutions such as the colleges at Oxford and Cambridge and at the newer seats of learning such as the Polytechnics. However, the teacher training institutions in the U.K. are not as well-staffed as the universities and have a student/teacher ratio of 10 to one.

The Irish, who have a less good ratio in the universities, have a better ratio than us when it comes to the importance of training teachers. In fact, the quality of Irish education improves as students climb the ladder.

If you want to study abroad, don't go to Italy, unless you're more interested in spaghetti than studying. They have 23 students to every professor in all third-level institutions. The U.S. is second worst with 17 to one.

Sources:
United Nations Educational, Scientific and Cultural Organization
+National statistical offices
0Basically primary schools
xBasically secondary schools

Ratio of students to teaching staff in second level schools

40

Secondary school pupils in Belgium had better be sure to do their homework every night. With one teacher for every seven students, each eager pupil stands a pretty good chance of getting asked questions in class.

35

It's a bit easier to get lost in the crowd if you go to school in Canada, Denmark or Holland. All those countries have a pupil-teacher ratio over twice as high as Belgium's.

Japan's overcrowding isn't as noticeable in its secondary school classrooms as it is in its city streets. There's one teacher for every 17 students.

30

The U.S. ratio is similar — 19 pupils to every teacher. That doesn't sound bad, till you realise it means only three minutes of the teacher's time for each pupil in a one-hour period. As any teacher knows, it can take that long just to get someone's attention.

25

In Spain, with a 39 to 1 ratio, the whole day could be spent on roll call.

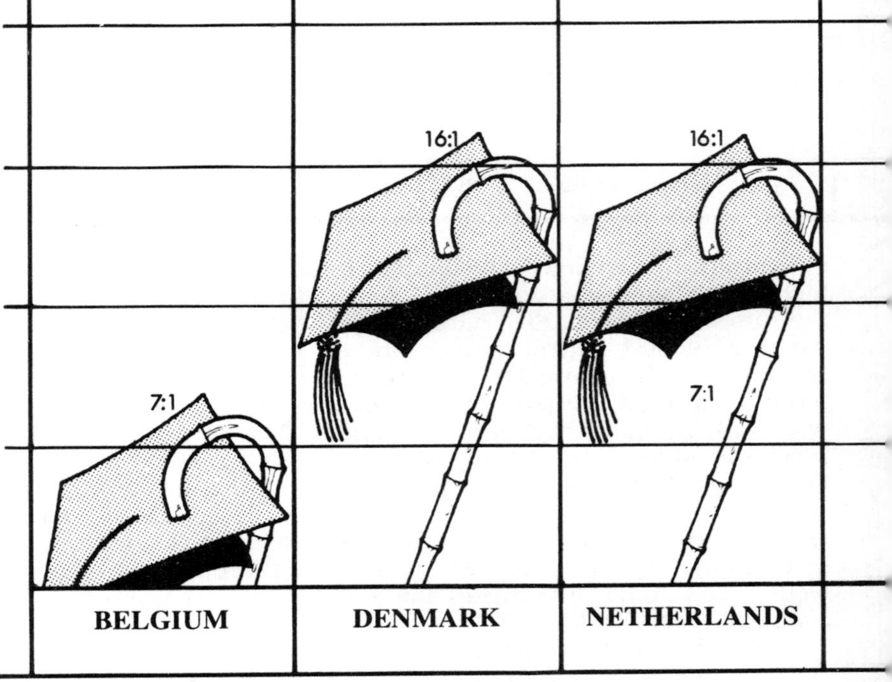

20

15   16:1   16:1

10

7:1   7:1

5

BELGIUM   **DENMARK**   **NETHERLANDS**

PUPILS: TEACHERS

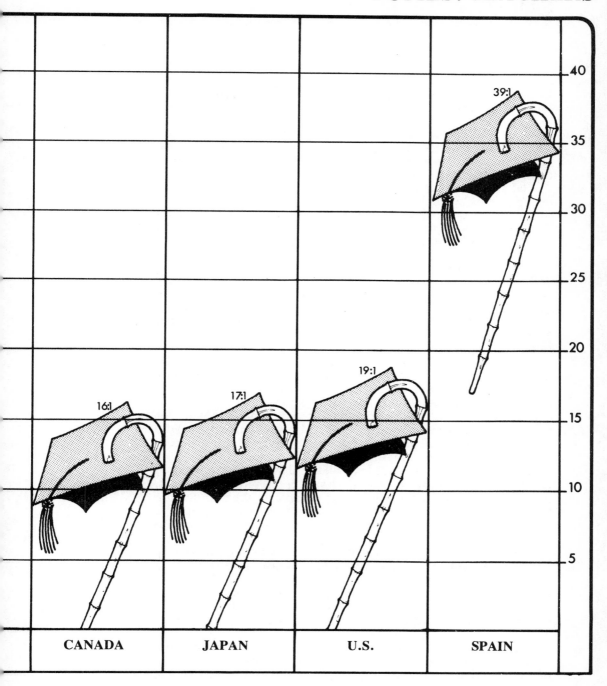

Crime

In this chapter, we take a look at many aspects of crime. What's the first thing you'd do if you were mugged in the street or found that your home had been broken into? The answer is obvious – call the police. How quickly that call was answered could depend a lot on what nationality you are. There's an enormous variation in the amount of police protection a citizen of France receives compared to, say, Holland.

And the size of a country's police force doesn't necessarily depend on how much protection those citizens need. As you'll see, there are countries where crime is virtually running rampant but there simply aren't enough policemen to match wits with the criminals. So in some cases, the old saying that "there's never a policeman around when you need one" is true.

But let's be fair to the boys on the beat. They're often faced with almost insurmountable problems. In urban areas in many countries, crime rates have soared sky-high, and they are often unfairly blamed on police inefficiency. There are many social factors beyond the control of the police force – like overcrowding in slums and unemployment among the young – that cause crime rates to be high.

Social factors also help determine what types of crime will be most common in certain countries. Obviously, a heavily urbanised, highly populated nation like Japan will have different problems from those of one like Switzerland.

So, whether you're a prospective master criminal looking for the ideal scene of the crime, or a peaceful law abiding citizen who wants to know where he's least likely to be burgled or assaulted, there's useful information for you here.

However, a word of warning on all the crime figures contained in the pages which follow. As we've said elsewhere in this book, our facts and figures can only be as reliable as the source data they're taken from. Crime statistics need to be viewed with the same suspicions the police use in questioning a suspect for several reasons.

The first of these is that an astonishing amount of crime simply goes unreported. There are a variety of reasons for this. One primary one is embarrassment. The man who's had his wallet lifted while visiting a prostitute, or the firm that's suffered embezzlement but doesn't want its reputation to suffer as well; in each of these cases more often than not the police are never called in.

Then there are the appalling crimes of assault on women, rape, wife beating and incest. In all these cases the hapless victim is often too embarrassed, humiliated and (often) frightened to come forward and complain.

Furthermore, much petty crime goes unreported. Many merchants don't prosecute all shoplifters. They build a stock loss factor into the price you pay instead. But these are thefts.

Finally in many parts of the world law enforcement authorities jiggle the figures to make things look good. Until recent years F.B.I. statistics were a good example. Agents were encouraged to concentrate on interstate car thefts (an easy crime to solve) and thereby make the overall "case solved" figure look good.

The first table in this chapter shows how many policemen there are in the countries covered – and also gives figures for the number of policemen per thousand population. There are some surprises – America, where they have so much violent crime, has fewer policemen per thousand population than most European countries. France, where the police, as a percentage of the population, outnumber anyone else's police force, has the highest ratio of police to citizens. The real crime indicator is in the next table. This shows crime rates in the countries we've covered. Here we have figures for the number of sex crimes committed per hour, and the number of larcenies, murders and attempted murders committed per hour. It's not surprising to find the U.S. easily tops the bill here. Over in the U.S., over 303 murders, attempted murders and attempted thefts are committed *every hour* around the clock.

The next table gives an indication of police efficiency. It gives the number of major and minor thefts reported and the percentage of them that are solved. Surprisingly, the U.S. fares comparatively well in solving minor theft offences – in spite of all those despair stories. So maybe their over-worked police are pretty efficient after all. Though Japan is the place where the police are really on the ball.

One of the major areas of crime in the U.S. is drug offences. But the U.K. has its share of the problem, as you'll see from the figures in the next table. And the accompanying text comes to some very interesting conclusions on the subject.

Though drug offences are bad, the worst crimes of all are surely rape and murder. Our table gives the gruesome figures here. At a glance one can see why the U.S. has such an appalling reputation for violence. But they're not alone: those West German *burghers* seem fairly violent too. On the brighter side, the U.K. is well down the list in sixth place. Not only can you see the number of murders and sex offences for the countries listed, but also how many the police manage to solve.

Lastly we have a table which deals with prisoners. It shows just how many there are as a percentage of the population.

	U.S.A.	CANADA
	1976	1976
Number of policemen (000s)	418	63.7
per 100,000 inhabitants[0]	211	280

No surprise to find that the home of Kojak has the most cops. The real story, however, is in the number of police per hundred thousand inhabitants. These figures give an indication of police capability to deal with problems that arise on the beat. Here the U.S. is way down the list – sixth, to be precise.

The figure for the U.K. shows that we are fifth from the bottom of the table. We only have 192 policemen per hundred thousand. This poor ratio undoubtedly has much to do with inadequate pay. Perhaps recent pay awards will stop the drift of our police out of the force.

France has proportionately the highest number of cops, way ahead of the rest of the field. In the home of gendarmes and "flics", they have more than 450 policemen to every hundred thousand inhabitants – nearly one to every 2,000 people. In spite of this, ten years ago in May, 1968, the Left Bank students took over Paris and paralysed the entire country for more than a week. Many experts believe that the repressive effect of having so many policemen around helped fan the flames of riot.

Austria comes second, with 328 policemen per hundred thousand people. A hidden factor here is that Austria's army is limited in size by treaty obligations. Given Austria's geographical position this means that those Austrian police are responsible for a comparatively long mileage of frontier with Eastern Europe.

The Italians almost equal the Austrians with 323 policemen per hundred thousand. This is a 1973 figure, and with Italy's recent record of political violence and urban terrorism it has risen considerably. Exact figures aren't yet available, but it's generally accepted that Italy now has the largest force – population-wise – in Europe. They certainly need it.

Italy is followed by Canada – home of the Mounties, the men who always get their man. This boast is no idle one – the Canadian police solve 83% of murders and attempted murders (see murders and sex crimes table).

The Belgians have 263 policemen per hundred thousand inhabitants – 13 more than they have in the U.S. (Not the impression you get from watching T.V. – where half the U.S. population seems to consist of policemen or private eyes.)

The sober, industrious Dutch have proportionately a very small police force – 148 per hundred thousand inhabitants. But this is a misleading figure. Recently Holland has been something of a target for terrorists. Splinter groups from the West German Baader-Meinhof gang have a habit of going into hiding in Holland, and the Dutch have their own terrorist problem. When South Moluccan terrorists take over a train or a school filled with children, it makes world headlines.

U.K.	AUSTRALIA	AUSTRIA	BELGIUM	DENMARK	FRANCE	(WEST) GERMANY	IRELAND	ITALY	JAPAN	NETHERLANDS	NORWAY	SPAIN	SWEDEN	SWITZERLAND
1977	1976	1973	1973	1973	1973	1973	1973	1973	1976	1973	1973	–	1976	1976
108.2	28.0	24.3[+]	25.2[+]	8.2[+]	233.0[+]	133.6[+]	6.5[+]	174.4[+]	231.1	19.9	5.5[+]	NA	14.1	13.00
192	203	328	263	164	452	219	218	323	208	148	137	NA	170	200

NA not available

Sources:
Government sources
[+]Euromonitor
[0]Heron House estimates

But what isn't always so apparent is that it isn't the police who deal with the problem in these cases. It's the army who surround the place, sit out the seige, and finally go in at the end. It's also the army who supervise many of the everyday anti-terrorist counter measures. This means that, strictly speaking, they are in fact taking over police duties in these fields. So although the Dutch police force in itself may be proportionately the smallest, many more people are actually involved in what are usually considered as policing duties.

For this reason, Norway really qualifies as the most under-policed nation. Here they have 137 policemen per hundred thousand inhabitants – and here in Norway they're lucky. They don't need the army to help them out with their work, because they haven't yet got a terrorist problem.

The next most underpoliced nations are also both Scandinavian. In Denmark they only have 164 policemen per hundred thousand inhabitants, and in Sweden they have 170. If you look at the tables on crime, you'll see there's comparatively little crime in Scandinavia. So perhaps those Scandinavians are just more law-abiding than the rest of us, although the experts have their own opinions on the matter. They say there's less need for police in Scandinavia and less crime because it's largely rural, with populations spread thinner on the ground than elsewhere. Another contributory factor is that Scandinavian cities are well planned and less crowded. (If you look at the table on green space in cities you'll see what they mean.)

One significant omission in our table is Spain. The figures are not available. Until recently, Spain was the only permanent police state in Western Europe, and they still don't want to talk about it.

Do cops curb crime? Our figures indicate that possibly they do, but not by very much. France and West Germany combined have about the same number of police as the U.S., but only half the population – and a lower crime rate.

In addition, in the U.K., there is an increase in the amount of work undertaken by private security organisations. Without such companies who provide night-watchmen and who supervise the delivery of money, our police would be more overworked than they are.

Serious crime occurs with such monotonous regularity that most rapes, robberies or murders aren't even mentioned in the newspapers in many cities of the world. The table shown here illustrates the extent of serious crime in the countries listed in two ways. The first line of figures is the number of serious sexual offences committed per hour. These include rape, attempted rape or serious molesting (except in the U.S. where only rape figures are given). The second line shows the hourly rate for other serious crimes. These include murder, attempted murder and large-scale robberies. Since the figures give the number of reported incidents on an hourly basis, multiply by 24 to arrive at a daily total and then by 365 for the annual figure.

Keep in mind, however, that these figures are only for *reported* crimes. Many crimes remain unreported because the victims are embarrassed or frightened of repercussions. In other cases, they feel it's hardly worth the bother: in some cities, an over-worked police force can't offer much more than sympathy which does little to catch the criminal or lower the crime rate.

The U.S. has the highest number for sex offences per hour of all the countries listed, with an average of 6.4 – almost one incident every ten minutes, day and night. And that is only the rape figures. Figures for all sex crimes would be considerably higher. West Germany comes second with 5.4. However, these

	U.S.A.	CANADA
	1975	1970
1. Sex offenses per hour[+]	6.4[x]	1.2
2. Major larceny[0], murders and attempted murders per hour[+]	426.6[‡]	1.4

figures (and the others on the table) take no account of differences in comparative populations.

Great Britain follows West Germany with a relatively low rate: 3.1 sex offences per hour, or about one every 20 minutes.

After Great Britain, the rates drop markedly. Australia and France report 1.5 offences per hour, only about half the British rate. Canada has 1.2 per hour. This rate is low – but so is Canada's population figure.

Those reputedly hot-blooded Italians and Spaniards certainly give the female tourist some embarrassing moments, but it's obviously all in fun – their sex crime rate is very low.

At the crime-free end of the scale, there are a handful of countries where people can walk the streets with virtually no danger of serious sexual offences occurring. In Ireland, on average, such an offence is reported only once every two days. Their rate is a miniscule 0.02 offences per hour. Also low are Norway (0.11 per hour), Denmark (0.23), Austria (0.25), Sweden (0.37). The Scandinavian countries which have a reputation with the rest of the world for freedom in sexual manners, movies and literature, also have

	GREAT BRITAIN**	AUSTRALIA ++	AUSTRIA	BELGIUM	DENMARK	FRANCE	(WEST)^00 GERMANY	IRELAND	ITALY	JAPAN	NETHERLANDS	NORWAY	SPAIN	SWEDEN	SWITZERLAND
	1974	1974	1974	–	1974	1974	1974	1970	1974		1974	1974	1970	1974	1974	–
	3.1	1.5	0.25	NA	0.23	1.5	5.4	0.02	0.75	1.3	0.86	0.11	0.49	0.37	NA
	65.2	15.3	10.3	NA	9.9	23.5	114.4	1.05	149.8‡‡	0.46	12.4	2.3	14.5‡‡	13.3	NA

x Rape only   ‡ Murders and non-negligent manslaughter, excludes attempted murders.
**Excludes Northern Ireland   ++ Includes Papua New Guinea   00 Includes West Berlin
‡‡ All larceny offenses.

exceptionally low rates of sexual violence. This contradicts the much-voiced opinion that an increase of sexual freedom leads to an increase in sex crimes.

The U.S. is undisputed leader in the remaining figures, which report hourly rates of serious crimes including murder, attempted murder, and major larcenies. The American rate, 426.6 per hour, is more than twice that of any other country. This means that, on average, a major crime is committed every hour in every city of any size throughout the U.S. – or six major crimes every hour in every state of the union.

While the U.S. has twice the crime rate of other countries on the table, it also has more than twice the population. Their population is 215 million – the closest contender is Japan with 108 million.

It looks as if crime is also big business in Italy – traditional home of the Mafia. Here, the rate of serious crimes reported is nearly 150 per hour. Obviously the new waves of urban terrorism are an important factor here – not so much in the figures themselves, but in the atmosphere of lawlessness which the (comparatively) few spectacular and well-reported cases help to create. But in these high Italian

Sources:
Heron House estimates based on figures from the Federal Bureau of Investigation and Interpol.
+ Reported cases
0 Robbery, burglary, etc.

figures there is a mitigating factor. The Italian rate includes all larcenies reported.

After the U.S. and Italy, West Germany weighs in with a still-substantial rate of 114.4 serious crimes an hour. Here again, there are problems with urban terrorism. You only need one Baader-Meinhof gang and the ensuing ripples of lawlessness spread throughout the entire country.

After the preceding high rates, Great Britain registers a rather moderate 65.2 per hour. However, even this modest average means a serious crime occurs more than once a minute, day and night.

Still, we British seem tranquil and pacific compared to Americans. In the minute or so it took to read this page, one major crime occurred in Great Britain, but in the same short time span, 50 Americans were attempting to murder or rob their fellow-men.

Crime certainly pays in the U.K. Two out of every three larcenies committed remain unsolved. To be fair to our British bobby, however, that is a much better rate than most police forces achieve. Only the Japanese police do better. Most countries have at least twice as many minor as opposed to major larcenies.

In the countries studied, there are generally twice as many minor larcenies as major thefts.

Austria and West Germany are interesting exceptions. There, crooks go for big-time robberies.

In the U.S., small-time crime is the order of the day. They have about 5.9 million minor robberies a year. That works out to about 28 for every thousand people in the country. There are also 17 or 18 major thefts for every thousand citizens.

	U.S.A.	CANADA
	1975	1970
1. Major larceny+ offenses (000s) known to the police	3717	11.6
per 100,000 population	1744	55
2. Major larceny offenses solved (000s)	NA	NA
as % of major larceny offenses known to police	NA	NA
	1975	–
3. Minor larceny⁰ offenses known to the police (000s)	5878	NA
per 100,000 population	2804	NA
4. Minor larceny offenses solved (000s)	1177	NA
as % of minor larceny offenses known to police	20.0	NA

Canada has only 11,600 major thefts for the year shown. Things may have become worse during the 70s. Even so, a rate of 55 major robberies per *hundred* thousand people makes Canada a much safer place to keep your goods than the U.S.

The Canadians don't keep a record of the percentage of the thefts that have been solved. Neither does the U.S. where major larcenies are concerned.

The U.S. has a lot of crime. But they don't come near the rate of thefts per unit of population found in Sweden and Denmark. One in every 25 Danes can expect to be robbed by a petty thief. In Sweden the rate is only slightly less.

When it comes to major larceny, Denmark has 17 major thefts per thousand people. In Sweden, the figure is roughly 14 for every thousand. Of the big jobs, only 15% are solved in Sweden, and 25% in Denmark. With the small, 82% get away in Denmark, and 84% in Sweden.

GREAT BRITAINˣ	AUSTRALIA⁺⁺	AUSTRIA	BELGIUM	DENMARK	FRANCE	(WEST) GERMANY**	IRELAND	ITALY	JAPAN	NETHERLANDS	NORWAY	SPAIN	SWEDEN	SWITZERLAND
1974	1974	1974	–	1974	1974	1974	1970	1974	1974	1974	1970	1974	1974	–
570.0	133.8	90.8	NA	86.8	204.6	999.9	9.3	⁺⁺1311	2.1	108.2	20.5	⁰⁰127.2	116.7	NA
1047	888	1214	NA	1723	387	1611	308	⁺⁺2355	1.94	802	524	⁰⁰360	1427	NA
188.8	26.5	26.7	NA	22.1	37.0	210.3	NA	NA	1.7	31.0	NA	NA	17.4	NA
33	20	29	NA	25	18	21	NA	NA	80	29	NA	NA	15	NA
1973	1974	1974	–	1974	1974	1974	–	–	1974	1974	–	–	1974	–
772	278.2	45.5	NA	200.3	919.9	833.3	NA	NA	1013	205.2	NA	NA	288.7	NA
1421	1847	609	NA	3976	1742	1343	NA	NA	920	1521	NA	NA	3530	NA
264.8	78.6	16.5	NA	35.7	172.0	348.1	NA	NA	517.7	34.5	NA	NA	47.2	NA
34.2	2.8	27.5	NA	17.8	18.6	41.7	NA	NA	51.0	16.8	NA	NA	16.0	NA

NA not available　　⁺⁺*All larceny offenses*　　⁰⁰*Includes all types of fraud*

To be fair, the rates of solving crimes of larceny are almost uniformly low except for those efficient Japanese who solve four out of five serious thefts. In fact, Japan is by far the most theft-free country on the table. There's only one major larceny for every 50,000 people. And only about one petty theft per hundred – not great, but better than anyone else. Our British bobbies are on their toes too. They solve 34.2% of all major larcenies committed – a pretty good average.

Source:
Interpol
USA: Federal Bureau of Investigation
⁺Robbery, burglary
⁰Theft, pickpocketing, shoplifting
ˣExcludes Northern Ireland
‡Includes Papua New Guinea
**Includes West Berlin

Nearly all human societies use drugs whether it's nicotine, alcohol, opium, hashish, aspirin, caffeine or peyote. According to anthropologists there's a remote Eskimo tribe which is a rare exception – they rely on deep breathing exercises instead.

All countries have some laws against narcotics. In most industrialised western countries, narcotics mean derivatives of the opium poppy – opium, morphine and heroin – as well as a number of other naturally occurring drugs. The term also covers a wide spectrum of man-made drugs developed for specific medical purposes. Illegal use or abuse of those drugs can also be a narcotics offence.

The laws vary all over the world. What is termed a "dangerous drug" or "controlled substance" in one country might well be exempt from penalties in another. So while using cannabis is still a crime in the U.K., the Dutch have decriminalised its use. Penalties for offences involving the drug will also vary.

The figures on the table give some idea of the rate of drug abuse in various countries. Exact comparisons can't be made since the definition of "narcotics offences" differs in each place.

In some countries, no figures are available. In the 14 where they are, the results are so uneven as to astonish even the most ardent fact-seeker. Rates vary between three offences per million people (Japan) and 1 in 356 people (the U.S.).

When it comes to narcotics offences, some nations are "have's" and others are

	U.S.A.	CANADA
	1975	1975
Narcotic arrests/convictions (000s):	601.4^{0}	60.0
per 100,000 inhabitants[+]	281.0	263.0

"have not's". The big three among the have's are the U.S., Sweden and Canada. One out of every 356 Americans is arrested every year for one offence or another against the drug laws. In Sweden the figure is 1 in 378, in Canada an almost identical 380. Australia follows with 1 in 869, then Denmark with 1 in 1,282. Norway (1 in 3,448) and the Netherlands (1 in 4,545) are next.

Japan, Ireland and Spain have the lowest number of narcotics arrests. All these countries have highly traditional, conservative cultures, which were largely unaffected by the turbulent "protest" movement of the 1960s.

The U.K. figure is also low – 16,000 arrests against over 600,000 in the U.S. This doesn't mean that British society is unusually drug-free. The U.K. figures on this table only show the number of people found guilty of drug offences and, strangely enough, such offences are usually to do with "soft" rather than "hard" drugs. We British are more inclined to treat addiction to "hard drugs" like heroin as a medical problem. Addicts are registered with the National Health Service, and they get their drugs

NARCOTIC OFFENCES

	U.K.	AUSTRALIA	AUSTRIA	BELGIUM	DENMARK	FRANCE	(WEST) GERMANY	IRELAND	ITALY	JAPAN	NETHERLANDS	NORWAY	SPAIN	SWEDEN	SWITZERLAND
	1976	1975	1976	–	1975	1976	–	1976	1976	1976	1975	1976	1976	1975	–
	$16.0^{‡}$	15.8^{0}	0.7	NA	3.9	3.8	NA	0.03	2.4	0.35^{x}	3.0^{x}	1.2^{x}	0.4	22.0	NA
	3.0	115.0	9.0	NA	78.0	7.0	NA	0.9	4.0	0.3^{+}	22.0^{+}	29.0	1.0	264.0	NA

^{0}All offenses known to the police connected with narcotics, including the sale of narcotics
^{x}People charged with any offenses contravening the narcotics laws
‡People found guilty of drug offenses   NA not available

Sources:
Government Departments of Justice
$^{+}$Heron House estimates

from the state free of charge. This may sound like condoning addiction, but it has at least two very practical results. The drug issue is out of the hands of criminals and policemen, and drug-related crime is minimised. In the U.K., there are fewer addicts who have to steal and traffic narcotics to support their habits.

The French underworld, especially that part of it centered around the port of Marseilles, is a major supplier of heroin to Europe and North America. Yet the figures for France are low.

Two main factors seem to be involved in the figures on the table: whether or not drugs are wanted by large segments of the population; and whether or not police and other enforcement agencies are aggressive in arresting and prosecuting drug offenders. In the U.S. there are millions of drug-users – and aggressive enforcement of the narcotics laws. As a result, figures for drug offences are high. The low figures for countries like France and Italy may be due to lack of interest in narcotics on the part of the French and Italians or lack of police interest.

If you're a fan of American T.V., it won't come as a surprise to learn that Americans murder each other at a faster rate than any other people shown. There are nearly ten murder attempts – successful and unsuccessful – for every hundred thousand people, or one for every ten thousand. That's the equivalent of one murder a year for every small town and neighbourhood in every large town and city. The real rate is even worse than that. Only murders for which somebody is arrested are counted. Unsolved murders, and murders that never get reported, send the true rate of murder sky high.

We in the U.K. manage to get along together much better than the Americans. Less than a third as many of us resort to killing each other. That puts us tenth in the murder league.

A big surprise on the murder table is the Netherlands, with its windmills, tulips and pink-cheeked population. Surprisingly, seven in every hundred thousand Dutchmen die at the hands of killers. In Holland, only one in ten murderers escapes arrest.

If you think somebody's after you, board the next flight for Japan. Statistically, at least, you'll get protection. The Japanese murder rate is 400% lower per hundred thousand population than in the

	U.S.A.	CANADA
	1975	1970
1. Murders and attempted murders known to the police	20,510[0]	690
per 100,000 population	9.6[0]	3.4
2. Solved murders and attempted murders	15,998[x0]	571
as % of cases known to the police	78[0x]	83
3. Sex offenses[+] known to the police	56,093[‡]	11,025
per 100,000 population	26.3[‡]	55.0
4. Solved sex offenses	28,725[‡x]	5951
as % of cases known to the police	51.2[‡]	54.0

U.S., and the efficient Japanese police solve all but four per cent of murder-related crimes.

The best places not to get murdered are Norway and Ireland – with probably the most rural economies of any countries on the chart. There must be some correlation between lots of elbow room and people living together peacefully.

Space to get away from it all doesn't seem to be the determining factor when it comes to sex offences. Those randy Aussies, with miles of lonely outback to roam in, top the table when it comes to sex crimes (including rape and "traffic in women" – also known as prostitution).

MURDERS AND SEX OFFENCES

	U.K.**	AUSTRALIA++	AUSTRIA	BELGIUM	DENMARK	FRANCE	(WEST) GERMANY 00	IRELAND	ITALY	JAPAN	NETHERLANDS	NORWAY	SPAIN	SWEDEN	SWITZERLAND
	1974	1974	1974	–	1974	1974	1974	1970	1974	1974	1974	1970	1974	1974	–
	1301	411	229	NA	102	1429	2771	24	1643	1912	964	6	233	275	NA
	2.4	2.7	3.1	NA	2.0	2.7	4.5	0.03	3.0	1.7	7.2	0.02	0.7	3.4	NA
	1162	374	216	NA	72	1151	2621	22	NA	1837	875	NA	NA	146	NA
	89	91	94	NA	71	81	95	92	NA	96	91	NA	NA	53	NA
	27028	13674	2274	NA	2068	13828	48075	261	6605	11338	7554	1047	4310	3313	NA
	49.7	90.8	30.4	NA	41.1	26.2	77.5	8.7	11.9	10.3	56.0	27.0	12.2	40.5	NA
	21016	8321	1942	NA	1071	10368	33994	216	NA	10428	3747	476	NA	1574	NA
	77.7	61.0	85.0	NA	52.0	75.0	71.0	83.0	NA	92.0	50.0	45.0	NA	48.0	NA

NA not available
0 Murder and non-negligent manslaughter   x Crimes cleared by arrest   ‡ Rape only

And the Aussies don't always have to pay a price for their crimes. Four out of every ten Australian sex offenders manage to get away scot-free.

After the Aussies, West Germany and Holland rank next in the number of sex crimes. Germany is second highest, with nearly 78 per hundred thousand. They have a pretty good success rate – in terms of solving crimes, that is. Almost three in four West German sex offenders get caught. Holland ranks third in sex offences – with about one sex crime for every two thousand people. Only half of Dutch sex offenders get caught. Tolerant attitudes towards prostitution may be partially the cause, and "traffic in women" is an offence that perhaps isn't taken very seriously.

The U.K. rate for sex crimes puts us fifth in the table.

Sources:
Interpol
USA figures from the FBI
+ Includes rape and traffic in women
** Excludes Northern Ireland
++ Includes Papua New Guinea
00 Includes West Berlin

In every country, there's a small minority that chooses to disobey society's rules. Some get away with it, but most are caught and many of those end up in jail. Our figures here show how the prisoner population varies from one country to another.

Where do we find the most prisoners? Both in terms of overall numbers and percentage of the population, more criminals are under lock and key in the U.S. than anywhere else. They have a total of nearly 400,000 jail-birds, that's 189 per hundred thousand of the population – over twice the figure for the U.K. and other countries.

Others with well filled jails are Canada, Germany and the U.K. When you recall that Australia was largely colonised by shady characters, you might expect to find quite a few jail-birds there. It's sixth in the table, but there are only 70 prisoners per hundred thousand of the population, so it's really a far more law-abiding country than the U.S.

Where do the largest number of people manage to keep out of trouble? They must be well-behaved in the Netherlands, or perhaps their police just turn a blind eye. The prison population is only 2,700 – just 21 out of every hundred thousand of the total population.

Dutch women are particularly law-abiding, they make up a smaller proportion of the prison population than in any other country: under two per cent. On the other hand, there are a lot of people awaiting trial in the Netherlands. Taken as a percentage of the number already in jail, the figure is higher in Holland than anywhere else except Italy. Nonetheless, even if all of them were eventually jailed, the Dutch prison population would still be the lowest.

There aren't many prisoners in Ireland either – only 35 in every hundred thousand of the population. It also has one of the lowest figures for people

	U.S.A.	CANADA
	1972	1974
1. Prisoners (000s)	393.7	20.7
per 100,000 inhabitants[+]	189	95
2. Female prisoners	28,112	702
as % of total prison population[+]	7.1	3.4
3. Male prisoners (000s)	365.6	20.0
	1973	1974
4. Prisoners under 21	61,794	5608
as % of prison population[+]	15.7	27.0
5. Accused persons in prison awaiting trial	NA	2533
as % of prison population[+]	NA	12.2

PRISONERS

U.K.	AUSTRALIA	AUSTRIA	BELGIUM	DENMARK	FRANCE	(WEST) GERMANY	IRELAND	ITALY	JAPAN	NETHERLANDS	NORWAY	SPAIN	SWEDEN	SWITZERLAND[0]
1974	1974	1974	1974	1974	1974	1974	1972	1972	1974	1972	1974	1972	1974	1975
41.7	9.2	7.8	5.6	2.7	27.1	50.5	1.1	27.8	46.0	2.7	1.5	13.8	3.5	4.5
75	70	104	58	54	52	81	35	51	43	21	39	40	43	69
1134	275	353	220	82	711	1364	26	1454	1059	49	35	715	94	800
2.7	2.9	4.5	3.9	3.0	2.6	2.7	2.8	5.2	2.3	1.7	2.3	5.2	2.7	17.7
40.6	9.0	7.4	5.4	2.6	26.4	49.2	1.0	26.4	45.0	2.7	1.5	13.1	3.4	3.7
1974	1974	1974	1974	1974	1974	1974	1972	1972	1972	1972	1974	1972	1974	1975
11912	1554	1007	611	497	4305	NA	428	4560	1161	983	344	1583	831	1000
28.5	16.8	12.9	10.9	18.3	15.9	NA	40.6	16.4	2.5	35.4	22.3	11.4	23.5	22.2
3541	1020	2223	1217	840	10731	15942	121	15116	8390	1311	460	5761	520	1200
8.4	11.0	28.5	21.7	31.0	39.6	31.6	11.5	54.4	18.2	47.2	29.8	41.7	14.7	26.6

NA not available

awaiting trial. But Ireland shares one striking characteristic with the Netherlands – they both have a high number of young offenders. In Ireland, people under 21 account for more than 40% of the jail population. The figure for the U.K. (28.5%) is much less dramatic, although there's plenty of unrest among our youth. The unrest does not necessarily mean that our young people actually break the law enough to go to jail. In Japan, there's the lowest number of young offenders, which is possibly because of the strict upbringing they receive.

Sources:
United Nations
+Heron House estimates
0Department of Justice

The figures for Italy are interesting. With only about 50 people in jail for every hundred thousand inhabitants, they seem much better behaved than us. But the number of Italians awaiting trial is over 54% of the number in jail, compared with our small proportion of 8.4%. Since these figures were made available, the level of crime in Italy has increased so dramatically that the courts can't keep up.

323

Murders and attempted murders known to the police

21,000		
17,500		
16,250		
15,000		
13,750		
12,500		
11,250		
10,000		
7,500		
6,250		
5,000		
3,750		
2,500		
1,250	1,301	1,429
10		
NORWAY	**U.K.**	**FRANCE**

The U.S. is famous for its fictional crime-busters like Kojak and Columbo. It needs all of them, fictional or otherwise, with over 20,000 known murders a year. That figure means about one American in every 10,000 will die at the hands of another.

No other country approaches that number of murders. West Germany is next, with 2,771 in one year. You probably associate Japan with tea-drinking ceremonies and flower-arranging. But the Japanese have the third-highest murder figure (1,912) which suggest that modern pressures are changing that gentle image.

Although there are only 964 murders committed in the Netherlands, the odds on a Dutchman being murdered are second only to those for an American. Seven in every hundred thousand Dutch are murdered.

6

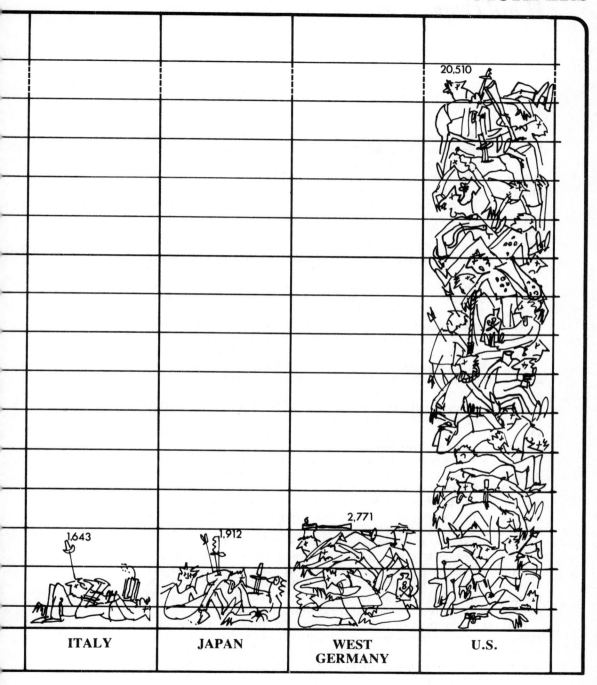

20,510

2,771

1,912

1,643

ITALY

JAPAN

WEST
GERMANY

U.S.

What People Use

Have you ever stopped to think how dependent you are on the hundreds of products that modern industry churns out? From the time you stumble out of bed and into the bathroom till you finally switch off the light and call it a day, you'll probably have used most of the products we talk about in the following tables.

As you'll see, nationality plays an important part in determining what people use and what they don't. Of course, most people, regardless of their nationality, wash themselves when they get up in the morning. And that involves soap, of course – or does it? Have a look at the section on soap and deodorants to find out who "the great unwashed" really are. While we're on the subject of personal hygiene, let's not forget another important area – your teeth. You might think absolutely everyone owns at least one toothbrush – and uses it. You'd be wrong. Turn to our table on toothbrushes to find out whose teeth are whiter than white – and whose aren't! (This section will be particularly interesting to dentists wondering where to set up shop.)

Then there's a product that only about half our readers have any use for – shaving equipment. According to the figures, there are some shaggy countries around the globe. You may be surprised when you find out which ones. Then there's the eternal male debate over electric shavers versus the old reliable razor. Do you shave wet or dry? This chapter will show you who agrees with your particular preference.

There's lots for the ladies here, too. You'll soon find out the countries to head for if you like the fresh natural look. You'll also see where the girls spend a lot of their time in front of the make-up mirror. But you'll have to decide for yourself if they do it because they actually need to or whether they're just showing an admirable interest in their appearance. There are some areas where even the most intrepid researchers fear to tread.

When it comes to giving nature a helping hand – or whatever – there is some enlightening information on bras. In which countries do most ladies wear them? And, where do most of the female population "let it all hang out"? We also have some eye-opening figures on – well, on eye-opening figures!

Still on the subject of female pulchritude, there is a table on a rather touchy subject – hair colourants. Do blondes really have more fun? It seems a lot of women think so. In some countries about half the women you meet may not be what they seem. Be sure to take a close look at the figures for the Scandinavian countries – some of those blonde Nordic beauties may be pulling the wool over our eyes!

Looking at hair care from the male point of view, can you still get a decent haircut for a reasonable price? Your answer will depend a lot on where you live.

While we're on the topic of hair, there's another vast range of products we rely on to keep us looking good – shampoos. Have a look at the table and see where the shiniest heads are to be found. And find out where they shun commercial hair cleaners and rely on water and old-fashioned elbow grease.

We also tackled the delicate subject of "white shoulders". You may assume there's an obvious connection between shampooing and dandruff – see if you're right.

Even though all the products mentioned so far are useful, you could probably live without them. But there's another product that a lot of people think of as an absolute necessity. That's the controversial weed, tobacco. Our section on smoking sheds some light on just how effective all those anti-smoking campaigns have been. You'll find out which countries you should head for if you're a clean air fan. This section demolishes some of the old stereotypes about who the heavy smokers are. It also points to some interesting trends in the major tobacco-producing nations of the world.

Finally, we come to what's probably the most necessary product of all – energy. Where would we be without oil and electricity? Who are the major consumers of the world's rapidly shrinking energy resources? These questions are especially important ones now, with countries like the U.K. facing a massive energy crisis and screaming headlines about the latest manoeuvres in the Arab world. It's a commonly known fact that North America is greedy when it comes to slicing up the energy pie – so where does that leave the rest of the world? Have a good look at this section and find out. These are just the kind of statistics that come in handy when settling those dinner-table arguments about economics.

In fact this whole chapter is full of facts – and some theories – about products that affect the lives of us all. Things that we take for granted but would be lost without.

When did you last take a bath? If you put that question to a lot of people, you might come up with some distressing answers. This week, last week, sometime, never... People in different countries vary in hygiene habits. Fortunately, at least in the countries surveyed, most people we spoke to did use soap.

If soap-buying is a measure of hygiene, the Scandinavians come out the cleanest of us all. You'd have a job finding a single household up in the Nordic countries without at least a bar or two of soap around the place. And if cleanliness *is* next to godliness, the Scandinavians should have nothing to fear when they're at Valhalla's gates. Unless, of course, they're not next to godliness at all – but are using that soap for waxing skis.

You'd think those bars of soap would be enough to keep them smelling sweet in Scandinavia. Well, they aren't. In Norway and Sweden, about 80% of women use deodorant regularly as extra security. In Denmark, the rate's lower – at 66%. But among Scandinavian men, only one in three bothers to use an anti-perspirant of some kind. Either there's a double standard, or Scandinavian men aren't as sweet as their gorgeous women.

Next in the soap league come Britain and Australia, where 99% of the population buy soap. That's enough to keep just about everyone well scrubbed, a toothbrush. According to the table, the population – about 560,000 people. That's equivalent to the population of Edinburgh. Or five times the number of a packed football crowd at Wembley.

Forty-one per cent of British men use deodorants, compared with 73% of women.

Except for Japan, it's always the women who use deodorants more frequently than the men.

Canada, Italy and the Netherlands come high on the table: 98% of them use soap. That leaves two per cent unclean. In Canada, that two per cent accounts for about 400,000 people. Maybe they're up in the backwoods of the far north, where it's too cold to take a bath. More Canadian men use deodorant than in any other country – twice as many as Dutch and Italian men. Of course in Italy you can always take a quick dip in the Mediterranean.

	U.S.A.	CANADA
	1977	1977
1. Households (%) who buy soap	97	98
	1977	1974
2. People who use a deodorant/ anti-perspirant regularly:		
men (%)	80	85
	1977	1974
women (%)	90	95

SOAP AND DEODORANTS

	U.K.	AUSTRALIA	AUSTRIA	BELGIUM	DENMARK	FRANCE	(WEST) GERMANY	IRELAND	ITALY	JAPAN	NETHERLANDS	NORWAY	SPAIN	SWEDEN	SWITZERLAND
	1977	1977	1977	1977	1977	1977	1977	1977	1977	—	1977	1977	1977	1977	1977
	99	99	83	91	100	91	84	72	98	NA	98	100	94	100	86
	1976	—	1976	1975	1975	1975	1975	—	1975	1977	1976	1975	1975	1976	1975
	41	NA	31	26	32	29	18	NA	43	25	45	35	39	37	34
	1976	—	1976	1975	1975	1975	1975	—	1975	1977	1975	1975	1976	1975	1975
	73	NA	51	73	66	70	74	NA	66	21	72	78	63	80	74

NA not available

You might expect the U.S., land of soap operas and soapbox derby's, to top the world in soap sales. Well, it isn't true. Comparatively they're dirty. Three per cent, or over six million people, don't touch the stuff. That's almost the population of Chicago or Los Angeles. But both the U.S. men and women make up for soap shortfalls when it comes to deodorants. Eighty per cent of the men and 90% of the women use antiperspirants regularly. That's nearly twice as many men as in the U.K.

Let's look at the bottom of the table. Where do soap salesmen have the toughest time? It's a different kind of bar they go for in Ireland. Mention bubbles to an Irishman and he's apt to think of a glass of Guinness stout. According to the study, one in every four Irish doesn't use soap at all. That's about 700,000 people – over 100,000 more than live in Dublin.

Sources:
Confidential industry sources

329

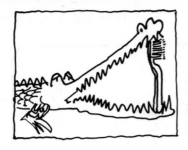

	U.S.A.	CANADA
	1977	1977
1. Toothbrushes bought per person per year	1.3	1.0
2. Toothbrushes sold per year (millions)	280	20
3. Households who buy toothpaste (%)	95	96

Never look a gift horse in the mouth. Why? Because you might well be put off by the state of its teeth.

But it's not just horses' teeth we care about. We have doctors to look after the health of our bodies. We have chiropodists to look after our feet. And we have dentists to look after our mouths. In fact, we spend millions every year at the dentist. Filling the cavities. Removing the plaque. Straightening the alignment. We are somewhat obsessed with the need for snow white, cover-girl choppers.

However, an ounce of prevention is better than a gram of novocaine. Toothbrushes! Toothpaste! The real route to healthy root canals is in our hands.

Who buys the most toothbrushes? And how many households use toothpaste among the countries on our table? As we all know, there's practically no such thing as 100% of anything. But what's surprising is the fact that a considerable percentage of people in each country don't brush their teeth. In the U.K., ten per cent of the households don't buy toothpaste. That adds up to about five million people. Maybe they're rubbing their teeth with baking soda. Or salt. Or old-fashioned tooth powder. But they'll never get that Michael Parkinson smile right if they keep stinting on their mouths like that.

In which countries do people take the best care of their teeth? First of all, before you buy toothpaste, it's useful to invest in a toothbrush. According to the table, the Australians buy more toothbrushes than anyone. Nearly two are sold for every person in the country every year. That means the average person buys a toothbrush every six months.

The pearly-smiling Aussies are way ahead of the rest of the field. The Japanese are next. They buy an average of one and a half toothbrushes every year – or one every eight months. The Swedes and the Americans buy an average of 1.3 toothbrushes a year – you need that extra third to get in where the ordinary bristles don't reach.

The U.K. is roughly average – one toothbrush bought every 15 months. If we follow our dentists' advice twice a day, our toothbrushes get worked out almost 1,000 times.

Who are the worst in the clean enamel club? The average Belgian buys a toothbrush less than once every three years. The Irish are just as bad. And the Italians, Spaniards and Austrians use the same brush for 30 months. The bristies must be worn down to stumps at the end of that kind of use.

TOOTHBRUSHES AND PASTE

U.K.	AUSTRALIA	AUSTRIA	BELGIUM	DENMARK	FRANCE	(WEST) GERMANY	IRELAND	ITALY	JAPAN	NETHERLANDS	NORWAY	SPAIN	SWEDEN	SWITZERLAND
1977	1977	1977	1977	1977	1977	1977	1977	1977	1977	1977	1977	1977	1977	1977
0.8	1.9	0.4	0.3	1.0	0.5	0.7	0.3	0.4	1.5	0.7	1.2	0.4	1.3	0.6
44	25	3	3	5	27	42	0.9	22	167	10	5	15	11	4
90	96	85	81	80	86	83	53	94	NA	86	75	87	95	90

NA not available

Getting a toothbrush operating is only half the battle. Now we need to know how many people use toothpaste. Here the Australians come out on top again – along with the Canadians. Ninety-six per cent of all Australian and Canadian households use toothpaste. We don't have data on what type of toothpaste they prefer – green, striped, plain, with fluoride or without – and neither do we know whether they press the tube from the bottom or the top.

In any case, the next countries on the list are the U.S. and Sweden. Here 95% of all households use toothpaste. The Italians are marginally less kissable, at 94%. Even though they don't change toothbrushes too often, they get high marks when it comes to paste.

Of all the countries studied, the Irish buy the least amount of toothpaste. Only slightly more than half of all Irish households use the stuff. No wonder the song goes "When Irish *eyes* are smiling...".

Sources:
Confidential industry sources

We are all familiar with the popular image of the Englishman shaving with soap and razor. But the common stereotype of an American male jumping into his Chevy, headed for work with an electric shaver plugged into his dashboard, is false. Three-quarters of all American men who shave do so with water! Curiously, of all the countries listed in our table almost the only places in the world where wet shaving is preferred are the English-speaking ones. Nowhere is there a lower percentage of electric shavers than in the U.S. What's more, bearded Uncle Sam keeps watch over virtually the hairiest male population in our studies. Only Canada – no doubt because of nippy weather, lumberjacks and hirsute trappers – has fewer shaved men. The proportion of shavers to non-shavers, of course, is still heavily on the side of the smoothies. But only 88% of Americans shave, compared to nearly 97% in both Belgium and Japan and 95% in the U.K.

Apparently there's little connection between whether a man shaves, and whether he prefers to shave with a wet blade.

It's true that in most places where shaving is practised by 95% or more, the preference is solidly in favour of the electric razor. In Norway, Sweden, West Germany and Belgium, shavers prefer electric about three to two. But in the U.K. and Italy – both big on shaving – the preference is heavily wet, 74% and 72% respectively. Perhaps our stiff upper lips simply can't get used to electric razors.

In about half the countries, the preference for electric is very pronounced – about 60% or a bit more. In three places – France, Spain and Japan men are split pretty evenly between wet and dry. But in the U.S., Canada, Australia, Ireland, the U.K. and Italy, there is a very strong preference for the old-fashioned razor blade. Many men say that the early morning philosophical speculation that accompanies every stroke of the blade is vital to the success of the day.

One peculiar fact shown on our table is that though Canadians prefer a blade, they do so to a much smaller degree than men in other non-electric countries. In fact, only 60% of Canadians use soap and blade, while three-quarters of the men in the other "wet" countries do so.

Maybe Canada's French-speaking population throws off the statistics by

	U.S.A.	CANADA
	1976	1976
1. Men (%) aged 15 and over who:		
shave regularly	88	87
do not shave regularly	12	13
2. Men (%) aged 15 and over who shave:		
wet	75	60
with an electric razor	25	40

	U.K.	AUSTRALIA	AUSTRIA	BELGIUM	DENMARK	FRANCE	(WEST) GERMANY	IRELAND	ITALY	JAPAN	NETHERLANDS	NORWAY	SPAIN	SWEDEN	SWITZERLAND
	1976	1976	1976	1976	1976	1976	1976	1976	1976	1976	1976	1976	1976	1976	1976
	95	90	91	97	90	94	95	92	96	97[+]	95	95	90	95	93
	5	10	9	3	10	6	5	8	4	3[+]	5	5	10	5	7
	72	72	38	40	40	54	41	72	74	51	33	40	48	40	38
	28	28	62	60	60	46	59	28	26	49	67	60	52	60	62

[+]Aged 18 and over

shaving dry out of respect for its ancestral country. However, Canada's shaving habits are remarkably like those of the Americans in terms of whether men shave or not. In fact, these two places are the only ones where more than ten per cent don't shave. Yet a full 40% of Canadian men who shave have been converted to electric razors, compared to only 25% of American men.

In conclusion, except for Italy, most of the old-fashioned shavers in the world speak English. Will the sun never set on hot foam and razor? Before any further deductions can be made, we need another set of statistics. Are non-shavers (like Bluebeard) more given to violence? Are English-speaking and Italian men tempting fate with every stroke of the razor? Or do we simply have more pimples? Or could it be that to shave or not to shave is decided at the whim of fashion as last year's beard gives way to this year's smooth look.

Sources:
Confidential industry sources

You can't tell from figures which countries have the most beautiful women. And a good thing too – long may beauty lie in the eye of the beholder. However, we can shed some light on what women use in an attempt to allure their opposite numbers.

What do they use most frequently in order to improve upon nature? This table shows that lipstick is the favourite in most countries. Overall, figures for perfume and toilet water are only marginally lower. (Maybe husbands around the world just can't think of anything else to buy their wives at Christmas.) In some countries, such as the U.S. and Switzerland, they're higher than for lipstick.

To a large extent the figures for lipstick are reflected in the use of other toiletries. However, there are notable exceptions. For example, Swiss ladies splash on perfume though they're only average users of lipstick. Eye make-up holds little attraction for them and rouge is positively shunned. Swiss women like to complement an outfit with an attractive smell and the fresh-faced look.

Canada is the country of the cupid's bow: 90% of all women wear lipstick. Canadians also wear twice as much eye-shadow as their European sisters. British women, along with the Dutch are Europe's most lavish lipstick-users.

The outdoor girls of Norway are lowest on the list: only 40% of them touch up their lips. Denmark comes next lowest, with a figure of just 43% who bother to colour their lips. And only three per cent more use eye make-up.

Despite their reputation for rose-like complexions, our English beauties slap on a lot of face-powder. Only Swedish women are more anxious to cover up – 43% of them use powder as against 39% of English women.

Overall, the Scandinavian trend is towards the natural look. But the Swedes are the odd women out.

Canadian and American women are the most anxious to improve their looks, with their British and Dutch sisters next.

France's figures are average right across the board. Parisian chic seems to take little more than a pot of foundation cream and a dash of lipstick. Clothes obviously play a more important part than make-up in creating that famous

	U.S.A.[+]	CANADA[+]
	1977	1974
Women (as % female population over 15) who use:		
make-up base	50	42
face powder	33	NA
lipstick	71	90
rouge; blusher	43	62
mascara	50	67
eye shadow	47	77
perfume; toilet water	85	82

U.K.	AUSTRALIA	AUSTRIA	BELGIUM	DENMARK	FRANCE	(WEST) GERMANY	IRELAND	ITALY	JAPAN	NETHERLANDS	NORWAY	SPAIN	SWEDEN	SWITZERLAND
1975	–	1975	1975	1975	1975	1975	–	1975	–	1975	1975	1975	1975	1975
38	NA	30	35	17	47	28	NA	31	NA	28	20	26	48	29
39	NA	29	32	17	28	22	NA	34	NA	29	22	32	43	23
74	NA	57	61	43	68	69	NA	48	NA	76	40	55	67	62
24	NA	21	18	17	17	18	NA	11	NA	14	31	15	14	11
42	NA	28	30	30	39	36	NA	36	NA	42	28	34	46	28
41	NA	26	31	16	37	34	NA	32	NA	43	19	46	44	26
68	NA	63	63	21	58	62	NA	43	NA	76	11	54	62	89

French look.

More Swiss women wear perfume than any other nationality. In Switzerland, 89% of all women use perfume or toilet water. Americans are second in the sweet-smelling league (85%), and Canadians third with 82%. These are the only exceptionally high figures on the table. The Netherlands are fourth (76%) and the U.K. fifth, with 68%. In France, home of Dior and Chanel, a mere 58% of ladies indulge.

The sporty Scandinavians care least what they smell like. In Norway, only 11% of women use perfume or toilet water. Their neighbours in Denmark come next, with 21%. Those sauna baths must give them confidence.

The figures for Japan, Australia and Ireland just aren't available – we'll never know what really puts the sparkle in those smiling Irish eyes.

Sources:
Euromonitor
+Confidential industry sources

	U.S.A.	CANADA
	1977	1977
Percentage of women who take bra size:		
A cup	15	22
B cup	44	42
C cup	28	22
D cup	10	7

In an effort to keep abreast of our times *The Book of Numbers* has obtained bra cup sizes on a multinational basis from confidential industry sources. We'd like to be able to say this accurately reflects the female physiognomy from country to country but it just isn't so for a variety of reasons.

In the first place these figures are based on those who buy brassières, not on the adult female population as a whole. In some countries – such as Mexico – owning a bra is restricted to those who can *afford* to lift and separate. As income increases pulchritude gives way to trim well-being and here the study was conducted amongst bra-owning women. Spain proves the point – 59% of women queried didn't specify their cup size. Modesty doubtless is part of the reason (as it is with the secretive Swiss) but another reason is quite simply many Spanish women don't wear bras. In the U.K., it's not so much whether you can afford a bra as whether it's fashionable to wear one.

These days, dresses are often strapless or backless and obviously designed to reveal what nature – rather than the lingerie department – has provided. Nevertheless, 96% of British women gave their bra cup size, so it seems they like that reassuring uplift, at least some of the time. Over half the British women surveyed wore a B cup; only two in ten took a C or D cup.

While all of the data on our chart was derived in the same year (1977), the research base isn't comparable in all countries so one has to be careful when drawing conclusions. Given that reservation here are some generalities that do hold true:
* American women are substantially bustier than Canadians – about four out of ten U.S. women wear C or D cups as opposed to three out of ten in Canada.
* American women are also bustier than British: 38 U.S. women take a C or D cup compared with 20 in the U.K.
* Italian women win on D cup ownership, but, overall, they're not the bustiest. That honour goes to the Danes where 60 out of every 100 women are C or D cuppers compared with 17 out of every 100 in Sweden.
* Japan is the hands down winner in the small cup sweepstakes.
* The French have a firm grip on B cups – very nearly seven out of ten are in this group.

BRA SIZES

U.K.	AUSTRALIA	MEXICO	BELGIUM	DENMARK	FRANCE	(WEST) GERMANY	SOUTH AFRICA	ITALY	JAPAN	NETHERLANDS	NORWAY	SPAIN	SWEDEN	SWITZERLAND
1977	1977	1977	1977	1977	1977	1977	1977	1977	1977	1977	1977	1977	1977	1977
22	23	41	25	4	14	17	26	11	65	32	NA	8	25	17
54	48	50	25	24	69	54	57	29	16	41	NA	13	54	37
17	21	5	16	42	14	23	16	30	NA	19	NA	13	15	12
3	7	4	NA	18	NA	NA	NA	19	NA	NA	NA	7	2	4

NA not available

Sources:

Confidential industry sources

If Anita Loos was right, gentlemen prefer blondes. And with that idea in mind, a lot of dark-haired ladies decide to change the colour of their hair. On the other hand, a lot of women see themselves as raven-haired beauties – and with the help of a few chemicals, they're able to achieve the desired effect.

From bleach to henna, highlighting to dyeing, hair colourants are big business as both women and men strive to improve on what nature gave them.

Here in the U.K. 23% of women use some sort of hair colourant. This is half the number of those who use them in the U.S. which has the highest percentage use on the table.

The only country to approach the U.S. figure is Denmark. It's generally assumed that there's an abundance of blondes in the Scandinavian countries. Maybe Danish ladies choose a darker colour to stand out from the crowd. Or maybe those natural looks aren't quite as natural as we've been led to believe.

After Denmark, there's a big drop to Austria and Sweden, where 30% of the women use a hair colourant to improve on their crowning glory. In Switzerland and Germany about a quarter of the women change the colour of their hair. Norway is next, equalling our 23% in the U.K.

That famous French chic obviously isn't the result of the use of chemicals. Only 20% of French women admit that they dye their hair.

	U.S.A.	CANADA
	1975	–
1. Women (%) who use colourants	45[+]	NA
	1977	1977
2. Average price of a man's haircut (UK £)	2.32[0]	2.32[0]
3. Average price of a shampoo and set (UK £)	6.38[0]	6.38[0]

At 14%, Belgium is right at the bottom of the table. Italy's figure is the same as France's. And Spain's 18% is even lower. It seems that the dark-haired ladies of the Mediterranean countries are quite happy to stay just the way they are. Or perhaps their menfolk don't have a yearning for blondes.

Although you may not feel you need to go as far as to change the colour of your hair, you doubtless get your hair done. The next figures show the average price a man pays for a haircut and a woman for a shampoo and set. If you look first at the figures for men, you'll see the Swiss have to pay the most – an average of £4.20. They're followed closely by the Swedes – £4.13. When Delilah sheared Samson's tresses, he lost his strength. Things aren't quite as bad for the men in these two countries, but when they have their hair cut, they certainly lose a lot of money.

Japan, where you have to pay £4.03 when you visit the barber, is third. France runs a not very close fourth: Frenchmen pay 87p less than the

U.K.	AUSTRALIA	AUSTRIA	BELGIUM	DENMARK	FRANCE	(WEST) GERMANY	IRELAND	ITALY	JAPAN	NETHERLANDS	NORWAY	SPAIN	SWEDEN	SWITZERLAND
1975	–	1975	1975	1975	1975	1975	–	1975	–	1975	1975	1975	1975	1975
23	NA	30	14	42	20	24	NA	20	NA	24	23	18	30	26
1977	–	1977	1977	1977	1977	1977	–	1977	1977	1977	1977	1977	1977	1977
2.99	NA	1.23	2.52	2.95	3.16	1.94	NA	2.31	4.03^0	2.31	3.08	0.94	4.13	4.20
2.99	NA	2.05	6.32	4.43	6.43	4.01	NA	4.62	5.92^0	4.29	6.04	2.55	6.21	8.46

NA not available

Japanese.

Spain is definitely the place to be if your hair needs a lot of attention. There, men only have to pay out 94p. for a short back and sides or the latest layered look. Here in the U.K., a man pays £2.99 on average so next time you're on holiday in Spain, pop into the local barber before you head home.

In general, women pay more for a hair-do than men – except in the U.K., where the average prices are identical for men and women. The Beatles and unisex salons have got a lot to answer for. Men in the U.K. certainly pay for a yearning for long hair and for sharing premises with the ladies.

In Switzerland, a woman pays an average of £8.46 for a shampoo and set – double what a man would pay for a haircut. At that price, there can't be all that many regular clients of Swiss hairdressers.

France is the next most costly country in the list. But even so, the French pay about a third less than the Swiss: £6.43.

Sources:
1 Euromonitor
2 & 3 Heron House conversions based on CBI *West European Living Costs,* 1977
+Industry estimates
|0Confidential industry sources

The U.S. and Canada are next. An American or Canadian woman saves only about five new pence. A Swedish woman pays £6.21 for a shampoo and set, and a Norwegian £6.04. Of the three Scandinavian countries, women in Denmark get the best deal. They pay an average of £4.43.

Austria's the place to go for ladies who want to save money on their hairdressing. There a shampoo and set costs just £2.05. Again the Spanish price is cheap: £2.55.

But remember the prices in the table are average. In Britain, for instance, there's a lot of difference in the prices charged in provincial towns and the larger cities. The prices charged in small-town beauty salons are obviously cheaper than those in big cities.

Who's got the squeaky-cleanest hair of all? The least dandruff? Three leading cosmetics firms provide the information for this earth-shattering study. The main conclusion? In the U.S., Canada, Japan and Sweden, at least 90% of all people use shampoo!

	U.S.A.	CANADA
1. Adults (%) who use shampoo	1977	1974
	90	95
2. Adults (%) with dandruff	1976	–
	14	NA

The Canadians top the list. Only five per cent of all Canadians *don't* use shampoo. This gives Canada the shiniest locks in the world. On average, there are twice as many shampoo-shunners in the U.S. as in Canada. (Perhaps, like Telly Savalas and Yul Brynner, they don't *need* shampoo.)

The U.K. comes seventh in the league of shampoo-users. Eighty-four per cent of us succumb to the admen's promises and try to put life, bounce and sunshine into our hair by washing it with their products. That doesn't, however, prevent an epidemic of dandruff. Twenty-four per cent of us still have to worry about our white-flecked shoulders.

There are some other odd contrasts in these tables. For instance, while 91% of Swedes prefer to use shampoo, only 76% of their Norwegian neighbours lather up. And in spite of their reputation for compulsive cleanliness, the West Germans have one of the lower shampoo rates, with only 79%. German frugality may overcome Germanic cleanliness, favouring good old-fashioned water, Castile soap and elbow-grease. Who knows? Maybe they wash their hair with beer (long used as a rinse).

In fact, for a fifth of the people in nearly half the places shown, Castile soap and elbow grease are the order of the day. But this is much more likely to be true on the European side of the Atlantic. In Ireland, for example, 22% decline to use shampoo. And they're not by any means the lowest on the table.

But before you jump to any snap conclusion about the Irish, take note: Ireland has less dandruff than any country on our table. In fact, Irish washing habits produce a population that is 90% dandruff-free. Maybe it's all that soft rainwater. Or the therapeutic benefits of Irish Mist.

Like the common cold, and the poor, dandruff is always with us. Unfortunately, there is no sure-fire cure for this perennial blight. Many have been suggested. The Romans used to drink garlic juice to cure the flaky problem.

These days, we stick to external solutions. And our remedies are no longer home-made. We can use one or all of several proprietary treatments. And they all come packaged in shampoos. But how effective is the shampoo solution against the unsightly flaky scalp?

We don't, unfortunately, have dandruff figures for those compulsive Canadian shampoo freaks, or for their equally eager Japanese friends. But in the other heavy shampoo-using countries – the

SHAMPOOS AND DANDRUFF

	U.K.	AUSTRALIA	AUSTRIA	BELGIUM	DENMARK	FRANCE	(WEST) GERMANY	IRELAND	ITALY	JAPAN	NETHERLANDS	NORWAY	SPAIN	SWEDEN	SWITZERLAND
	1977	–	1976	1976	1976	1976	1976	1976	1976	1977	1976	1976	1976	1976	1976
	84	NA	80	82	87	80	79	78	72	93	89	76	77	91	82
	1976	–	1976	1976	1976	1976	1976	1976	1976	–	1976	1976	1976	1976	1976
	24	NA	21	13	15	30	13	10	21	NA	15	14	20	15	19

U.S. and Sweden – the dandruff rates are relatively low. The fact is, however, there's no clear-cut correlation between using shampoo and avoiding dandruff.

The Irish are not alone in being able to neglect shampoo without the risk of itchy pates. The Germans also have low dandruff rates. They have the same proportion of afflicted flakers (13%) as the Belgians, who also place low in the shampoo stakes.

But shampooing habits in countries with moderate dandruff rates (around 15%) vacillate wildly. Of these middling types, over one in four is a non-shampooer. Yet the keen, clean Swedes shampoo at a rate of 91% with almost the

Sources:
Confidential industry sources

same flake figure as Norway.

When we turn the coin over, this discrepancy doesn't quite apply. The scurfier extremes do fall pretty low on the shampoo table. The worst afflicted places are those where 20% or more of the people admit to a dandruff problem.

Austria has 20% non-shampooers and 21% dandruff sufferers. Spain, with 23% non-soapers, has 20% with itchy scalps. In Italy, only 72% lather up, while 21% scratch. In France, a fifth of all the French surveyed decline to use shampoo.

Today, smoking is a world-wide pleasure – or vice, depending on your point of view. We've based the figures for men and women on regular smokers, who smoke three or more cigarettes a day. Our total figures include people who smoke regularly and also occasional smokers (fewer than three cigarettes a day).

Some of the most interesting figures relate to the U.K. and the U.S. The English and their colonies were among the first smoking enthusiasts in the West. Today, the British and North Americans are the most health-conscious smokers in the world. The government health warnings that appear on every cigarette packet in both countries must have some effect: both the U.K. and the U.S. are nearer the bottom than the top of the table. Just over a third (37%) of their populations smoke. At 41%, Canada has a higher proportion of smokers. Something else stands out in the U.K. figures: the figures for men and women smokers are exactly equal. In the States, only 34% of women claim to be regular smokers. This trend has been changing over the last ten years. Before tobacco is given up altogether, it's likely that the equality found in the U.K. will apply in the U.S. as well.

America is the world's largest single producer of tobacco. This makes it all the more ironic that Americans have lately taken such a puritanical line on smoking. Tobacco is a multi-billion dollar business. Many European governments, as well as the Japanese government, have state monopolies on the manufacture and sale of tobacco products.

The French have always had a reputation for being great smokers. The cliché is Jean-Paul Belmondo, looking surly with a Gauloise hanging suggestively from his mouth. And it was a Frenchman – Jean Nicot – who gave his name to the weed in the first place. It's true that the French smoke more than Britons or Americans – but, even so, less than half their population (43%) smoke today. The figure is the same for their neighbours in Germany. Though French men (57%) have the edge on the German *Herren* (51%). In both these countries, the women lag far behind. Could it be that the ladies of France and Germany are more sensible – or less liberated?

In Japan, it's illegal to smoke before the age of 20. So it comes as quite a shock to see that the Japanese have the second highest number of smokers on our table – 44% of the total population smoke.

But if you look more closely at the figures, you'll see that the sexes are more dramatically split than anywhere else. Although an astronomical three out of

Smokers as % of population:[+]	U.S.A.[o]	CANADA[o]
	1977	1976
men	41	45
women	34	38
total	37	41

U.K.^x	AUSTRALIA^x	AUSTRIA^‡	BELGIUM^‡	DENMARK^‡	FRANCE^‡	(WEST) GERMANY^x	IRELAND^‡	ITALY^‡	JAPAN**	NETHERLANDS^‡	NORWAY^‡	SPAIN^‡	SWEDEN^‡	SWITZERLAND^‡
1976	1977	1976	1977	1974	1975	1976	1977	1977	1977	1974	1977	1974	1977	1977
37	34	49	42	47	57	51	42	45	75	65	46	62	40	37
37	30	26	23	39	30	30	33	19	15	50	39	16	35	28
37	32	37	32	42	43	43	37	32	44	57	42	37	38	33

every four Japanese men lights up, fewer than one in six Japanese women smoke.

It's surprising to see that the thrifty Dutch lead in the smoking stakes. Holland is far ahead of all the other countries. In fact, it's the only place where well over half (57%) of all adults smoke.

On the whole, more men smoke than women. But Dutch women aren't at all slow to light up with a higher proportion of smoking women than anywhere else in the world. One out of two Dutch ladies lights up at least three times a day.

As far as women smokers are concerned, Holland is followed – though not very closely – by Norway, Denmark and Canada, where four out of ten women smoke. The Spanish ladies – like the Japanese – aren't too keen on the taste of tobacco. So if you offer a pack of cigarettes around a group of ladies, you're more than three times as likely to have them accepted in Holland as in Spain, and twice as likely in Australia or Canada as in Japan. However, if you're in the company of a few Japanese gentlemen, you'd better carry plenty of cigarettes with you: they'll go twice as

Sources:

Confidential industry sources

+ Includes smokers who roll their own cigarettes

o Aged 18 and over

x Aged 16 and over

‡ Aged 15 and over

** Aged 20 and over

fast as they would in the U.K. Male smokers from Holland and Spain are the only people who can hope to keep up with Japanese men.

Overall, the figures on smoking suggest that it's very hard to generalise. The heaviest smokers – the Dutch and Japanese – are completely different as far as the smoking habits of their women are concerned. And the places where smoking is least common seem to be wildly different in character. What do Australia, Belgium and Italy have in common? A non-smoker would be happier in all three – under a third of the population smokes.

Keeping the world's energy stoked up is a crucial problem. Coal, oil and natural gas are nature's resources – and they run out if exploited too greedily. Industrial nations have all begun to use a newer source of energy – nuclear power, but it isn't without its dangers. In many countries, heavy breeder reactors have met with protest.

	U.S.A.	CANADA
	1976	1976
1. Consumption per person of oil equivalent+ (tons)	8.9	9.4
	1975	1975
2. Consumption per person of electricity (kw hours^0)	9.3	11.8

We all know the chaos that shortages can cause. When the Arab sheikhs stopped the oil flow in 1974, and paralysed industry they threw governments into disorder. Millions of Europeans faced powercuts and long waits at the petrol station.

In Britain, an attempt was made to curb petrol consumption by legally limiting car speeds. A government advertising campaign urged electricity users to "Switch it Off".

The by-products of the oil industry are just as basic to our daily lives as petrol or fuel oils. Our homes and gardens are full of them: paints, detergents, plastic goods, pesticides.

To estimate energy consumption around the world, the tables use a measurement of a ton of oil equivalent per head. This includes oil, natural gas, solid fuels, water power and nuclear power.

North America is the leading energy user. In 1976, Canada consumed 9.4 tons of oil equivalent per person and the U.S. 8.9 tons per person – nearly twice as much as any other country. In terms of rich natural resources, both Canada and the U.S. are among the world's best-endowed countries. They produce most of their own fuel requirements. But to maintain industrial power in world markets and a high standard of living, they also import vast quantities of the world's energy supplies, which are then exported as manufactured consumer goods.

Spain, the least industrialised nation in Europe, is at the very bottom of the table. In 1976, each Spaniard got through just two tons of oil equivalent. Ireland and Italy followed with 2.5 and 2.7 tons per person respectively. These countries have remained largely agricultural producers. With few natural energy resources of their own, the road to massive industrialisation is a long and expensive one. In the future, there's a possibility of oil off Ireland's Atlantic coast and this could change the picture.

In 1976 Sweden and Norway used more tons of oil equivalent than any other European country. The Swedes used 6.7 tons per person and the Norwegians 5.7, which places them third and fourth overall in the table. The Scandinavians are among the most highly developed industrial European nations and need a lot of power. They were pioneers in the

ENERGY CONSUMPTION

	U.K.	AUSTRALIA	AUSTRIA	BELGIUM	DENMARK	FRANCE	(WEST) GERMANY	IRELAND	ITALY	JAPAN	NETHERLANDS	NORWAY	SPAIN	SWEDEN	SWITZERLAND
	1976	1976	1976	1976	1976	1976	1976	1976	1976	1976	1976	1976	1976	1976	1976
	4.1	5.2	3.6	5.0	4.2	3.7	4.7	2.5	2.7	3.4	5.2	5.7	2.0	6.7	3.9
	1975	1975	1975	1975	1975	1975	1975	1975	1975	1975	1975	1975	1975	1975	1975
	4.9	5.2	4.1	4.3	3.5	3.4	5.1	2.6	2.7	4.2	4.0	17.8	2.3	9.6	5.3

Sources:
1 Organization for Economic Co-operation and Development
2 Euromonitor
+Includes oil, natural gas, solid fuels, water power, nuclear power.
0Unit of energy: 1 kilowatt hour = 1000 watts for an hour

development of hydro-electric power. And with the North Sea full of Scandinavian oil-rigs, it looks as if they can look forward to an energetic future. Australia is fifth on the table with a consumption of 5.2 tons per person. Yet the country has few natural energy resources of its own. Australians have coal and hydro-electric systems, yes – but oil and natural gas, no. What they do have – and what the rest of the world wants and needs – is vast deposits of essential minerals like lead, zinc, copper and gold. The export of these buys the crude oil and petroleum products which – as with most countries in the world – are Australia's largest single import and essential to its industries and high living standards.

The Norwegians are way ahead in the industrial use of electricity. They used 17.8 kwh per person in 1975. Norway has an extensive hydro-electric system, which has been the basis of their industrial prosperity. Canada comes next with 11.8 kwh per person and. Sweden third with 9.6.

It's interesting that not only does Sweden keep its own lights burning but those of neighbouring Denmark as well.

Since the Danes have no natural energy resources and must import all their fuels, they have to be economical.

The U.K. has natural resources in coal and Natural Gas but relies upon imported Oil for industry and home consumption. However, our North Sea oil wells, which are now in production, will reduce the level of imports in the next few years.

The Dutch – high energy consumers with a rate of 5.2 tons per person – are more thrifty when it comes to the industrial use of electricity. In 1975 they used only four kilowatt hours per person. In the energy crisis of 1974, when the rest of Europe was still intent on their family outings, the Dutch banned weekend pleasure motoring and stayed at home.

Consumption per person of oil equivalent (tons)

Who are the greediest countries when it comes to consuming the world's energy? North America takes the cake here. Canada tops the list — each Canadian consumes almost ten tons of oil equivalent. No doubt a lot of it goes toward keeping warm during the long cold winters. The U.S. is next. The 8.9 tons per person consumed annually help keep all those television sets, blow-dryers and electric toothbrushes running.

Two Northern nations follow. Sweden and Norway, like Canada, use up a lot of energy just keeping cosy in the winter months. An "odd couple", tiny Belgium and vast Australia, both use about five tons of oil equivalent per person. But there the similarities end.

Consumption is lowest in Spain. The climate is warm and sunny — who wants to consume energy when there's all that free-flowing wine?

| 5.0 | 5.2 |
| 2.0 | | |

SPAIN **BELGIUM** **AUSTRALIA**

ENERGY CONSUMPTION

NORWAY	SWEDEN	U.S.	CANADA
5.7	6.7	8.9	9.4

What People Think

In this chapter we're going to look at a wide variety of topics, from a standpoint that probably caused the last argument you got into – personal opinion. Are you in favour of Women's Lib? Do you think most ads are misleading drivel? What do you think about the possibility of a Third World War? These are just three of the topics covered.

Our first sections deal with the most vital area of all – your personal life. Are you the cheerful type who looks on the bright side? Or do you find life trying with all those unpaid bills and the rising cost of living a constant worry? You'll be able to see who shares your approach, and whether yours is the majority or minority view.

Then we move into an area of really universal concern – job satisfaction. Let's face it, everybody has gripes about their daily grind. But when you come down to it, a surprising number of people get real satisfaction from their work. That's especially true in certain countries. So have a look at these tables to find out who escapes those Monday morning blues, and who's the most likely to be muttering ''Thank God it's Friday'' at the end of the week.

But not all jobs are nine to five – ask any housewife. We've devoted part of this chapter to the ladies, taking a look at a very controversial issue – the role of women in society. Obviously, both men and women have strong opinions on this, so we look at polls of both sexes in many countries for a comprehensive view. But bear in mind that these are opinions. What people feel about things isn't necessarily a true reflection of the way things *are*.

Laws have been passed in several countries to prevent sexual discrimination in job opportunities. In which countries do people feel that equality is a reality? As you'll see, things are changing rapidly in some societies, while others have remained backward. Where would *you* say that the women are still chained to the sink? Check it out against the table.

That leads us to a particularly argument-provoking section – whether husbands help with household work. Just to make things interesting we show the answers to the same question for both halves of interviewed couples.

Then we move into the larger issue of the community as a whole. Do you feel safe in your neighbourhood? What about those frightening headlines on crime and civil disorder? Are they accurate reflections of the way *you* feel about life today?

Some say the only solution to the social tensions confronting society is revolutionary extremism. Others favour less drastic reform. Still others think global conflict is the only possible conclusion to it all. In this section, you'll see which views prevail in various nations.

We also take an interesting look at the travel business. Does everyone long for a holiday in the sun? It may come as a surprise to learn that not everyone is tempted to even peek at what goes on in other countries. Some people really do prefer to stay safely at home.

Then we have opinions on the opinion-makers. In other words, do you think you're being well-informed by the media, or just led down the garden path?

Everybody regards their own views as important. The sector of society which probably cares most about its opinions is youth. Why? They're just starting out in life, they're filled with idealism and ambition, and they're going to make things work their way.

The last part of this chapter is devoted entirely to the opinions of youth. What they think about the world, what they think about themselves, what they think is wrong with society and how it ought to be changed.

First we profile young people's ideas on the neccessary ingredients for happiness – whether they think it's more important to be healthy or rich or intelligent and so on. The next tables deal with satisfaction with society, and then, more practically, how to deal with dissatisfaction in society.

After this, the questions become more specific and searching. There are tables giving figures on attitudes to government. We also look at the opinions of today's youth concerning women's role in society.

Finally, young people tell us about their own problems. What do they consider to be their most vital problems? Finding a job? Deciding which career to take up? Whether they should sleep with their boy/girlfriend? What are their attitudes to education? How much do they feel it determines their future?

By the time you've finished this chapter, you should have a pretty good idea of what all sorts of people think about all sorts of things. We're willing to bet it'll change a few of your own opinions – even if they're only opinions about what people think.

The top figures on the table are satisfaction ratings on standard of living. As such, they represent our countries' answers to the most common question, "How are you doing?"

The answer in Scandinavia is "very well" with over eight out of ten citizens expressing some degree of overall satisfaction and one out of three saying they're very satisfied compared with a low seven per cent in France, eight per cent in Japan.

Among English-speaking countries the overall satisfaction order is Australia, U.S., Canada and U.K. So younger societies with room to expand do seem to have special attractions.

Looked at the other way round Italy and France are both countries where just under six out of ten citizens are unhappy with their present way of life.

A family's standard of living is clearly affected by the number of children sired. That makes the figures shown for desired number of children by country surprising. The concept of zero population growth has a long way to go in many of our countries. In Australia, for example, three out of ten respondents want four or more children. While back here at home about one out of eight British is in this category.

Quite clearly there appears to be a correlation between desired children and available space. This cuts both ways. 68% of U.K. citizens and 65% of those in Benelux opt for one or two child families.

Japan is an odd man out here. The country is cramped yet 56% want three or more children.

The Germans' interest in keeping family size down is also notable. It shows in their overall population decline (see population table).

Overall it's surprising how few people want to have just one child or remain childless. Given today's widespread use of planned parenthood methods, this is a realistic alternative. Clearly it often doesn't happen by choice.

	U.S.A.	CANADA
1. Satisfaction with standard of living:[+]		
very satisfied (%)	26	24
satisfied (%)	20	24
quite satisfied (%)	13	19
dissatisfied (%)	30	26
very dissatisfied (%)	10	6
2. Ideal number of children:[0]		
1 (%)	3	3
2 (%)	52	44
3 (%)	19	22
4 - 7 or more (%)	18	22
0 (%)	2	2
don't know/no answer (%)	6	7

U.K.	AUSTRALIA	BENELUX	SCANDINAVIA	FRANCE	(WEST) GERMANY	EEC	ITALY	JAPAN
16	31	19	33	7	18	15	14	8
23	26	17	32	15	26	19	12	19
18	15	26	19	19	21	19	15	14
36	22	28	15	47	33	38	45	32
6	6	6	1	11	2	9	13	7
4	2	6	3	3	7	5	7	*
64	38	59	55	53	67	60	60	41
14	25	18	31	33	17	21	21	41
13	29	5	6	4	5	7	8	15
3	2	7	3	2	2	3	1	*
2	4	5	2	3	2	3	3	3

*0.5 or less

Source:
Gallup International affiliated institutes in the countries concerned
The poll was conducted during winter 1974-75 and spring 1976.
[+]The original question used a mountain card and asked: "Considering everything, how satisfied or dissatisfied are you with your standard of living? Just point to the step that comes closest to how satisfied or dissatisfied you feel. Remember the higher the step the more satisfied, the lower the step the more dissatisfied!" We have interpreted the mountain cards so that levels 1 - 2 = very satisfied; level 3 = satisfied; level 4 = quite satisfied; levels 5 - 7 = dissatisfied; levels 8 - 11 = very dissatisfied. Totals do not add up to 100 because there is a small proportion of "don't knows" and "no answers".
[0]The original question asked: "If a young married couple could have as many or as few children as they wanted during their lifetime, what number would you, yourself, suggest?"

The Japanese are the biggest worriers when it comes to facing those monthly bills. Even though Japan's economy is booming, 41% are afraid of not having enough yen in their pay cheque to pay for what they need. And well over one family in ten says it can't afford adequate nourishment.

If they suddenly had a lot more money, the majority of Japanese would invest it. Only five per cent would spent the extra cash on food and other necessities.

In Italy, things seem to be a bit easier on the pocket, yet three people in ten worry about household bills most of the time. Almost a third of all Italians – more than in any other country – would spend a windfall on improving their housing.

The U.S. comes third on the list of worriers. More than a quarter of all Americans worry about paying their bills. And about twice as many people in the U.S. as anywhere else (except Italy and Japan) are occasionally unable to buy food. They're the worst off nation when it comes to health care too – 3 out of every 20 of them just can't pay the ever-increasing costs of medical attention and hospitalisation.

In the U.K., fortunately, our comprehensive national health service means that only one per cent of us ever worries about health bills. In fact, a calm 43% of us almost never worries about anything – though eight per cent of us have sometimes been unable to afford food. But over one in four of us feel that we would have to spend a sudden windfall on essentials – the highest figure on the table.

The easy-going Australians worry least about family finances – 57% of them hardly give the matter a thought, even with inflation running high.

	U.S.A.	CANADA
1. People who worry about money:[+]		
all and most of the time (%)	26	15
some of the time (%)	36	30
almost never (%)	38	51
2. Economic deprivation:[0]		
people (%) who have been unable to buy food	14	6
people (%) who have been unable to pay for medical/health care	15	4
3. What people would do with extra money:[x]		
save; invest in business, farming (%)	32	40
buy essentials (%)	21	10
buy new house; repair present house; move to new home or community (%)	21	23
travel; buy non-essentials (%)	28	48
pay bills (%)	13	5

MONEY WORRIES

U.K.	AUSTRALIA	BENELUX	SCANDINAVIA	FRANCE	(WEST) GERMANY	E.E.C.	ITALY	JAPAN
22	12	9	17	22	10	19	29	41
34	31	34	40	37	33	36	39	44
43	57	56	43	35	55	43	32	14
8	4	4	6	6	7	8	15	14
1	4	6	2	8	1	5	9	5
33	46	32	23	32	41	34	27	61
26	5	11	9	24	16	19	16	5
22	16	22	28	28	20	24	32	17
37	34	29	43	29	34	31	23	20
2	6	*	5	*	2	1	1	*

*0.5 or less

Source:
Gallup International affiliated institutes in the countries concerned
The poll was conducted during winter 1974-75 and spring 1976. The questions asked:
[+]"How often do you worry that your total family income will not be enough to meet your family's expenses and bills?"
[0]"Have there been times during the last year when you did not have enough money to buy the food your family needed? When you did not have enough money to pay for medical health care?"
[x]"Suppose you had more money, say twice as much as you have now. What would you do with it? Totals do not add up to 100 as the answers used in this table have been extracted from more extensive questionnaires.

Everyone has family problems. They worry about not having enough money, or losing their jobs; about illness in the family or trouble with the children.

Americans are the biggest worriers of all. They score a total of 85% for all the problems on the table. We in the U.K., on the other hand, only have a total of 66%. Family finance is the biggest headache for us and for the Americans – though ten per cent fewer of the British worry about it. But we're not as secure as the West Germans. Their booming economy means that only 12% of them worry about making money ends meet. For the entire E.E.C., the figure is 18% of the population who think they have money problems. One reason for the high U.S. figure is the increasing instability of their economy, and the fact that they're protected by fewer social welfare and health care programmes.

The French are the next biggest worriers (79%), followed by the Japanese (77%) and the Italians (75%) each. The Scandinavians have the fewest worries (51%), with Benelux citizens (54%) and the Australians (55%) just above.

When it comes to family illness, the West Germans, who are lowest in financial worries (12%), take top position (24%), closely followed by the French with 20%. Health care costs are extremely high in both countries. There clearly aren't many health worries for us in the U.K. (nine per cent), or for the

	U.S.A.	CANADA
1. People (%) who think their most important family problem is:[+] finance	40	25
illness in the family	13	10
rearing/disciplining/communicating with children	8	5
unemployment	8	5
family (social/in-laws)	5	4
shortage of food/clothing; others	11	18

Scandinavians and the Japanese, both with eight per cent. It obviously helps if you have a ubiquitous health service, or a large corporation to care for you.

What about problems with children? The Italians come top with 23%. It figures. With all the kidnappings and bombings in Italy, you'd probably worry too. The Japanese figure is also high (19%); perhaps this is one of the problems caused by the country's emergence from centuries of a different family tradition. Sixteen per cent of the French have cause for concern over their children. Americans (eight per cent), have few worries about their offspring. It's even less of a worry for parents over here. Only seven per cent of us in the U.K. found that it was difficult to raise children. Only Scandinavian parents (six per cent) and Canadian parents (five per cent) seemed to have fewer problems with their kids. Only three per cent of people in Japan and nine per cent in France worry about unemployment. Even in Italy – the country with the highest figure for unem-

FAMILY PROBLEMS

U.K.	AUSTRALIA	BENELUX	SCANDINAVIA	FRANCE	(WEST) GERMANY	E.E.C.	ITALY	JAPAN
30	23	13	14	19	12	18	17	30
9	10	14	8	20	24	17	15	8
7	7	11	6	16	10	13	23	19
6	3	4	4	9	10	8	11	3
6	3	3	2	5	7	5	7	2
8	9	9	17	10	1	6	2	15

Source:
Gallup International affiliated institutes in the countries concerned
The poll was conducted during winter 1974-75 and spring 1976.
+The question asked: "What is the most important problem facing your family at this time?"
Totals do not add up because of a high proportion of "no problems" and "no answers".

ployment worries – it's only 11% and for the whole of the E.E.C., it's eight per cent.

Japan and Australia have the lowest figure here: only three per cent worry about unemployment. People in the U.K. – which is supposed to have a very bad unemployment problem – don't seem very concerned either: our figure is only six per cent. Perhaps our extensive system of unemployment benefit has finally banished the worries that used to be associated with losing your job.

What about all those mother-in-law jokes? As you can see from the table, in Japan and Scandinavia (two per cent) and Benelux countries and Australia (three per cent), the in-law problem doesn't really exist. The U.K. figure for this was

six per cent. And in West Germany and Italy – the countries with the highest figures – only seven per cent admit to family difficulties.

Many people in other countries worry more about feeding and clothing their families than we do. For example, in Canada and Scandinavia, nearly one person in five has difficulty in balancing his budget. Japan is third here with 15%, followed by the U.S. with 11%.

Italy comes second lowest in this category: only two per cent worry about food and clothes. Only West Germany is lower (one per cent.) Perhaps this reflects the Italians' priorities – with so much to worry about in society at large, they don't have time to worry about things closer to home – except their children.

Worriers are distinctly in the minority these days. In most countries, two out of three people say they take life as it comes and don't spend a lot of time fretting about it. However, this generally cheery state of affairs doesn't prevail equally throughout the world. People living in English-speaking countries – the U.S., Canada, Australia and the U.K. – appear to take life pretty much as it comes. Contrary to all popular mythology, the gloomiest domain is France: a startling 50% of the population questioned admitted that they're habitual worriers. This pervasive dejection sends the average worrying quotient of the E.E.C. up to 39% – several points higher than the scores of most individual member countries.

If Gallic insouciance is at an unexpected low, the dark clouds don't stop at the French border. The Italians are second among the worriers in the survey. In the land once known for *la dolce vita*, 45% of people questioned confessed that they worry a lot.

The second category on this table shows whether people find their lives interesting or dull. It correlates with, and perhaps helps to explain, the fretful state of affairs among the French and the Italians. In a sharp blow to conventional stereotypes, about a third of the Italians questioned described life as fairly or very

	U.S.A.	CANADA
1. People (%) who worry a lot[+]	34	36
take life as it comes	64	62
don't know/no answer	2	2
2. People (%) who think their life is:[0]		
very interesting	36	30
fairly interesting	52	59
fairly dull	8	9
very dull	4	2
don't know/no answer	*	*

dull, and a quarter of the French felt the same way. Perhaps we must say *au revoir* to the much-vaunted Latin *joie de vivre*.

The other spot on the globe where people find life notably lack-lustre is Japan; again, a quarter of the citizens find life fairly dull; and the same number confess that they worry a lot. But a further 27% either didn't know or wouldn't tell whether they worry a lot. (In most other countries, only one to five per cent didn't answer the question.)

In the U.K. and Australia the level of worriers goes up a bit to over 30% – but the number of people who find life interesting also takes a mighty leap upwards. Eighty-two per cent of us are enthusiastic about life and only 18% find it a dull old round. But the Australians are

U.K.	AUSTRALIA	BENELUX	E.E.C.	FRANCE	(WEST) GERMANY	ITALY	JAPAN	SCANDINAVIA
31	35	42	39	50	31	45	25	34
65	63	57	56	45	62	51	48	64
4	2	1	5	5	7	4	27	2
28	30	30	20	14	19	14	6	26
54	62	55	56	60	57	53	67	65
16	6	11	20	22	19	28	25	8
2	2	2	3	3	2	5	2	1
0	0	2	1	1	3	*	*	*

*0.5 or less

even more grateful to be alive – only eight per cent of them suffer from the blues.

Clearly, the overwhelming majority of people everywhere do find life fairly interesting. The teasing differences – the enigmas – lie at the extreme ends of the spectrum. While approximately the same number of people both in the U.K. and in Italy find their lives fairly interesting, fully twice as many of we British find our lot in life very interesting. And similarly, more than twice as many Italians find things very dull.

The last hurrah definitely comes from the world's younger countries. With unmatched zest, a third of Americans, Canadians and the Australians describe their lives as very interesting indeed.

Source:
Gallup International affiliated institutes in the countries concerned
The poll was conducted during winter 1974-75 and spring 1976.
[+]The question asked: "Would you say you worry a lot or that you take life as it comes?"
[0]The question asked: "Would you say that your life at this time is very interesting, fairly interesting, fairly dull or very dull?"

BEER TESTER

The Scandinavians are at the top of the league when it comes to enjoying work. Fifty-five per cent find nothing at all to complain about in their jobs. Perhaps it's all that clean fresh air and modern working conditions. In Scandinavia, employers try to keep their workers happy by varying the dull routine of the assembly line. This seems to work. Only three per cent of the entire work force are dissatisfied and frustrated with their jobs.

	U.S.A.	CANADA
1. Satisfaction with employment:[+]		
very satisfied (%)	49	46
satisfied (%)	46	45
dissatisfied (%)	5	9
2. Choice of occupation if re-starting working life:[0]		
present job (%)	51	46
different job (%)	41	42

Job satisfaction in Scandinavia isn't unanimous. Forty-eight per cent of workers say they'd choose a different kind of job altogether if they had a chance.

Australia is the next most contented country. Exactly half of all Aussies say things are just great at work down under. They seem to enjoy their work in that energetic, thrusting country and say that they have no complaints.

Over in the U.S., most of them also appear to whistle while they work. Only one out of 20 is a grumbler – the other 19 are either satisfied or very satisfied with their jobs. More than half of all Americans want to stay in their present jobs, and only 41% would choose a different one if they were starting out all over again – a rather surprising statistic for that mobile, job-hopping society.

In the U.K. only three per cent are grumblers – which seems at first sight rather surprising. The other 97% are satisfied or very satisfied with their jobs.

The odd thing is though, while 45% of us would choose the same job if we started out all over again, the same number would opt for a completely different way of earning a living if they had the chance. There are obviously different levels of "satisfaction" involved.

Other happy fellows are the Canadians and workers in the Benelux countries.

Who has the longest faces when it comes to clocking in on Monday morning? The French are in the lead. Only one Frenchman in four shows enthusiasm for his work. Japan and Italy, where the figure is one out of three, tie for second place. The highest percentages of people stuck in jobs they hate are in France and Canada (nine per cent), with Australia (eight per cent) coming second and Italy (seven per cent) third.

Only 28% of workers in Japan would choose the same job over again. Despite this sign of discontent, the number of workers who'd opt for another line of work is small – a mere 38%.

U.K.	AUSTRALIA	BENELUX	SCANDINAVIA	FRANCE	(WEST) GERMANY	E.E.C.	ITALY	JAPAN
49	50	46	55	25	44	41	33	33
48	42	50	42	66	55	55	60	64
3	8	4	3	9	1	4	7	3
45	54	57	46	48	53	50	50	28
45	40	40	48	48	39	43	43	38

Source:
Gallup International affiliated institutes in the countries concerned
The poll was conducted during winter 1974-75 and spring 1976.
[+]The original question used a mountain card and asked: "How satisfied are you with your present work, that is, your main employment? If you are extremely satisfied point to the top of the mountain, if very dissatisfied point to the bottom. If you are satisfied with some parts of it and dissatisfied with others, point to some step that comes closest to how satisfied you are with your job. Remember, the higher the step the more satisfied you are, the lower the step the more dissatisfied." We have interpreted the mountain card so that levels 1-3 = very satisfied; levels 4-8 = satisfied; levels 9-11 = dissatisfied.
[0]The question asked: "If you had the chance to start your working life over again would you choose the same kind of work you are doing now or not?"
Totals do not add up to 100 because there is a small proportion of "don't knows" and "no answers".

Is woman's role in the world really changing? Or are things much the way they've always been – despite all the talk of women's liberation?

In the U.K., 82% of us feel that the position of women has changed. The government has passed legislation to guarantee equal opportunities for women and not a day passes but that we hear more debate and comment on the subject. Yet we're not in the fore of the women's liberation.

People (%) who think the part played by women in their country is changing:[+]	U.S.A. 1976	CANADA 1976
a great deal	63	45
a fair amount	28	42
not much	6	11
not at all	1	1
don't know	2	1

It seems that woman's role in society is changing most in the U.S. – according to 63% of American men and women.

There's no doubt that America has a more flexible society than many European nations. Life styles have changed enormously – and frequently – in the last 200 years. That means women's life styles too. Compare this with a country like Spain or Italy, where a woman's place has been in the home for centuries. In the U.S., women simply have a less rigid tradition to fight against when they choose to play a more dominant role in society.

U.S. women also have easier access to higher education than many Europeans. That means they're better equipped to take jobs outside the home and so make their influence felt.

The next highest figure is for the Scandinavian countries. In Denmark, Sweden and Norway, nearly half of the population feels that the role a woman plays in society is changing rapidly. If we include those who think things have changed a fair amount, that's 80%. Only one per cent feel things haven't changed at all, the identical figure to the U.S. and U.K.. The Scandinavian countries like the British and Americans have an advanced social outlook. Their well-known liberal views obviously include an openness to social change that's reflected in these figures. They also have a high standard of living – and a healthy economy certainly creates more opportunities for women who want to get out of the home and play a more active role in society.

Canada is next: 45% think that things are changing a great deal for women, and nearly as many more think that a fair amount of change is taking place. At a total of 87%, the Canadian figures are close to the American ones. Of course, in many ways the two countries are alike. One difference might be found in the

WOMAN'S ROLE

	U.K.	AUSTRALIA	BENELUX	FRANCE	INDIA	ITALY	JAPAN	SCANDINAVIA
	1976	1976	1976	1976	1976	1976	1976	1976
	39	33	31	38	17	32	23	47
	43	49	43	41	29	50	48	33
	13	14	21	16	18	9	17	18
	1	2	2	1	6	2	4	1
	4	2	3	4	30	7	8	1

majority of French-speaking Canadians. A clue is that in France only 38% of the population thinks their women's roles have felt the winds of change. But French women play a greater part in the running of society than is often realised.

India is one of the few countries which has had a woman prime minister – Mrs. Indira Gandhi led India for over five years. Yet only 17% of Indian citizens think she changed women's role in society to any great extent. Six per cent don't think things have changed at all – a higher percentage than in any other country. That's understandable, since a large percentage of the population still lives in villages, where the traditional way of life is slow to change. Isolation from the influences that affect Western society may account for the very large percentage (30%) of Indians who don't really know which way things are going regarding women's roles.

Japan is also quite low in the table.

Source:
Gallup International affiliated institutes in the countries concerned

+The question asked: "Do you think that the part played by women in your country is changing a great deal, a fair amount, not much, or not at all?"

Less than a quarter of the population thinks that any dramatic changes have taken place, and four per cent are quite sure they haven't. That's surprising when you consider the massive changes that have occurred in Japanese society since the end of World War II. The powerful Western influences on the Japanese haven't managed yet to alter the female role of graceful flower-arranger and tea-pourer in favour of something more dynamic. The Japanese have traditionally felt that a woman's place is in the home...and it looks as though that's where she's going to stay.

	U.S.A.	CANADA
	1976	1976
1. People (%) who agree women have equal job opportunities with men[+]	48	44
disagree	48	50
don't know	4	6
2. People who agree women should have equal job opportunities with men[0]	39	42
disagree	8	7
don't know	1	1

From the early suffragettes to women's lib, the campaigners for equal rights for women have made themselves a powerful force.

Here in Britain it's illegal to discriminate against women in employment. In fact, it's illegal even to advertise a job as specifically requiring a man. But is this enough? Do women in most countries have equal job opportunities or not? That's what this table tries to answer. Those who answered "no" were then asked "Do you think women should have equal opportunities or not?"

It's quite a surprise to find the Italians at the top of the league (at least in their own eyes) for equal job opportunities. Fifty-five per cent of those questioned thought that women had equal opportunities. It could be just wishful thinking. Or maybe Italy's fiery women's lib movement has achieved some of its goals. If you look into those Italian figures more closely it seems that the latter is the case.

The U.S. is next in line when it comes to satisfaction with the way things are going for women. Forty-eight per cent of Americans thought women have equal job opportunities with men.

Australia (third on the table), like Canada, is a relatively new society, where traditional roles are changing rapidly and forty-six per cent thought that women have equal job opportunities.

The Australian figure of 46% is closely followed by the one for India – 45%.

Many upper caste Indian women are encouraged to enter the professions. Even so, it's surprising that the figure for India is higher than for Canada (44%), where we tend to think women are more "liberated".

Where do most people think women get a raw deal? In West Germany, 64% of those polled thought they don't get an even break.

The Japanese agree with the Germans. Sixty-two per cent are convinced that women don't get equal job opportunities. Japan has traditionally always been a male-oriented society. Things are changing, though. This may in part be due to the strong influence that the U.S. has had on Japanese culture since the end of World War II. The women's lib movement there is fanatical and strong – and it needs to be. Although Japanese business methods are extremely up-to-date, many of the old

362

EQUALITY AT WORK

AUSTRALIA	INDIA	BENELUX	U.K.	FRANCE	(WEST) GERMANY	SCANDINAVIA	ITALY	JAPAN
1976	1976	1976	1976	1976	1976	1976	1976	1976
46	45	40	38	37	31	36	55	22
49	34	58	54	58	64	60	36	62
5	21	2	8	5	5	4	9	16
40	20	48	41	51	57	55	27	34
9	11	8	8	5	6	4	8	22
*	3	2	5	2	1	1	1	6

*Less than 1%

Source:
Gallup International affiliated institutes in the countries concerned
+The question asked: "Do you think that women in your country have equal job opportunities with men?"
0The question asked (of respondents who answered 'no' to the previous question): "Do you feel that women should have equal job opportunities with men?" Figures are expressed as percentages of total original sample.

social attitudes are taking a long, long time to die. A large number of Japanese – 16% – said that they didn't know whether women have equal job opportunities or not. The same sort of reaction exists in India. 21% said they didn't know.

Scandinavians rank third (60%) in believing that men retain the advantage when it comes to job opportunity – though Scandinavian women have seemingly been treated as equals for many years. Maybe it was easy enough to get the basic equalities, but Swedish women (like their German neighbours) now demand *complete* equality in *every* field.

People who thought women didn't get equal job opportunities were also asked whether they thought they should. The most "yes" answers came from countries where a high proportion of people thought women weren't getting a square deal. West Germany heads the list with 57%, and Scandinavia comes a close second with 55%. France is third with 51%, and the Benelux group is fifth (48%). (Here in Britain, the vast majority (41%) of those who thought women don't have equal job opportunities felt they should).

363

Compared to the U.S., Europe is the stronghold of male chauvinism. In all the eight nations surveyed, more men would place their confidence in a member of their own sex representing them in Parliament than in a woman. In the U.S., only a small minority (ten per cent) of men would not vote for their party candidate if a woman were nominated. The European preference was most marked in Germany. Over half the German men questioned favour male politicians. (In fact, the Germans are at the top of the list in feeling a woman's place is in the home.)

In the U.K., although about a third of men and women would prefer a man to represent them in Parliament, over a half of all of us think the sex of our M.P. is irrelevant.

	U.S.A.	U.K.
	1976	1975
Reactions to voting for a woman if she were running for Parliament[+]/ Congress[0]:		
would prefer a man (%)		
men	10	37
women	8	31
would prefer a woman (%)		
men	88	5
women	89	12
no difference (%)		
men	–	53
women	–	52

Perhaps European men are being falsely accused. Across the board almost as many European women prefer a man to represent them in government, reflecting a conservative attitude in many families. Yet even in a male-oriented society like Italy's, almost one man in ten says he'd *prefer* a woman to represent him in parliament. This just might reflect the Italians' utter disdain of their government. Or it may be a healthy reflection of growing egalitarian values. Almost 40% of Italian males surveyed agree that there is no difference between a male and a female when it comes to representation in government – so much for Italian chauvinism.

Some European women are raising their voices and helping to change values in their societies. In Italy, 15% of the women surveyed feel more enthusiastic about having a female represent them in parliament than a man. Their Irish sisters are even more outspoken: 24% prefer to see one of their own sex in a position of political power. It seems that a female backlash has taken place in countries where women have been traditionally the most domesticated.

The Danes have the most liberal attitude of all the European countries on

WOMEN IN POLITICS

BELGIUM	DENMARK	FRANCE	(WEST) GERMANY	IRELAND	ITALY	NETHERLANDS
1975	1975	1975	1975	1975	1975	1975
42	20	35	53	42	47	28
35	15	28	37	33	41	23
4	4	7	2	10	9	3
8	6	9	8	24	15	4
46	70	51	41	45	38	57
48	73	56	50	40	40	59

— Question not asked

Sources:
+Commission of the European Communities, 1975
The original question asked: "In general, would you have more confidence in a man or a woman as your representative in Parliament?"
0The Gallup Organization Inc, 1976
The original question asked in the U.S. was: "If your party nominated a woman to run for Congress from your district, would you vote for her if she were qualified for the job?" The response "yes" corresponds here to "would prefer a woman", "no" to "would prefer a man". Totals do not add up to 100 because there is a small proportion of "don't knows".

the table. The great majority of both Danish men and women don't think sex is an important issue in choosing a political representative.

The Americans are even more liberated. The vast majority of American men and women wouldn't hesitate to vote for a female Congressional candidate. In fact, an almost identical number of both sexes agrees that they would vote for a woman for a Congressional seat as long as she were as qualified as a man, which indeed does seem a good way to cast your vote.

"A woman's work is seldom done – by her husband" (to rephrase an old saying). Although many men would probably tell a different tale.

The table on the right shows that European couples just don't agree on the touchy subject of how often a husband lends his wife a helping hand. In every country, however many men there are who think they help around the house, fewer women would agree with them.

In the U.K., 65% of men thought that they helped round the house "frequently" or "occasionally". Only 54.5% of their wives agreed. It seems highly likely that 10.5% of our married couples spend a lot of time arguing about division of labour.

	U.K.
Husbands who help in household work (men's/women's responses): [+]	
frequently	
male (%)	35.9
female (%)	22.7
occasionally	
male (%)	29.1
female (%)	31.8
never	
male (%)	9.2
female (%)	15.3
no answer	
male (%)	0.6
female (%)	1.3

Italian men obviously believe that housework is a woman's area. Only 9.6% of them claim they help round the house "frequently" and only 8.1% of their wives agree. Over 35% of them say they wield a duster "occasionally", but seeing as just 21% of Italian women would endorse this statement, they must overestimate their usefulness.

Men in Holland are the most industrious when it comes to housework – both in their own eyes and in the eyes of their wives. Over 75% claim to help round the house "frequently" or "occasionally" and over 63% of Dutch women would agree. But that still leaves 12.6% of husbands in the Netherlands who think they work harder than they actually do. Or perhaps there's a lot of Dutch ladies who don't fully appreciate their men.

The smallest disparity in the figures comes in Ireland: 46.2% of Irishmen think they're helpful round the house and 41.6% of the wives agree. However, when you consider that apart from the Italians, the Irish men are the most unhelpful in the table, it doesn't necessarily mean that there are fewer argu-

HELPFUL HUSBANDS

BELGIUM	DENMARK	FRANCE	(WEST) GERMANY	IRELAND	ITALY	NETHERLANDS
26.3	—	30.1	20.5	17.8	9.6	33.3
18.1	28.4	19.6	14.8	14.0	8.1	21.2
30.0	—	34.5	40.9	28.4	35.3	42.4
26.0	25.7	31.4	35.7	27.6	21.0	41.9
11.7	—	8.0	9.3	9.0	19.6	4.7
20.0	15.7	19.2	15.1	13.0	29.6	14.2
1.0	—	3.8	1.3	0.2	0.4	1.0
1.0	3.3	0.5	0.4	0.8	1.5	1.1

—Question not asked

Source:
Commission of the European Communities
[+]The question asked (of married women): "Does your husband ever help you with the household work? If yes, frequently or occasionally?" Married men were asked: "Do you help with household work? If yes, frequently or occasionally?"
Percentages do not add up to 100 as the proportion of unmarried respondents has been omitted.

ments about whether or not husbands in Ireland are pulling their weight.

It's just as well we don't have the U.S. figures on the table. In general, Europe doesn't have the many hours of sport on T.V. that act as an excuse for idle American husbands.

Whatever the country, it looks as if the men are doing work round the house of which their wives are unaware.

How frightened are we by crime? And to what degree are our fears justified?

The U.K. has the highest figure in Europe (nine per cent) for people who've suffered an attempted or successful break-in within the last five years.

In the U.S., 17% of those questioned had suffered from a burglary or an attempted one in the last five years: the highest rate in the table.

The U.S. also has the highest figure for stolen property, with over 26% of the people questioned reporting a theft.

Australia is second in both categories and Canada is third – well below the U.S. rates.

Scandinavia comes fourth in the figures for stolen property (18%), just above the U.K. (16%).

Only three per cent of Japanese have confronted a burglary or an attempt at it, and they're second lowest for stolen property. Italy at second lowest in the housebreaking category is a real surprise; the same holds true for physical assaults. You're three or four times more apt to be physically assaulted in the English speaking world than in Italy. Only Benelux has lower assault figures.

However, in all the countries on the table, assault is low – no more than five out of a hundred people need fear being attacked during any given year.

Surprisingly, the U.K. is well ahead of all our Common Market counterparts in

	U.S.A.	CANADA
1. People who, during the last five years, have[+]:		
had their home broken into or an attempt made (%)	17	12
had property stolen from themselves or their family (%)	26	21
been physically assaulted (%)	4	5
2. Fear of walking alone at night within about a mile of home[o]:		
afraid (%)	41	31
not afraid (%)	56	66
don't know/no answer (%)	4	3

this category. A careful analysis of the underlying facts and figures would show that this is primarily a Northern Irish phenomenon – there are 21 deaths per hundred thousand people in Northern Ireland.

Astonishingly, fear of crime doesn't seem to be related very closely to the actuality. The greatest fear of crime is in the Benelux countries. There, 46% of the people questioned are afraid of being physically assaulted, yet the actual number of assaults is lower than in any other country on the table.

West Germany is next highest: 45% of Germans are afraid of being physically assaulted.

Finally, how far has society really progressed when, in all save two countries, at least a third of the population are frightened of areas within 20 minutes of their own front door?

U.K.	AUSTRALIA	BENELUX	SCANDINAVIA	FRANCE	(WEST) GERMANY	EEC	ITALY	JAPAN
9	14	7	8	8	5	7	4	3
16	25	2	18	11	8	11	11	7
5	5	*	4	3	2	3	1	2
34	33	46	26	36	45	39	35	33
62	64	53	71	58	51	57	60	63
4	3	1	3	6	4	4	5	4

*0.5 or less

Source:
Gallup International affiliated institutes in the countries concerned
The poll was conducted during winter 1974-75 and spring 1976.

+ The original question asked: "During the last five years, have any of these happened to you: Had your home broken into or had an attempt made? Had money or property stolen from you or any other household members? Been personally physically assaulted?"

0 The original question asked: "Is there any place around here — that is within 20 minutes walk — where you would be afraid to walk alone at night for fear of being physically attacked?"

If you believe in revolution, you're an extremist. If you believe in defending your society against "all subversive forces", you're also an extremist. Statistically, you belong on the extreme end of popular opinion, where revolutionaries and reactionaries begin to look alike.

Revolutionary extremists tend to be left-wing and are mostly active in right-wing or democratic societies. Those in Western countries who believe that society must be changed through revolution incline towards anarchy or extreme Marxism of some kind. Statistics aren't available for Russia. Presumably, revolutionary Soviets are fighting to move their society further to the right. While in China, considered by some to be the most left-wing country in the world, some "revolutionaries" are pressing to move the government further left. At this point, "left" and "right" as terms of political demarcation become meaningless.

This table gives a general picture of the political views of eight European countries. Predictably, in the democratic West, moderation takes the day. In 1977, over half those polled in all countries except West Germany voted for improvement through social change. However, fewer people voted for moderation than seven years previously. In West Germany, 30% fewer people voted for social change than in 1970, throwing

	U.K.
1. People (%) who thought society should be changed by revolution, in:[+]	
1977	6
1970	7^x
2. People (%) who thought society should be defended against subversion, in:[0]	
1977	28
1970	25^x

their vote to the right in favour of defending society against "subversive forces".

In the U.K. there was almost no hardening of attitudes between 1976 and 1977. In fact, fewer people were in favour of revolution and there was only a three per cent increase in those who thought that there was a danger of subversion.

Italians are responding to their economic and political crisis by turning the other way: ten per cent of those polled thought that revolution was the only answer – three per cent more than in 1970.

Except for Belgium, France and above all Italy, revolutionary zeal waned in the surveyed countries during the same period. So did confidence in the power of gradual social change. The Dutch lost faith by 20% between 1970 and 1977; turning to a conservative standpoint by a dramatic 22%.

BELGIUM	DENMARK	FRANCE	(WEST) GERMANY	IRELAND	ITALY	NETHERLANDS
4	3	8	2	7	10	4
3	4^x	5	2	7	7	6
19	40	22	50	26	27	37
14	38^x	12	20	23	11	15

^{x}1976

Source:
Euro-Barometre 8, 1978
Polls were mainly conducted in February/March 1970 and October/November 1977.
Respondents were shown a card and told: "On this card are basic kinds of attitudes vis-à-vis the society we live in. Please choose the one which best describes your own opinion." The statements on the card read:
+"The entire way our society is organized must be radically changed by revolutionary action."
0"Our present society must be valiantly defended against all subversive forces."

How much real civil tension is there in Europe today? Is it so bad that there's a possibility of the breakdown of civil order within the next ten years?

This table shows that most Europeans would answer "no" to these questions – although there are exceptions.

Over 63% of Italians think that civil disorder in the next ten years is either "certain", "very possible" or "possible" – almost ten per cent of them think there's no doubt that it'll happen. But then there's already civil disorder there at the moment: the *Brigate Rosse* and other terrorist groups have nearly paralysed the country on several occasions.

The Netherlands is the second most pessimistic country: over 37% of the people polled think that the breakdown of civil order is "very possible" in the next ten years and the combined figure for "certain", "very possible" and "possible" is only two per cent behind that of the Italians. The Netherlands has recently experienced savage attacks by South Moluccan terrorists in their attempt to gain independence from formerly Dutch Indonesia.

Overall, France is the next most pessimistic country. In 1968 Parisian students took over the capital – something that hadn't happened in Western Europe for decades. In the Pyrenees, the Basque nationalists are also waging a sporadic war for independence.

	U.K.
	1977
People (%) who think that, in the next ten years, civil disorder is:[+]	
certain	8.5
very possible	26.4
possible	15.9
unlikely	28.2
impossible	16.3
don't know/no answer	4.6

The U.K. isn't without its problems: Northern Ireland is being torn apart in a tragic civil war. Racial tension between whites and immigrants from India, Pakistan, Africa and the Caribbean also flares up from time to time. Consequently, only 44.5% of us are prepared to say that civil disorder is unlikely or impossible.

The Irish are more confident than we are that they won't have to face the breakdown of order in the next ten years: just over 53% think it's "unlikely" or "impossible". The Danes aren't far behind them at 53%. In fact only 36% of the Danes think there is any risk at all of civil disorder – the lowest figure on the table.

Today's world is changing and complex, so it might be sensible to assess not *whether* various countries will experience civil disorder, but how often it will occur – and whether or not people think that it will be contained.

CIVIL DISORDER

BELGIUM	DENMARK	FRANCE	(WEST) GERMANY	IRELAND	ITALY	NETHERLANDS
1977	1977	1977	1977	1977	1977	1977
3.1	5.6	3.8	3.7	4.8	9.7	4.0
28.9	14.9	27.8	27.8	19.6	39.3	37.2
14.7	15.5	21.7	17.5	15.1	14.5	20.3
25.0	30.3	30.1	41.0	25.8	26.3	32.2
14.5	22.8	6.1	5.2	27.8	7.7	4.6
13.8	10.7	10.4	4.7	6.7	2.7	1.9

Source:

Euro Barometre 8, 1978

[+]People were shown a card upon which were printed the numbers 100, 90, 80, down to 0 and asked: "Using this scale, could you indicate to what extent you think there is a danger over the next ten years of an increase in tensions in (your country) leading to actual civil disorder." We have interpreted the card so that level 100 = certain; levels 90-60 = very possible; level 50 = possible; levels 40-10 = unlikely and level 0 = impossible.

War is no longer the romantic charge of the light brigade. It's a grisly, depersonalised prospect – the fingers on the buttons, the intercontinental ballistic missiles, the nuclear devastation of major cities, megadeaths and massive radiation levels – possibly even the extinction of life on the planet. We all know what lies in store for us if the world goes to war again. Even the most hawkish politicians no longer advocate it, recognising mega – madness for what it is.

But the dangers of war still remain too apparent. The superpowers continue a billion-dollar and billion-rouble arms race. And the arsenals of the world grow larger and more explosive every day. The world is also full of hot spots where the clash of interests between the superpowers may escalate into a major confrontation.

How long can we be sure that there will be no fatal misunderstandings? No major blunder that'll take us all past the point of no return?

From the look of the figures in the table, most Europeans are mildly optimistic about world stability. Nearly half the people questioned in the U.K., Ireland and Denmark feel it's impossible that there will be a world war in the next ten years. The Danes are the most optimistic – nearly 48% of them think war is an impossibility.

Italy, West Germany and France also vote in favour of world sanity. Of all

	U.K.
	1977
People (%) who think that, in the next ten years, world war is:[+]	
certain	3.6
very possible	8.5
possible	10.0
unlikely	27.0
impossible	44.6
don't know/no answer	6.2

Italians polled three-quarters think that a Third World War in the next ten years is either unlikely or impossible. Yet 23% of the Italians feel war is certain, very possible or possible – a few points more than the West Germans.

The West Germans seem surprisingly optimistic, considering their vulnerable geographical position. If war were to break out between East and West, the Germans would be on the front line. But they've got more troops stationed in their country than any other in Western Europe (with large treaty contingents from the U.S., the U.K. and France). Perhaps this accounts for their feeling of optimism and security.

The second highest figure in the unlikely category is for the Netherlands. More than 45% of the Dutch are confident that there will be no war.

The French are also optimistic. Roughly two out of three Frenchmen

BELGIUM	DENMARK	FRANCE	(WEST) GERMANY	IRELAND	ITALY	NETHERLANDS
1977	1977	1977	1977	1977	1977	1977
3.3	3.2	2.3	1.8	2.6	2.7	1.9
14.2	5.6	10.1	10.3	10.1	11.2	14.5
11.5	9.5	14.2	9.3	10.7	8.8	19.0
29.7	19.1	33.3	49.5	25.2	35.7	45.1
26.9	47.9	28.1	21.0	46.1	38.6	16.5
14.5	14.6	11.9	8.0	5.1	3.0	2.9

surveyed doubt that war is in the offing. The Belgians are less positive; over half vote for world stability. But a quarter recognise the possibility of a holocaust and a sixth don't know.

Where are the most people who suspect there'll be a world war in the next ten years? Even in the most pessimistic countries the figures are low. In the Netherlands and Belgium just over 14% are fairly certain there'll be a world war within the next decade.

In all the other countries that gloomy view is shared by an average of about ten per cent of people. The striking exception is Denmark, where only one person in 20 thinks disaster is lurking just around the corner.

How many are "absolutely certain" there will be a world war in the next ten years? In the U.K. almost four per cent are talking of the end of civilization, the gloomiest record on the table.

Source:
Euro-Barometre 8 1978
[+]People were shown a card upon which were printed the numbers 100, 90, 80, down to 0 and asked: "Here is a sort of scale. Would you, with the help of this card, tell me how you assess the chances of a world war breaking out within the next ten years." We have interpreted the card so that level 100 = certain; level 90-60 = very possible; level 50 = possible; levels 40-10 = unlikely and level 0 = impossible

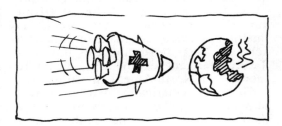

If you ever want to open a travel agency, open it in Scandinavia or France – potentially the most profitable countries for such an undertaking. Inside every Scandinavian there is a Viking trying to escape. Given time, opportunity and money, 84% of Norwegians, Swedes and Danes would choose to travel to other countries – particularly the U.S.

The French (also 84%) have the same trans-Atlantic inclinations. Spain is their second choice. Maybe they are hoping to avoid all the Canadian tourists who choose France.

The average Australian is travel-mad. Eighty-three per cent want to travel, preferably to the U.K. or U.S. Most nationalities pick the U.S. as their destination.

Canadians come next: 79% here want to travel – the fourth highest figure on the table. The choice of countries is a direct reflection of their origins. Eighteen per cent want to visit France and 14% the U.K. The higher figure for France isn't surprising, especially since French is taught as a second language in Canadian schools. Their third choice is the U.S. Twelve per cent feel the urge to explore more of the North American continent.

How do Americans themselves feel about travelling? Only 68% want to leave their own shores – the ninth highest figure on the table. Thirty-one per cent don't feel any wanderlust at all. That's the next highest number of stay-at-homes after the Japanese. Of those who want to travel, the majority (13%) pick the U.K.

The next most popular choice is Italy (ten per cent). No doubt this is because so many Americans are of Italian ancestry. Canada is their least favourite destination.

Twenty-one per cent of the British put the U.S. in first place as the nation they would most like to visit. Many of us would also like to visit our emigrated

	U.S.A.	CANADA
1. People (%) who want to travel[+]	68	79
don't want to travel	31	18
2. People (%) who would like to travel to:[0]		
USA	–	12
France	9	18
UK	13	14
Italy	10	4
Spain	5	3
USSR	4	3
Switzerland	5	5
Canada	2	–
Australia	5	9
Japan	8	3

U.K.	AUSTRALIA	BENELUX	E.E.C.	FRANCE	(WEST) GERMANY	ITALY	JAPAN	SCANDINAVIA
74	83	70	74	84	67	73	42	84
24	15	30	24	16	30	23	49	14
21	21	9	20	16	25	22	13	23
6	4	8	7	—	8	12	4	8
—	26	2	4	2	4	11	2	12
6	5	8	6	8	9	—	6	9
5	3	13	9	11	11	7	NA	13
5	3	3	6	8	5	8	1	9
7	3	9	5	2	5	5	7	6
12	4	2	6	5	5	3	2	3
14	—	2	5	1	3	1	1	5
7	11	2	5	8	4	2	—	7

NA not available   —*Respondents' own country*

Source:
Gallup International affiliated institutes in the countries concerned
The survey was conducted during winter 1974-75 and spring 1976.

[+]The question asked: "If you had the time, money and opportunity, would you like to travel to other countries of the world or not?" Totals do not add up to 100 because of a proportion of "don't knows" and "no answers".

[0]The question asked (of respondents who answered 'yes' to the previous question): "To which countries would you like to travel?" Answers have been extracted from a more extensive questionnaire.

friends and relatives in Australia. Canada is our third most popular choice.

A mere five per cent of those polled in the U.K. choose Spain as the country they'd most like to visit. Rather surprising when you consider the number of package tours to the Costa del Sol.

Most Japanese are anti-travel. Only 42% of them want to go anywhere.

We live in a world of information. The media bombard us with news all the time – but are we getting the information we want? As consumers, we're faced with dozens of decisions every week. Which products, goods or services to buy...how to handle money, credit and taxes... whether to take a holiday now or later. How much help do we get from the media with these matters, and do we believe what they tell us?

Fifty-three per cent of West Germans and 50% of the British approve of the consumer information given in their newspapers, and approbation of T.V. is even higher – 63% for the West Germans and 57% for the British.

The French are the most distrustful of newspaper information. Fifty-nine per cent thought it either poor or worthless. The Irish (56%), the Italians (53%) and the Belgians (48%) come next. In general, all nationalities feel television gives better consumer information than newspapers.

The table also shows the number of people who feel they have been cheated recently or sold merchandise that wasn't quite what they'd been led to expect. The U.K., has one of the highest numbers of consumers who've felt cheated: 47%. Italy is highest of all: 53% claim to have been cheated at one time or another. Forty-five per cent of the Dutch and 41% of the French also say they've been cheated or deceived on a purchase. By contrast, only 21% of Americans say they've been bilked.

Better consumer information from the media would cut down the number of customers (an average of over 35% for all the countries surveyed) who succumb to misleading or deceptive advertising and sales practices.

	U.S.A.
	1976
1. Quality of consumer information provided by newspapers: [+]	
good (%)	NA
poor (%)	NA
none at all (%)	NA
television	
good (%)	NA
poor (%)	NA
none at all (%)	NA
2. People (%) who have been cheated over a purchase [0]	21 [+]
3. Reactions to being cheated:	
did nothing; didn't mention any action (%)	NA
complained to sales assistant/asked for a replacement/refund (%)	NA
took legal action (%)	NA

U.K.	BELGIUM	DENMARK	FRANCE	(WEST) GERMANY	IRELAND	ITALY	NETHERLANDS
1975	1975	1975	1975	1975	1975	1975	1975
50	22	42	25	53	38	27	31
34	21	23	35	24	37	30	27
7	27	12	24	14	19	23	19
57	47	63	46	63	35	32	49
29	16	11	29	19	35	29	25
6	13	4	13	7	22	25	8
47	22	25	41	28	38	53	45
32	45	31	24	32	44	35	70
53	40	54	50	42	38	33	27
1	*	2	1	2	1	1	*

NA not available   *Under 0.5

Sources:
1 European Consumers, Brussels. 2 Europe: European Consumers, Brussels USA:
Public Opinion Index, ORC
[+]The question asked: "Do you think that the following information media provide good.
poor or no information at all for consumers?"
[0]The question asked (US): "Have you or your family been cheated or deceived in regard
to any product or service that you have purchased during the past year or so?"
   The question asked (Europe): "People sometimes say that they feel they have been cheated
when buying something. Can you remember you or your husband/wife buying something
with which you were not wholly satisfied this year or last year? If yes, did you take any
action? What did you do?"

Is death a justified penalty for certain types of crimes? Society has been faced with this question since the beginning of history. In most places, at one time or another, the answer has been "yes". How do you feel about the death penalty? Are some crimes so horrible that only death is a justifiable punishment? Or does the commandment "Thou shalt not Kill" always apply? Our first table shows you how others in the U.K. feel about these questions.

The first figures give the total response and are then broken down according to the sex of the respondents. The overwhelming majority of people questioned think that the death penalty is justified in certain circumstances: over four out of five people take this viewpoint. It's the men who feel most strongly about it – 84% of them think capital punishment is a justified response to some crimes. Only 78% of the women agree. Perhaps this isn't too surprising. Think of all those old movies with the condemned murderer's mother insisting tearfully, "He was always such a good boy". Twenty per cent of women feel that the death penalty is never justified whatever the circumstances, whereas only 14% of British men agreed with them.

The 81% who do advocate the death penalty were then asked which crimes they thought deserved it. By far the most (96%) feel that the murder of a policeman should be punished in this way. Policemen are symbols of society's desire to enforce law and order. When one is murdered, society demands revenge. The position is especially understandable here in the U.K. where policemen are unarmed.

Throughout history, murder has generally been considered the worst crime possible. So it's not too surprising that nine out of ten feel the murder of an ordinary citizen is the next crime deserving capital punishment. But we can look at it another way. These figures also mean that ten per cent of the people who favour the death penalty think it should be applied to crimes other than murder.

Perhaps they think capital punishment should be reserved for some of the new forms of crime that the 20th century has brought with it: terrorism, for example. What should be done with political extremists who plant a bomb in a crowded pub or store? In the U.K., 84% of the believers in the death penalty think it should be used against convicted terrorists.

These first three crimes, all of which involve killing other people, are the ones which Britons think should be punishable by death. Fewer people feel as strongly about crimes that don't necessarily involve murder. Hijacking for instance. Even though it is a crime involving risk to the lives of many innocent people, only 67% of those questioned thought hijackers deserved to die for it. But that's still more than the 62% who think violent rape is deserving of the death penalty. Strangely, this crime, which involves direct physical aggression against one helpless person, isn't considered as dreadful as the more impersonal political crimes.

There's a considerable drop in the figures when it comes to robbery. This crime doesn't involve direct aggression against anyone, nor does it necessarily endanger human life. However, just under half of those who believe in the death penalty think it should be applied to convicted thieves although the figures don't reveal whether these people feel that the type of robbery is significant. Do they think a bank robber should be hung, but not someone who holds up a small shop? Or is the value of the thief's target a factor to be considered in deciding on his punishment?

	1977	1977	1977	U.K. 1977
	—	All	Men	Women
1. People (%) who think the death penalty is justified:[+]				
ever	—	81	84	78
never	—	17	14	20
don't know	—	2	2	2
2. People, who approve of death penalty, (%) who believe the death penalty is a suitable punishment for:[0] murder of a policeman	96	—	—	—
murder	90	—	—	—
terrorism	84	—	—	—
hijacking a plane	67	—	—	—
violent rape	62	—	—	—
violent robbery	49	—	—	—
kidnapping for ransom	44	—	—	—

Source:
Market and Opinion Research International, London, on behalf of the Daily Express
[+]The question asked: "Do you think the death penalty is ever justified or not?"
[0]The question asked (of respondents who answered that the death penalty was sometimes justified): "I would like you to tell me whether or not you feel the death penalty would be a suitable punishment for those who commit each of the following crimes (shown on table, above)."

The lowest response was for the punishment of kidnapping. Only 44% think that holding someone for ransom deserves death yet kidnapping is a crime that involves a great deal of emotional suffering, both for victims and their families.

The death penalty is no longer used in the U.K. But a lot of people believe it's a suitable punishment for many types of crime. It looks as if the old adage, "an eye for an eye, a tooth for a tooth", isn't out of date yet.

Today's youth are strongly idealistic. They believe that happiness is a matter of being healthy – and that wealth and wisdom are far less important than honesty and kindness.

Unfortunately, we don't have figures for U.S. youth. There are some general trends which they would probably fall in with, but there are also some surprising national differences. So just what young Americans believe must remain speculative.

However, in all the countries on the table they are unanimous on one point: health tops the list in every country – and in Austria 91% of the youngsters mention it. Even in Finland where it is named least often, health is singled out by 53%.

Kindness is almost equally important to the Finns – 51% include it as an ingredient of happiness. The Dutch also put a premium value on kindness. Forty-seven per cent of them considered it as an ideal for happiness. These two countries are way ahead of the rest of the field here. The Spanish are the next highest, with a figure of 30% – then the U.K., along with the English Canadians, scoring a slightly above average 26%.

However, in the majority of countries honesty was considered the second essential to happiness after health. Though whether this was because the young people questioned thought that a bad conscience would make them unhappy, or whether they simply thought that a society which made a practice of dishonesty would be an unhappy one to live in, isn't disclosed. Even so, more than half the Greeks, Italians, Spanish and Canadians questioned thought that honesty was an essential for happiness.

A big surprise – especially considering that those questioned were young people between the ages of 15 and 25 – is to find how unimportant humour is in almost everyone's ideals for happiness. It looks as if only we British – who traditionally pride ourselves on our sense of humour – actually consider it to be an essential ingredient of a happy life. Here 56% of the young people questioned considered that a sense of humour is one of their ideals for happiness.

However, if the British enjoy a hearty laugh, the Greeks on the other hand seem

Young+ people (%) whose ideal for happiness is:	CANADA	French
to be healthy	58	61
to be honest	53	60
to be kind	22	11
to have a sense of humor	31	31
to be intelligent and clever	28	36
to be practical and handy	22	37
to be rich	11	10
to be beautiful/handsome	6	5
to be modern	5	4
to be slim	3	2

YOUTH: IDEALS FOR HAPPINESS

English	U.K.	AUSTRIA	FINLAND	GREECE	ITALY	NETHERLANDS	SPAIN
57	69	91	53	68	60	67	68
51	41	42	36	60	59	38	50
26	26	17	51	18	17	47	30
31	56	25	21	8	15	21	44
24	22	43	15	35	45	37	66
17	18	26	14	20	23	14	20
12	14	13	7	11	16	6	9
6	4	7	4	12	6	13	4
5	6	5	3	8	8	5	8
4	5	6	3	2	2	3	2

Source:
McCann-Erickson Youth Study, 1976-77
+Male and female, 15-25

to take their happiness very seriously indeed. Only eight per cent of them considered a sense of humour as one of the ideals for happiness.

Intelligence is uniformly downgraded almost everywhere. In general, only about a third of young people mention it, except in Spain where 66% rate it high.

Being practical is moderately essential to happiness – mostly to the Austrians (26%). The Dutch and the Finns (both 14%) mention it least.

The old adage that money doesn't buy happiness gets a ringing endorsement from young people. The Italians (16%) include money most often, but across the board just about ten per cent of respondents mention being rich as one of the qualities that make for happiness.

Other ideals that are often equated with worldly success – being beautiful or handsome, slim, modern – also seem to be comparatively unimportant to the younger generation. In most cases, fewer than ten per cent bother to single out these qualities, though a few more young Dutch have dreams of being beautiful. The Greeks, Italians and Spanish put most value on being modern (eight per cent). And Austria puts the highest value on being slim: six per cent of the young Austrians believe that happiness is a matter of shedding poundage. This concern isn't too surprising in a country where whipped cream pastries are part of the national heritage.

The young are usually the most discontented in any society. In this report, young people, between 18 and 24, were offered a choice of four means to express dissatisfaction.

Swedish youth appear to be the most docile. More than half of them would vent their displeasure only at the polling booth. More than half the young people in the U.K. and Japan also believe in the power of the vote, with the West Germans only a few percentage points behind.

The most activist youth is over in the U.S., where 54% believe that active, legal protest is the most effective instrument for social change. The not-so-peaceful Swiss come next on the activist list, followed by the U.K., Sweden, France and Japan. The fewest demonstrators are in Brazil.

Young French and West Germans are the leading advocates of violence, with six out of every hundred believing that only terrorism can effect reforms. The Swiss are next with five percent, and the U.S., India and Japan follow with four per cent. The most peaceful nations are the U.K. and Sweden (two per cent) and Brazil (one per cent). The Brazilians are probably more disgusted with their soc-iety than their figure suggests – 40% of their young people regard dropping out of society as the best solution to national problems. The fed-up French are next,

	U.S.A.	U.K.
	1973	1973
1. Measures young[+] people would take if they were dissatisfied with society:[0]		
use their vote (%)	36	54
take active, but legal, measures (%)	54	37
resort to violence and/or other illegal measures (%)	4	2
drop out from society (%)	5	4
2. Reasons why young[x] people would vote, rather than take more active measures to change society:[‡]		
individuals are ineffective (%)	30	30
people in the proper position should cope (%)	46	39
other things are more important (%)	22	28

with 17% wanting out, followed by the perennially discontented Swiss, with 12%. Fewer young people are "dropping out" from society as a cure for dis-satisfaction than in most other countries.

The young people who said they'd restrict their protest to voting were asked why they chose not to take more active measures. Almost three-quarters of the Japanese and over half the West Ger-mans felt that the individual was inef-fective in trying to change society. The most frequent answer for British, Swedes, Americans, Brazilians, Indians and French was that reform should be left to people in the proper positions.

YOUTH: CHANGING SOCIETY

BRAZIL	FRANCE	(WEST) GERMANY	INDIA	JAPAN	SWEDEN	SWITZERLAND
1973	1973	1973	1973	1973	1973	1973
41	34	49	44	54	55	37
18	37	33	42	37	38	46
1	6	6	4	4	2	5
40	17	4	8	5	4	12
30	16	53	33	73	18	32
48	44	31	45	9	57	30
21	32	11	20	18	23	38

Source:
Gallup International affiliated institutes in the countries concerned
+Male and female, 18-24
0The question asked: "Suppose you were dissatisfied with society what attitudes do you think you would take: "I will use my voting right but nothing more." "I will actively resort to a variety of measures such as petitions, letters of complaint, demonstrations, strikes etc. so long as the means are permitted by the law." "I will resort to violence and/or other illegal measures if necessary." "I will become a drop-out from society."
XMale and female, 18-24; respondents who said they would use their voting right but nothing more in answer to the previous question.
‡The question asked: "Why don't you take more active measures as well? Please choose one answer from: "The problems involved are beyond the reach of individuals." "The affairs of society should be handled by persons in the proper position." "There are other things which are more important to me."
Totals do not add up to 100 as there is a proportion of "don't knows" and "no answers".

Everybody's welfare depends on the good of the nation. But the good of the nation also depends on the welfare of the individual. Which should come first is a question of national temperament. And the type of government in power.

Just over two-thirds of the young people in Britain think that the government benefits the nation at the expense of the individual. Also, almost nine out of ten questioned think that money reigns supreme in our grossly materialistic society – a high figure, but lower than the figures for the U.S., France and Switzerland. On the other hand, only 42% think that a person's future is decided by family background.

Americans are traditional believers in "rugged individualism". Almost three out of every four young Americans feel that the emphasis on the national good undermines the well-being of the individual.

Nine out of every ten Americans think money reigns supreme in our society. At the same time, the majority feel that their success depends on their own hard work, not on family background.

Traditionally, the West Germans think differently. And the thoughts of young Germans are true to form. Only 44% of them think their individual benefits take back seat to those of the nation. A full 49% approve of the nation taking priority.

		U.S.A.	BRAZIL
1.	"The Government emphasizes benefits to the nation at the cost of the individual"+		
	young people who agree (%)	74	55
	young people who disagree (%)	25	44
2.	"Money reigns supreme in this grossly materialistic society"⁰		
	young people who agree (%)	88	77
	young people who disagree (%)	11	21
3.	"A person's future is decided by his or her family background"ˣ		
	young people who agree (%)	48	61
	young people who disagree (%)	51	37

In industrial West Germany, young people aren't at all cynical about the importance of money. Only 16 per cent reject the idea that money reigns supreme in their society.

Again, when it comes to the importance of family background, West Germany is odd man out. Nearly three-quarters of all young Germans think a person's success depends largely on the status and wealth of his parents. This is almost twice as high as the percentage in Sweden and France. It's over ten per cent higher than in Brazil.

The figures for France are equally interesting. Mammon reigns in the land of luxury and good food. Despite an old guard network, French youth seem more optimistic about the chances of a person making it on his own than any people but

YOUTH: ATTITUDES TO GOVERNMENT

U.K.	INDIA	FRANCE	(WEST) GERMANY	JAPAN	SWEDEN	SWITZERLAND
68	71	66	44	88	68	63
28	27	13	49	11	27	35
86	86	90	78	84	82	90
12	13	5	16	16	15	10
42	52	38	72	48	34	57
57	46	53	23	51	63	42

Source:
Gallup International affiliated institutes in the countries concerned. The poll was conducted in 1973.
In the survey young people (male and female, 18-24 years) were told "I am going to read some statements about our national government or society. For each one please tell me if you think it's 'true' or 'false' " The statements:
+"The government is placing too much emphasis on the benefits of the nation as a whole at the cost of individuals."
0"In the present grossly materialistic society, money reigns supreme."
x"Man's future is often virtually predetermined by his father's (mother's) profession as well as his family background."
The percentages don't add up to 100 because of the "don't knows" and "no answers".

the Swedes. Only 38% think you really need a Proustian pedigree to get on in life. Two-thirds of French youth is also on the side of the individual.

The Japanese figures are surprisingly like those for the U.S. Even more young Japanese resent government interference than in America. Eighty-four per cent believe in the cult of yen. But most surprisingly their views on the importance of a good background are exactly the same as in the U.S..

Young+ people (%) who agree:	CANADA	French	English
"A woman's place is in the home"	20	31	16
"If a couple both earn money, both should share the housework"	93	90	94
"Men and women should be paid the same for the same job"	92	91	92
"Washing clothes should still be done by women"	28	46	22
"Cooking should still be done by women"	30	52	23

From this table you can't tell at a glance which country has the most macho males or subservient females because the questions were answered by both men and women. However, we may be able to approach the subject from another direction.

We could ask in which country the macho males would receive most support from dutiful wives.

The answer is Greece, where 80% believe that housework should be shared between men and women if both work, and that women should be paid the same for doing the same job. But there's almost a complete reversal when it comes to the nitty gritty of who should do the washing and cooking. Here, very few believe in sharing the work. By 75% and 72% respectively, the agreement is that women should do both.

The country in which young women appear to be on the most equal footing is Sweden. Their extremely liberated views make youth in the U.K. look, in comparison, like male chauvinist pigs. Only 11% of young Swedes think that a woman's place is in the home whereas over a quarter of British youth think this. Likewise, fewer of us are prepared to share the housework, and 13% of us still do not believe in equal pay.

The Finns are consistent supporters of sex equality with only 22% of young people believing that a woman's place is in the home. They're also enthusiastic about sharing the housework (93%) and equal pay (90%). An extremely low percentage of Finns – 13% – believe cooking is solely a woman's job and even fewer – seven percent – think that only she should do the washing.

For a more traditional view of a woman's place, turn to the Austrians, a whopping 57% of whom think women belong in the home. Most of them believe she should do the washing (62%) and the cooking (65%), which puts them almost on a par with the Greeks. The macho spirit isn't confined to the Latins.

Only 22% of young Spaniards believe a woman belongs at home, and 97% think she should be paid the same money for the same job. They're also fairly liberal about the washing and the cooking, with only about 36% feeling those are solely women's jobs.

The Italians run neck and neck with

YOUTH: A WOMAN'S PLACE

U.K.	AUSTRIA	FINLAND	FRANCE	GREECE	ITALY	NETHERLANDS	SPAIN	SWEDEN
26	57	22	42	39	30	29	22	11
84	96	93	86	80	83	96	92	97
87	87	90	93	81	87	95	97	97
46	62	7	47	75	50	45	35	—
44	65	13	26	72	37	30	37	—

—Question not asked

our own youth, save that just 37% of Italians versus 44% of us believe cooking is a woman's job. It seems that the mother's days of ruling the kitchen are numbered – perhaps to her relief.

Turning to France, we find a country that's not over-enthusiastic about equality: 42% of young people believe that a woman should stay at home.

But there's an interesting 21% gap between whether cooking and washing are woman's sole province. Forty-seven percent vote for women doing the washing, while only 26% feel the same way about cooking.

Overall, one interesting fact emerges: around 90% of youth believe a couple should share the housework and get equal pay for equal work.

Perhaps there's a better future for women's rights than some of our tables indicate.

Source:
McCann-Erickson Youth Study, 1976-77
[+]Male and female, 15-25

Work takes up a huge chunk of the waking day for most adults in the world. So how do young people today feel about setting out on a lifetime of it? Why are they going to do it? Forget all about the dignity of labour, self-expression and mutual self-help. The majority of 18 to 24 year olds in the countries listed see work simply as a way of getting money to pay all those bills.

	U.S.A.	U.K.
	1973	1973
Young[+] people's reasons for working:[0] to earn money (%)	59	80
to be a responsible member of society (%)	11	4
to find self-fulfilment (%)	30	14

Top of the list come the young people in the U.K. and France. A massive 80% in both countries are under no illusions as to why they are spending their time in offices or factories – it's to get money. They sweat for eight hours a day to get enough pounds and francs to buy the things they need or just fancy. They don't expect to get satisfaction from their jobs so they hope that their wages will allow them to get it in their private lives.

Our figures are identical with those of the French in all categories. Only four per cent believe that work is about keeping the wheels of society rolling and that everybody contributes to the effort. The remaining 14% are the ones who see work as an end in itself. These young people are not looking for any reward other than the satisfaction that their jobs provide. It's a fairly safe bet that the young people who opted for this answer are the lucky ones who are not involved in repetitive, mechanical work. Their jobs obviously give them room to stretch themselves and use their abilities to the full.

The youths in two other Western European countries, West Germany and Sweden, have roughly the same attitudes. The major difference between Swedish young men and women and their French and British counterparts is that twice as many (eight per cent) feel that work is their way of being a responsible member of society. An even higher percentage (11%) of German youth gave that as their reason for working.

By far the highest percentage of people who believed that work was their way of helping society is found, not unsurprisingly, in Yugoslavia. Twenty-three per cent of young Yugoslavs have obviously been deeply influenced by the Communist doctrines that prevail in their country. When one considers the enormous emphasis placed by Communist countries on the importance of the State and the common cause, it's surprising that this figure is not higher. It would be interesting to see the figures for Red China or Russia. As it is, the large majority of Yugoslavs still put their own desire to possess money as their priority. They also returned the lowest number of responses in the "self-fulfilment" column – just eight per cent.

There's a high rate of idealism in India. There is a great emphasis on spiritual

YOUTH: MEANING OF WORK

	BRAZIL	FRANCE	(WEST) GERMANY	INDIA	JAPAN	SWEDEN	SWITZERLAND	YUGOSLAVIA
	1973	1973	1973	1973	1973	1973	1973	1973
	43	80	70	56	55	75	63	69
	13	4	11	21	11	8	13	23
	42	14	15	23	34	15	23	8

rather than material matters in Eastern philosophies and religions, and it shows in this table. Even in the face of the daily fight for the basics in order to stay alive, a low 56% of Indians believe that their working day is spent only in getting the daily bread. Forty-four per cent of them have their minds on other things. Twenty-one per cent forget their own needs and concentrate on working for the general good, while 23% find self-fulfilment in their work.

The Eastern attitude to worldly goods even applies to industrialised, competitive Japan. They may be challenging the markets of the world with cars, motorbikes and electrical goods, but the people who make and sell these products aren't just in it for financial reward. Only 55% of Japanese workers felt that they put the effort in just to earn money. The other 45% had idealistic reasons. Much of the Japanese people's way of thinking and acting is influenced by Shinto, which until recently closely identified religion and state. This may explain the 11% who see work as a duty towards their society. The 34% of Japanese who chose "self-fulfilment" as their reason for work is the second highest response in this category.

Source:
Gallup Opinion Index, 100
+Male and female, 18-24
0The question asked: "Why do you think man works? Please choose one answer closest to your feelings from the following: to earn money; to do his duty as a member of society; to find self-fulfilment."
Totals do not always add up to 100 because of a proportion of "don't knows" and "no answers".

Perhaps the possession of wealth eases the desire to earn more by working. Only 59% of Americans and 63% of Swiss put this as their main priority. A high 30% of Americans worked in order to fulfil themselves and 23% of Swiss workers felt the same.

The most amazing figures come from Brazil. Only 43% of Brazilians listed earning money as their reason for working. Fifty-five per cent of them felt that the other two reasons were more important. One thing seems to be pretty certain. Young Brazilians must get much more satisfaction from the time they spend at work than the youth of Britain and France.

Previous younger generations often had little voice in selecting their careers. Circumstances, social class and parental choice usually dictated which occupations the child could pursue. Today, society is more liberal and its values less absolute. This means greater freedom for young people – freedom to do as they please, freedom to choose what they do. However, this freedom also places a much heavier burden on the individual. Young people have more say in deciding the nature and quality of their lives. They are also aware that these choices can be vitally important. This table is an indication not only of the difficulties that face young people, but also how seriously they take them.

Freedom isn't everything. Unemployment figures (see table on unemployment) show there aren't enough jobs to go round. In some countries the situation is extremely bad, and, in spite of government-sponsored employment programmes, the unemployment problem obviously hits young people hard. They're just starting out in life and are very concerned about the opportunities which will or will not be available to them. The first question on the table shows how economic problems have affected the attitudes of young people today. Eighty-six per cent of young Spaniards think finding a job is a problem.

Young people (%) who think it's a problem:	CANADA	French	English
finding a job	43	52	40
deciding which job or career to take up	43	46	42
deciding whether to get qualifications or go for a job	30	33	29
deciding whether to live with their girl/boyfriend	6	8	6
deciding whether to sleep with their girl/boyfriend	5	8	4

Italy, where 64% of the young people are worried about getting a job, comes next, followed by France with 59%.

Unemployment is high in Spain and Italy, so these figures are understandable. But why France? The fact that 59% of young French people are undecided about a career reflects increasing discrimination. They don't want any old job but one that will be interesting and well paid. Lack of opportunity in their chosen field is a bitter blow.

When jobs are scarce, dissatisfaction and rebellion are likely to follow. More than half the young people in French Canada, for instance, are worried about finding a job. No wonder there's so much agitation for Quebec Libre (independent French Canada).

Two other problems the young people have to cope with are whether to live with

YOUTH: PROBLEMS

U.K.	FINLAND	FRANCE	GREECE	ITALY	NETHERLANDS	SPAIN	SWEDEN
27	29	59	46	64	28	86	30
28	42	59	–	49	32	53	40
28	29	38	49	36	19	41	25
6	8	15	24	22	6	28	13
5	6	10	25	19	6	25	–

—Question not asked

Source:
McCann-Erickson Youth Study,
1976-77

their girl or boyfriend and whether to sleep with them. These are tough decisions for young people who live in morally rigid societies. For instance, in Roman Catholic Spain more than one out of four young people say that deciding whether to live with their girlfriend (or boyfriend) is a major problem; and exactly a quarter are worried about whether to sleep with her or him. These are the highest figures for both these categories, though the Greek figure is also one in four for the second question. In Greek society, the Orthodox church is still a ruling force, just as the Roman Catholic Church is in Italy and Spain. Catholic influence is also still fairly strong in France – and the French figures are accordingly high.

However, these questions appear to pose hardly any problem in the other countries listed. Less than ten per cent considered that sleeping with or living with your girlfriend (or boyfriend) is a problem. If you polled their parents you'd get a very different result.

The young people in the U.K. score lower than all the other countries in nearly every category. This could indicate that they feel that there are fewer stresses and problems and that all the ingredients for a happy life are readily available. This complacency, however, could lead to a lack of push in recognising and overcoming the challenges of life.

We British, with our long tradition of public school education and our continuing controversy over grammar and comprehensive schools, are not as snobbish as we're made out to be. Forty per cent of our young people think that a person's actual qualifications are more important than his "old school tie". However, the number of British youth who think that higher education is a waste of time is revealing. Nearly 20% think it's best to go to work as soon as possible. Perhaps this is where our class structure shows through. Our working-class youths aren't always encouraged to seek a higher education. School teachers and parents often steer them into some form of job or apprenticeship at an early age to learn a wage-earning skill. It is the middle-class British parents who are more apt to push for college or university. There's a significantly high proportion (17%) of youngsters who'd like to go on with academic studies, but who are forced to go out and earn a living.

Most young Americans believe an education is more than reading, writing and arithmetic. It's a way to change social and economic status – and to find a good marriage mate. Only the Swiss place more emphasis on the prestige of schools than Americans do. As many as 70% of young Americans surveyed believe that the school name embossed

	U.S.A.	U.K.
1. Young[+] people who agree a school's social prestige influences job opportunities and future[0]	70	59
disagree	29	40
2. Young[+] people (%) who think higher education:[x]		
improves chances of a good job and marriage	33	32
is a waste of time	5	19
3. Young[+] people (%) who would like to stay on at school but have to earn a living[x]	10	17

on a diploma means just as much as one's qualifications.

In conservative Switzerland, the emphasis on higher education shows up markedly. Nearly half the young Swiss go to college to advance their career and marriage prospects. And three-quarters think the name of your school is the name of the game.

The French agree with the Americans and Swiss in the value they place on prestige in education.

School prestige counts for the least in West Germany and Sweden. In Germany, 40% think it's important, in Sweden, only 31%. In fact, two out of every three Swedes disagree with the idea that the prestige of schools is important to future and job.

Young Germans don't all see college as a practical matter. Only 23% go for career

YOUTH: ATTITUDES TO EDUCATION

BRAZIL	FRANCE	(WEST) GERMANY	INDIA	JAPAN	SWEDEN	SWITZERLAND	YUGOSLAVIA
56	70	40	63	63	31	74	67
42	20	47	34	35	66	24	33
26	21	23	36	25	28	45	18
2	7	14	4	11	11	10	3
18	24	10	9	10	15	12	13

Source:
Gallup International affiliated institutes in the countries concerned.
The survey was conducted in 1973.
⁺Male and female, 18-24
⁰The original question asked whether or not the following statement applied to the experience of the young people questioned: "It is accepted by most people that regardless of your qualifications the social prestige at the school you graduate from will influence your job and future."
ˣThe original question asked which of the following statements came closest to the view of higher education of the people questioned: "I have to go to a higher school in order to improve my chances of obtaining a good job and marriage." "It's a waste of time staying on at school or college. I should go out to work as young as possible." "I would like to have stayed on but I have to earn a living." Other options were also offered.

reasons. And 14% think it's better to go out and get a job.

Ironically, the places where college is most valued are also those where the fewest go for practical reasons. In Brazil, India and Yugoslavia, very few people think higher education is a frivolous matter. But only a quarter of Brazilian youth go on for career reasons. And in spite of the fact that 67% of surveyed Yugoslavians think college is worth the effort, only 18% actually think they'll be advancing themselves by more education.

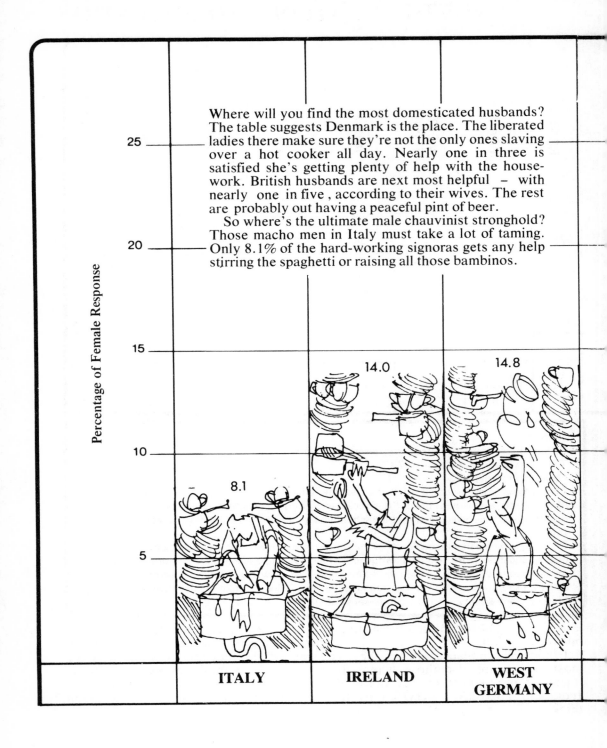

Where will you find the most domesticated husbands? The table suggests Denmark is the place. The liberated ladies there make sure they're not the only ones slaving over a hot cooker all day. Nearly one in three is satisfied she's getting plenty of help with the house-work. British husbands are next most helpful – with nearly one in five, according to their wives. The rest are probably out having a peaceful pint of beer.

So where's the ultimate male chauvinist stronghold? Those macho men in Italy must take a lot of taming. Only 8.1% of the hard-working signoras gets any help stirring the spaghetti or raising all those bambinos.

Percentage of Female Response

25

20

15

14.0

14.8

10

8.1

5

ITALY **IRELAND** **WEST GERMANY**

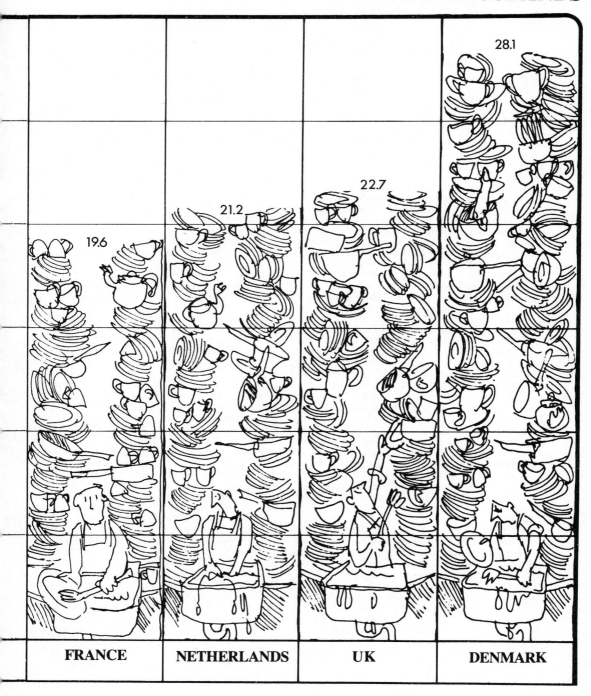

BIBLIOGRAPHY

People
18/19: United Nations; data compiled from document number ESA/P/WP.55, May 28th 1975, "Selected World Demographic Indicators by Country 1950-2000", by the Population Division, Department of Economic and Social Affairs, © United Nations Secretariat, reproduced with their permission. Population Reference Bureau Inc., Washington, D.C.; data calculated by Heron House from 1977 estimates. © Population Reference Bureau Inc., Washington, D.C., reproduced with their permission. 20/21: United Nations; data compiled from medium variant, document number ESA/P/WP.60, February 25th 1976, "Population by Sex and Age for Regions and Countries, 1950-2000", as assessed by the Population Division, Department of Economic and Social Affairs, © United Nations Secretariat, reproduced with their permission. 22/23: United Nations; data compiled from document number ESA/P/WP.56, October 6th 1975, "Single Year Population Estimates and Projections for Major Areas, Regions and Countries in the World, 1950-2000", as assessed by the Population Division, Department of Economic and Social Affairs, © United Nations Secretariat, reproduced with their permission. 24/25: United Nations, data compiled from document number ESA/P/WP.58, November 21st 1975, "Trends and Prospects in the Population of Urban Agglomerations, 1950-2000", as assessed in 1973-1975, by the Population Division, Department of Economic and Social Affairs © United Nations Secretariat, reproduced with their permission. 26/27: United Nations, data compiled from document number ESA/P/WP.55, May 28th 1975, "Selected World Demographic Indicators by Country 1950-2000", by the Population Division, Department of Economic and Social Affairs, © United Nations Secretariat, reproduced with their permission. Population Reference Bureau Inc., Washington, D.C.; data calculated by Heron House using birth and death rates estimated for 1977, © Population Reference Bureau Inc., Washington, D.C., reproduced with their permission.

The Land
32/33: Food and Agriculture Organization, data compiled from *F.A.O. Production Yearbook 1976*, Volume 30, © Food and Agriculture Organization, reproduced with their permission. © 34/35: Food and Agriculture Organization; data compiled from *F.A.O. Production Yearbook 1976*, Volume 30, © Food and Agriculture Organization, reproduced with their permission. 36/37: I.U.C.N.; data compiled from "1975 United Nations List of National Parks and Equivalent Reserves", published 1975, number 33, © I.U.C.N. (International union for Conservation of Nature and National Resources), reproduced with their permission. Vision, data compiled from *Vision* magazine, July/August 1973 issue, © Vision, reproduced with

their permission.
Man and the Environment
42/43: "Tables of Temperature, Relative Humidity, Precipitation and Sunshine for the World", H.M.S.O.; data compiled from Parts 1 and 3 and reproduced with the permission of the Controller of Her Majesty's Stationery Office. *World Survey of Climatology*, Elsevier Scientific Publishing Company, reproduced with their permission. 44/45: National Oceanic and Atmospheric Administration, U.S. Department of Commerce. 46/47: World Health Organization, data compiled from "Air Quality in Selected Urban Areas 1973-4", published by W.H.O., Geneva, 1976, reproduced with their permission. 48/49: Overseas Development Council, data compiled from various documents, published by the Overseas Development Council, Washington, D.C. and reproduced with their permission.

People and Money
54/55: United Nations, data compiled from United Nations "Monthly Bulletins" and from "National Accounts", reproduced with their permission. 56/57: *World Bank Atlas 1976*, published by the World Bank, reproduced with their permission. Organization for Economic Co-operation and Development, data compiled from *O.E.C.D. Financial Statistics*, number 11, 1977 and from "Quarterly National Accounts Bulletins", published in Paris in 1977 by O.E.C.D., reproduced with their permission. 58/59: International Monetary Fund, data compiled from "International Financial Statistics" for the years concerned, published by the I.M.F. 60/61: Organization for Economic Co-operation and Development, data compiled from the *O.E.C.D. Financial Statistics Tables* and from "Quarterly National Accounts Bulletins" for the years concerned, reproduced with their permission. 62/63: Bank of England and International Monetary Fund, data compiled from "The Bank of England Quarterly Bulletins" and the "International Financial Statistics Tables", published by the I.M.F. for the years concerned. 64/65: Government sources, data compiled from the various statistical offices concerned and conversions by Heron House. 66/67: Market Research Surveys and Consolidated Gold Fields Ltd., London. Data compiled from their publication "Gold 1977" and reproduced with their permission. 68/69: Aston Martin Lagonda Ltd., Rolls-Royce Motors Ltd. and Panther Cars Ltd., data provided by Public Relations Departments of these three companies. Sojuzpushnina, Moscow, data provided by their Public Relations Department. All data reproduced with permission of the companies concerned. 70/71: United Nations, data compiled from document number ECSC/CIRC/PAC/39, June 30th 1978, "Schedule of Post Adjustment Classifications" by International Civil Service Commission. 72/73: International Labor Organization, data compiled from Part III

of the "Bulletin of Labor Statistics 1977", 2nd Quarter, © International Labor Organization 1977, converted by Heron House, reproduced with their permission. 74/75: Euromonitor, data compiled from *European Marketing Data and Statistics,* volume 14, 1977/8, London and from © *International Marketing Data Statistics,* volume 3, 1977/8, © Euromonitor Publications Ltd., London, reproduced with their permission. 76/77: *Survey of Europe Today,* Reader's Digest 1970. Heron House estimates based on I.L.O. figures; Heron House has interpreted the Table 2b, chapter 1 of the *International Labor Organization Statistics 1977* to match the categories found in *Survey of Europe Today* Reader's Digest, 1970.

Where the Money Goes
82/83: Europa Publications Ltd, data compiled from *The Europa Yearbook, A World Survey 1977,* published by Europa Publications Ltd., London, reproduced with their permission. 84/85: Organization for Economic Co-operation and Development, data compiled from "Development Co-operation, 1977 Review", published by O.E.C.D., Paris, November 1977, reproduced with their permission.

The Taxman Cometh
90/91: Organization for Economic Co-operation and Development, data compiled from *Revenue Statistics of O.E.C.D. Member Countries 1965-1974,* © O.E.C.D. 1976, reproduced with their permission. 92/93: Organization for Economic Co-operation and Development, data compiled from *Revenue Statistics of O.E.C.D. Member Countries 1965-1974,* © O.E.C.D. 1976, reproduced with their permission. 94/95: Inland Revenue, London. 96/97: Official government documents, data compiled and converted by Heron House based on government data made available to Heron House. 98/99: Organization for Economic Co-operation and Development, data compiled from *Revenue Statistics of O.E.C.D. Member Countries 1965-1974,* © O.E.C.D. 1976, reproduced with their permission. Confidential industry estimates, conversions by Heron House.

At Work
104/105: Organization for Economic Co-operation and Development, data compiled from *Labor Force Statistics 1964-1975,* published in Paris in 1977, © O.E.C.D. reproduced with their permission. 106/107: Organization for Economic Co-operation and Development, data compiled from *Labor Force Statistics 1964-1975,* published in Paris in 1977, © O.E.C.D., reproduced with their permission. 108/109: Organization for Economic Co-operation and Development, data compiled from *Labour Force Statistics 1964-1975,* published in Paris in 1977, © O.E.C.D., reproduced with their permission. Organization for Economic Co-operation and Development, January 1978 reprinted from the "O.E.C.D. Observer", data number 90, January 1978. 110/

111: International Labor Organization, data compiled from *The Yearbook of Labor Statistics 1977,* © I.L.O. 1977, reproduced with their permission. Heron House estimates were based on I.L.O. figures and converted into a different set of units. 112/113: British Steel Corporation, data compiled from *International Steel Statistics 1976* and converted into the US/UK system of units by Heron House, together with B.S.C.'s permission. 114/115: Union Bank of Switzerland, data compiled from *Prices and Earnings around the Globe,* published in October 1976. 116/117: International Labor Organization, data compiled from *The Yearbook of Labor Statistics 1977,* © International Labor Organization 1977, reproduced with their permission. Heron House estimates based on I.L.O. figures and converted into a different set of units. 118/119: International Labor Organization, data compiled from *The Yearbook of Labor Statistics 1977,* © International Labor Organization 1977, reproduced with their permission. 120/121: International Labor Organization, data compiled from *The Yearbook of Labor Statistics 1977,* © International Labor Organization, 1977, reproduced with their permission. 122/123: International Labor Organization, data compiled from *The Yearbook of Labour Statistics 1977.* © International Labor Organization 1977, reproduced with their permission. 124/125: International Labor Organization, data compiled from *The Yearbook of Labor Statistics 1977,* © International Labor Organization 1977, reproduced with their permission.

At Home
130/131: Euromonitor, data compiled from *European Marketing Data and Statistics,* volume 14 1977/8, and from *International Marketing Data and Statistics,* volume 3, 1977/8, © Euromonitor Publications Ltd., London and reproduced with their permission. United Nations, data compiled from the *United Nations 1976 Statistical Yearbook,* 28th issue, New York, 1977, © United Nations, reproduced with their permission. 132/133: Euromonitor, data compiled from *European Marketing Data and Statistics,* volume 14, 1977/8, and from *International Marketing Data and Statistics,* London, volume 3, 1977/8, © Euromonitor Publications Ltd., and reproduced with their permission. National Statistical offices, data compiled by Heron House based on documents sent by the statistical offices concerned. 134/135: National statistical offices, data compiled by Heron House based on various publications provided by the statistical offices. Union Bank of Switzerland, data compiled from *Prices and Earnings around the Globe,* published in October 1976 in Switzerland. 136/137: National statistical offices, data compiled by Heron House based on documents sent by the statistical offices concerned. Euromonitor, data compiled from *Euro-*

pean *Marketing Data and Statistics,* volume 14, 1977/8, and from International *Marketing Data and Statistics,* volume 3, 1977/8, © Euromonitor Publications Ltd., London, reproduced with their permission.

At Play

142/143: Euromonitor, data compiled from *Consumer Europe 1977* © Euromonitor Publications Ltd., London, reproduced with their permission. Gallup Organization Inc., data reproduced from the "Gallup Opinion Index of March 1974", report number 105, reproduced with their permission. 144/145: Organization for Economic Co-operation and Development, data reprinted from "O.E.C.D. Observer", number 91, March 1978. Heron House estimates, data compiled by Heron House based on *U.N.E.S.C.O. 1975/76 Statistical Yearbook.* 146/147: United Nations, data compiled from *United Nations 1976 Statistical Yearbook,* 28th issue, New York, 1977, © United Nations, reproduced with their permission. United Nations Educational, Scientific, and Cultural Organization, data extracted from the *U.N.E.S.C.O. Statistical Yearbook 1976,* table 14.1, © U.N.E.S.C.O. 1977, reproduced with their permission. 148/149: International Olympic Committee Archives, data compiled from a selection of documents provided by the National Olympic Committee in London. 150/151: Golf Digest Inc., data reprinted from the February, 1975 issue of Golf Digest Inc., U.S.A., © 1975 Golf Digest Inc., U.S.A., reproduced with their permission. 152/153: Euromonitor, data compiled from *European Marketing Data and Statistics,* volume 14, 1977/8, © Euromonitor Publications Ltd., London reproduced with their permission. Government trade offices, data supplied by the respective offices. *Gambling in America,* by the Commission on the Review of the National Policy Toward Gambling, Washington, D.C. 1976. 154/155: Euromonitor, data compiled from *European Marketing Data and Statistics,* volume 13, 1976/7, © Euromonitor Publications Ltd., London, reproduced with their permission. Gallup Organization Inc., data provided by the Gallup Organization Inc., in Princeton, New Jersey, U.S.A., reproduced with their permission. *Leisure Survey of the Secretary of State* published in 1976 by the Secretary of State, Ottawa, Canada. 156/157: United Nations Educational, Scientific, and Cultural Organization, data compiled from *U.N.E.S.C.O. Statistical Yearbook 1976,* ©U.N.E.S.C.O. 1977, reproduced with their permission. 158/159: United Nations Educational, Scientific, and Cultural Organization, data compiled from the *U.N.E.S.C.O. Statistical Yearbook 1975,* © U.N.E.S.C.O. 1976, reproduced with their permission. 160/161: United Nations Educational, Scientific, and Cultural Organization, data com-

piled from *U.N.E.S.C.O. Statistical Yearbook 1975,* © U.N.E.S.C.O. 1976, reproduced with their permission. Library Association, data compiled by the Library Association, London. 162/163: E.M.I. Ltd., data compiled from *World Record Markets 1976* published by E.M.I., reproduced with their permission. 164/165: Euromonitor, data compiled from *European Marketing Data and Statistics,* volume 14, 1977/8, © Euromonitor Publications Ltd., London, reproduced with their permission. Industry estimates. The Australian Tourist Board. 166/167: Organization for Economic Co-operation and Development, data compiled from "Tourism Policy and International Tourism in O.E.C.D. Countries", Paris 1978, © O.E.C.D., reproduced with their permission.

Underway

172/173: International Road Federation, data compiled from *World Road Statistics 1972-1976,* 1977 edition, Geneva, Switzerland. Society for Motor Manufacturers and Traders, London, data supplied by the Society and reproduced with their permission. 174/175: Union Bank of Switzerland, data compiled from *Prices and Earnings around the Globe,* October 1976, reproduced with their permission. 176/177: International Road Federation, data compiled from *World Road Statistics 1972-1976,* 1977 edition, Geneva, Switzerland, data converted by Heron House. 178/179: Automobile Association, London. Department of Motor Transport, Australia. Embassies. Data supplied upon request. 180/181: International Road Federation, data compiled from *World Road Statistics 1972-1976,* 1977 edition, Geneva, Switzerland. 182/183: International Civil Aviation Organization, data compiled from *Civil Aviation Statistics of the World - 1976,* published by I.C.A.O. in Montreal, Canada, reproduced with their permission. 184/185: International Civil Aviation Organization, data compiled from "Digest of Statistics", number 218-A, series T, number 36, "Airline Traffic", volume 1, published by I.C.A.O. in Montreal, Canada, reproduced with their permission. Heron House conversions. 186/187: International Civil Aviation Organization, data compiled from "Digest of Statistics", number 218-A, series T, number 36, Airline Traffic, volume 1, published by I.C.A.O. in Montreal, Canada, reproduced with their permission. 188/189: *Destination Disaster* by Paul Eddy, Elaine Potter and Bruce Page, © Times Newspapers Ltd 1976, reproduced with their permission. *Flight International* magazine, data compiled from January 24th 1976, January 22nd 1977 and January 21st 1978 issues of *Flight International,*©*Flight International,* reproduced with their permission. 190/191: *Flight International* magazine, data compiled from January 21st 1971, January 20th 1972, January 18th 1973, January

17th 1974, January 23rd 1975, January 24th 1976, January 22nd 1977 issues of *Flight International,* © *Flight International,* reproduced with their permission. 192/193: *Sunday Times* files, London. 194/195: *Aerospace International* magazine, data compiled from "Security in the Air" by Chris Eliot in February/March 1978 issue of *Aerospace International,* published by Mönch Verlag, Germany, reproduced with their permission. *Flight International* magazine, data compiled from January 21st 1971, January 20th 1972, January 18th 1973, January 17th 1974, January 23rd 1975, January 24th 1976, January 22nd 1977 issues of *Flight International,* © *Flight International,* reproduced with their permission.

Diet
200/201: Euromonitor, data compiled from consumer Europe 1977, © Euromonitor publications ltd, reproduced with their permission. 202/203: Euromonitor, data compiled from *European Marketing Data and Statistics,* volume 14, 1977/8, and from *International Marketing Data and Statistics,* volume 3, 1977/8 © Euromonitor Publications Ltd., London, reproduced with their permission. U.S. Department of Agriculture, data provided by the Department for reproduction. 204/205: Brewers' Society, data compiled from the *1976 Brewers' Society Statistical Handbook,* London, England, reproduced with their permission. *The World Atlas of Wine,* revised edition by Hugh Johnson published by Mitchell Beazley Publishers, © 1977, data compiled from the above and reproduced with their permission. Heron House conversions.

Health: Mind and Body
210/211: World Health Organization, data compiled from *World Health Statistics Annual, 1977,* © W.H.O. 1977, reproduced with their permission. 212/213: Organization for Economic Co-operation and Development, data compiled from "O.E.C.D. Public Expenditure on Health," July 1977, © O.E.C.D. 1977, reproduced with their permission. 214/215: Euromonitor, data compiled from *Consumer Europe 1977,* London, © Euromonitor Publications Ltd., London, 1977, reproduced with their permission. Confidential industry source. *The Japanese Statistics Yearbook 1975.* 216/217: Heron House estimates based on World Health Organization figures, data compiled from *World Health Statistics Annual 1977,* © W.H.O. 1977, reproduced with their permission. 218/219: Commission on Narcotic Drugs, data compiled from "Drug Abuse: Extent, Pattern and Trends", Commission on Narcotic Drugs 5th special session. "Annual Abstract of Congression", data compiled from from "1976 C.I.S. Annual Abstract of Congression Drug Abuse". *Man Alive,* B.B.C. Television, April 4th 1978, reproduced with their permission. National statistical offices, data compiled by Heron House based on documents provided by the various government offices. United Nations, data compiled from documents provided by the 29th International Congress on Alcoholism and Drug Dependence 1970. 220/221: World Health Organization, data compiled from from "World Health Statistics Report", volume 30, number 3, 1977, © W.H.O. reproduced with their permission, Heron House estimates. Centre for Disease Control, data compiled from documents provided by the Centre for Disease Control, U.S. Department of Health, Education and Welfare. H.M.S.O., data compiled from "On the State of Public Health", © H.M.S.O. 1976, reproduced with the permission of the Controller of Her Britannic Majesty's Stationery Office. Office of Health Economics, data compiled from documents provided by the Office of Health Economics, Paris. Department of Health, data compiled from documents provided by the Department of Health, Dublin, Eire. 222/223: Addiction Research Foundation, Toronto, data compiled from "The Epidemiology of Alcoholism: The Elusive Nature of the Problem, Estimating the Prevalence of Excessive Alcohol Use and Alcohol - Related Mortality, Current Trends and Issue of Prevention", by J. de Lint, reproduced by permission of the author. Government sources, data compiled by Heron House based on documents provided by the various government offices. 224/225: World Health Organization, data compiled from *World Health Statistics Annual 1977,* © W.H.O. 1977, reproduced with their permission. 226/227: World Health Organization, data compiled from *World Health Statistics Annual 1977,* © W.H.O. 1977, reproduced with their permission. National statistical offices, data compiled by Heron House based on documents provided by various government offices. 228/229: National statistical offices, relevant Commissions, British Pregnancy Advisory Service and International Planned Parenthood Federation files, data compiled from documents provided by the above. 230/231: World Health Organization, data compiled from *World Health Statistics Annual 1977,* © W.H.O. 1977, reproduced with their permission. 232/233: World Health Organization, data compiled from *World Health Statistics Annual 1977,* © W.H.O. 1977, reproduced with their permission. 234/235: World Health Organization, data compiled from *World Health Statistics Annual 1977,* © W.H.O. 1977, reproduced with their permission. 236/237: World Health Organization, data compiled from "World Health Statistics Report", volume 28, number 6, 1975, © W.H.O. 1975, reproduced with their permission. 238/239: Department of Health and Social Security, London. World Health Organization, data compiled from *World Health Statistics Annual 1977,* © W.H.O. 1977, reproduced

with their permission. *Canada Yearbook 1976-77.*

Sex

244/245: "The Virginia Slims American Women's Opinion Poll", The Roper Organization Inc., 1974 (for U.S.A.). *Honey* magazine, May and June issues 1977, survey conducted by the Schlachman Research Organization Ltd., London (for U.K.). *Cleo* magazine, August and September 1974 issues, survey conducted by Roy Morgan Research Centre Pty. Ltd. (for Australia). "Allensbacher Berichten 1970", Institut für Demoskopie Allensbach. All data reproduced with their permission. 248/251: *Cleo* magazine, August and September 1974 issues, survey conducted by Roy Morgan Research Centre Pty. Ltd. (for Australia), reproduced with their permission. "Family Planning Perspectives", volume 9, number 2, 1977, from "Sexual and Contraceptive Experience of Young Unmarried Women in the U.S.A., 1976-71" by Melvin Zelnik Ph.D. and John F. Kantner Ph.D. (for U.S.A.), reproduced with their permission. "Riksforbundet für Sexuel Upplysning", Stockholm (for Sweden). *The Sexual Behaviour of Young Adults* by Michael Schofield, published by Allen Lane, 1973, © Michael Schofield 1973, reproduced with the permission of Penguin Books Ltd. (for U.K.). 252/255: "The Virginia Slims American Women's Opinion Poll", The Roper Organization Inc., 1974 (for U.S.A.). *Rapport sur le Comportement Sexuel des Français* by Dr Pierre Simon, published by Rene Julliard, Pierre Charron, 1972, survey conducted by *l'Institut Français d'Opinion Publique* (for France). *Weekend* magazine, Toronto, December 3rd 1977 issue (for Canada). *Honey* magazine, May and June 1977, survey conducted by the Schlachman Research Organization Ltd., London (for U.K.). All data reproduced with their permission. 256/257: Gallup International Affiliated Institutes in the countries concerned, survey carried out in 1973 sampling attitudes of young people aged 18-24, data compiled from the "Gallup Opinion Index" U.S.A., report number 100, October 1973, reproduced with their permission. A list of the participating institutes is given under Chapter 18 bibliography. 258/259: "Family Planning Perspectives", volume 9, number 2, 1977 from "Sexual and Contraceptive Experience of Young Unmarried Women in the U.S.A., 1976-71", by Melvin Zelnick Ph.D. and John F. Kantner, Ph.D. (for U.S.A.), *Honey* magazine, May and June 1977 issues, survey conducted by the Schlachman Research Organ-.ization Ltd., London (for U.K.). "Sexualiteit in Nederland", published in *Margriet*, a women's magazine, 1968. All data reproduced with their permission. 260/263: "Sexual Experience, Birth Control Usage and Sources of Sex Education among Unmarried University Students", by Dr.

M. Barrett and Dr. M. Fitz-Earle, 1974 (for Canada). "Riksforbundet für Sexual Upplysning", Stockholm (for Sweden). *Rapport sur le Comportement Sexuel de Français* by Dr. Pierre Simon, published by Rene Julliard, Pierre Charron, 1972, survey conducted by *l'Institut Français d'Opinion Publique* (for France). "Family Planning Perspective", volume 9, number 2, 1977, from "Sexual and Contraceptibe Experience of Young Unmarried Women in the U.S.A., 1967-77", by Melvin Zelnik Ph.D. and John F. Kantner Ph.D. (for U.S.A.). *Honey* magazine, May and June 1977 issues, survey conducted by the Schlachman Research Organization Ltd., London (for U.K.). "Sexualiteit in Nederland", published in the *Margriet,* magazine 1968. All data reproduced with their permission. 264/267: "Family Planning Perspectives", volume 9, number 2, 1977 from "Sexual and Contraceptive Experience of Young Unmarried Women in the U.S.A. 1976-71", by Melvin Zelnik Ph.D. and John F. Kantner Ph.D. (for U.S.A.), reproduced with their permission. *L'Express,* September 5th-11th 1977 (for France), reproduced with their permission. Michael Schofield, *The Sexual Behaviour of Young Adults,* published by Allen Lane, 1973, © Michael Schofield, 1973, reproduced with the permission of Penguin Books Ltd. (for U.K.). "Riksforbundet fur Sexuel Upplysning", Stockholm (for Sweden). *Honey* magazine, May and June 1977 issues, survey conducted by the Schlachman Research Organization Ltd., London (for U.K.). *Cleo* magazine, August and September issues, survey conducted by Roy Morgan Research Centre Pty. Ltd. (for Australia). All data reproduced with their permission. 270/271: United Nations, data compiled from document ESA/P/WP.59, January 30th 1976, "Up-dated Study of Urban-Rural Differences in the Marital Status Composition of the Population", prepared by the Department of Economic and Social Affairs, © United Nations, reproduced with their permission. 272/273: United Nations, data compiled from the *United Nations 1976 Statistical Yearbook,* 28th issue, New York 1977, © United Nations, reproduced with their permission. 274/275: *Sex and Marriage in England Today* by Geoffrey Gorer, published by Thomas Nelson & Sons Ltd., London, 1971 (for U.K.), reproduced with their permission. Survey conducted by Isopublic, Zurich, on behalf of *Weltwoche* (for Switzerland), reproduced with their permission. The Canadian Gallup Poll (for Canada). The National Opinion Research Centre of the University of Chicago (for U.S.A.). "Sexualiteit in Nederland" published in *Margriet* magazine, 1968 (for Netherlands). 276/277: *Sex and Marriage in England Today* by Geoffrey Gorer, published by Thomas Nelson & Sons Ltd., London, 1971 (for U.K.).

The Sex Survey of Australian Women by Prof. Robert E. Bell, published by Sun Books, 1974 (for Australia). "Sexualiteit in Nederland", published in *Margriet* magazine, 1968 (for Netherlands). All data reproduced with their permission. 278/279: *Weekend* magazine, Toronto, December 3rd and 17th 1977 issues and The Canadian Gallup Poll Ltd. (for Canada). The Gallup Organization Inc., U.S.A. and The National Opinion Research Center of the University of Chicago (for U.S.A.). Social Surveys (Gallup Poll) Ltd. (for U.K.). "Sexualiteit in Nederland", published in *Margriet* magazine, 1968 (for Netherlands).

Citizenship
286/287: Government documents, data compiled from documents provided by various government offices. Election Research Center, Washington, D.C. 288/289: Thomas P. Kane, Europe's Guest Workers, Intercom, January 1978 (Population Reference Bureau, Washington DC). National statistical offices, data compiled by Heron House based on material provided by statistical offices. 290/291: National statistical offices. Government departments, data compiled from documents provided by various government offices. Embassies, data provided by various embassies in London. 292/293: International Institute for Strategic Studies, London, data compiled from *The Military Balance 1977-78,*© International Institute for Strategic Studies, 1977, reproduced with their permission. 294/295: "Amnesty International Report", 1977, data extracted from the Amnesty International Report 1977, reproduced with their permission.

Education
300/301: Organization for Economic Co-operation and Development, data compiled from *The Educational Situation in OECD Countries,* © O.E.C.D. reproduced with their permission. 302/303: Organization for Economic Co-operation and Development, data compiled from *The Educational Situation in O.E.C.D. Countries,* © O.E.C.D. reproduced with their permission. United Nations Educational, Scientific, and Cultural Organization, data compiled from *U.N.E.S.C.O. Statistical Yearbook 1976,* © U.N.E.S.C.O. 1977, reproduced with their permission. 304/305: United Nations Educational, Scientific, and Cultural Organization, data compiled from *U.N.E.S.C.O. Statistical Yearbook 1976,* © U.N.E.S.C.O. 1977, reproduced with their permission. 306/307: United Nations Educational, Scientific, and Cultural Organization, data compiled from *U.N.E.S.C.O. Statistical Yearbook 1976,* © U.N.E.S.C.O. 1977, reproduced with their kind permission. Euromonitor, data compiled from *European Marketing Data and Statistics 1977/78,* © Euromonitor Publications Ltd., London, reproduced with their permission. Government statistical offices.

Crime
312/313: Euromonitor, data compiled from *European Marketing Data and Statistics,* volume 12, 1975/76, © Euromonitor Publications Ltd., London, reproduced with their permission. 314/315: Data calculated by Heron House using data provided by the Federal Bureau of Investigation and by the International Criminal Police Organization, Paris. 316/317: Interpol, data compiled from *International Crime Statistics 1975,* published by the International Criminal Police Organization, Paris, reproduced with their permission. F.B.I., data supplied by the Federal Bureau of Investigation. 318/319: Government Departments of Justice, data supplied by the Department of Justice of the countries concerned. 320/321: Interpol, data compiled from *International Crime Statistics 1975,* published by the International Criminal Police Organization, Paris, reproduced with their permission. F.B.I. data supplied by the Federal Bureau of Investigation. 322/323: United Nations, data extracted from document numberA/CONF/56/6, New York 1975, presented to the Fifth United Nations Congress on the Prevention of Crime and the Treatment of Offenders.

What People Use
328/329: Confidential industry sources. 330/331: Confidential industry sources. 332/333: Confidential industry sources. 334/335: Euromonitor, data compiled from *Consumer Europe 1976,* © Euromonitor Publications Ltd., London, reproduced with their permission. Confidential industry sources. 336/337: Confidential industry source. 338/339: Euromonitor, data compiled from *Consumer Europe 1977,* © Euromonitor Publications Ltd., London, reproduced with their permission. Confederation of British Industry, data based on *Western European Living Costs 1977,* © Confederation of British Industry, reproduced with their permission. Confidential industry sources. 340/341: Confidential industry sources. 342/343: Confidential industry sources. 344/345: Organization for Economic Co-operation and Development, data reproduced from the *O.E.C.D. Observer,* number 91, March 1978, © O.E.C.D. reproduced with their permission. Euromonitor, data compiled from *European Marketing Data and Statistics 1977/78,* volume 14 and from *International Marketing Data and Statistics, 1977/78,* volume 3, © Euromonitor Publications Ltd., London, reproduced with their permission.

What People Think
350/359: Gallup International Affiliated Institutes in the countries concerned, the surveys were carried out during the fall and winter of 1974/5 and during the spring of 1976, reproduced with their permission. The list of participating institutes is given hereafter. 360/363: Gallup International Affiliated Institutes in the countries

concerned, data compiled from "Women in America", U.S.A. Gallup Opinion index 128, March 1976, reproduced with their permission. The list of participating institutes is given hereafter. 364/365: Commission of the European Communities, data compiled from *European Men and Women*, December 1975, Brussels, Gallup International Affiliated Institutes in the countries concerned carried out the survey in May 1975, reproduced with their permission. Gallup Organization Inc., data compiled from the "Gallup Opinion Index", U.S.A. number 128, March 1976, reproduced with their permission. 366/367: Commission of the European Communities, data reproduced with their permission. 402/403: Gallup International Affiliated Institutes in the countries concerned, the surveys were carried out during the fall and winter of 1974/5 and during the spring of 1976, reproduced with their permission. The list of participating institutes is given hereafter. 370/371: Commission of the European Communities, data compiled from *European Consumers 1976*, Brussels, survey carried out in 1975 by Gallup International Affiliated Institutes in the countries concerned, reproduced with their permission. Opinion Research Corporation, data compiled from "O.R.C. Public Opinion Index", May 1976, survey carried out in 1976, reproduced with their permission. 372/373: Commission of the European Communities, data compiled from *Euro-Barometer* number 8, January 1978, published by the Commission of the European Communities, Brussels, reproduced with their permission, survey carried out by the Gallup International Affiliated Institutes in the countries concerned. The list of participating institutes is given hereafter. 374/375: Commission of the European Communities, data compiled from *European Consumers 1976*, Brussels, survey carried out in 1975 by Gallup International Affiliated Institutes in the countries concerned, reproduced with their permission. Opinion Research Corporation, data compiled from "O.R.C. Public Opinion Index", May 1976, survey carried out in 1976, reproduced with their permission. 378/379: Commission of the European Communities, data compiled from *European Consumers, 1976*, Brussels, survey carried out in 1975 by Gallup International Affiliated Institutes in the countries concerned, reproduced with their permission. Opinion Research Corporation, data compiled from "O.R.C. Public Opinion Index", May 1976, survey carried out in 1976, reproduced with their permission. 380/381: Market and Opinion Research International, London; research conducted on behalf of the Daily Express in November 1977, reproduced with the permission of MORI. 482/483: "McCann-Erickson Youth Study", 1976/77, the survey was carried out between June 1976 and April 1977, the coverage

was 10 to 25-year-olds except in Scandinavia and Spain where it was 15 to 25, the survey includes Japan and Brazil and is still being extended, reproduced with their permission. 384/387: Gallup International Affiliated Institutes in the countries concerned, the surveys were carried out in 1973 sampling attitudes of young people aged 18 to 24, data compiled from the "Gallup Opinion Index", U.S.A., Report number 100, October 1973, reproduced with their permission. The list of participating institutes is given hereafter. 388/389: "McCann-Erickson Youth Study", 1976/77, the survey was carried out between June 1976 and April 1977, the coverage was 10 to 25-year-olds except in Scandinavia and Spain where it was 15 to 25, the survey includes Japan and Brazil and is still being extended, reproduced with their permission. 390/391: Gallup International Affiliated Institutes in the countries concerned, the survey was carried out in 1973 sampling attitudes of young people aged 18 to 24, data compiled from the "Gallup Opinion Index", U.S.A., Report number 100, October 1973, reproduced with their permission. The list of participating institutes is given hereafter. 392/393: "McCann-Erickson Youth Study", the survey was carried out between June 1976 and April 1977, the coverage was 10 to 25-year-olds except in Scandinavia and Spain where it was 15 to 25, the survey includes Japan and Brazil and is still being extended, reproduced with their permission. 394/395: Gallup International Affiliated Institutes in the countries concerned, the survey was carried out in 1973 sampling attitudes of young people aged 18 to 24, data compiled from the "Gallup Opinion Index", U.S.A., Report number 100, October 1973, reproduced with their permission. The list of participating institutes is given hereafter.

List of all companies, members of Gallup International Affiliated Research Institutes, who have participated in the surveys quoted above:
Australia: The Roy Morgan Research Centre Pty., Ltd., Melbourne
Brazil: Instituto Gallup de Opiniao Publica, Sao Paulo
Canada: The Canadian Gallup Poll Ltd., Toronto
Denmark: Gallup Markedsanalyse A/S, Copenhagen
Finland: Suomen Gallup O/Y, Helsinki
France: I.F.O.P.-E.T.M.A.R., Paris
Germany: E.M.N.I.D. Institut Gmbh & Co, Bielefeld
India: The Indian Institute of Public Opinion Ltd., Delhi
Ireland: Irish Marketing Surveys Ltd., Dublin
Italy: D.O.X.A., Milan
Japan: Nippon Research Center Ltd., Tokyo
Netherlands: N.I.P.O. het Nederlands Instituut voor de Publieke Opinieen het Marktonderoek

B.V., Amsterdam
Norway: Norsk Opinionsinstitutt A-S, Oslo
Sweden: Swedish Institute of Public Opinion
Research Ltd., Stockholm
Switzerland: I.S.O.P.U.B.L.I.C., Zurich
United Kingdom: Social Surveys (Gallup Poll)
Ltd., London
United States: The Gallup Organization Inc.,
Princeton, New Jersey
Yugoslavia: Zavod za Trzisna Istrazivanga
(Z.I.T.), Belgrade

INDEX

BE A
BOOK OF NUMBERS
CONTRIBUTOR

fill in the questionnaire which follows

QUESTIONNAIRE

Be a Book of Numbers Contributor

While we have scoured the world's resources to come up with the answers contained in this book, there are many other questions we would have liked to answer. This is particularly true since we plan to publish a new *Book of Numbers* in about two years time.

One way to get additional answers is to commission our own research. We are doing that.

Another important way is to ask readers of the first edition for their answers to vital questions. The questionnaire which follows does so.

If you would be good enough to fill in this questionnaire and mail it to us, we would be ever so grateful. If you complete the questionnaire you have the opportunity to be listed as a research contributor in your country's next edition of the work.

As you can appreciate it is difficult to pose questions in an unbiased manner. We have certainly attempted to do so and have no prejudgements on the topics raised below. However, the editors of Heron House take full responsibility for the questions and the way these have been posed. This has not been done in consultation with the many excellent research organisations which have assisted us.

It is possible that in some cases you will disagree with the choices you can circle under a question. Many of these are, after all, complicated topics. In this case we would suggest that you disregard the question rather than circle a view which doesn't accurately reflect your own.

You can complete the questionnaire in one of two ways: either by cutting out the appropriate pages or by taking a separate sheet of paper and writing your answers on it, and then sending them to us.

Thus, for example, if you believe on question 2, that your life is much more satisfactory than it was five years ago, you would put 2 a) on a piece of paper; if on question 5 you believe that police protection is excellent you would list this as 5a)i).

It is vital that you also give us the personal data about yourself at the end of the questionnaire so that we can compare response by sex, age, geographic location, etc. We guarantee that all returned questionnaires will be kept in strict confidence. If you want to be listed as a contributor in the next edition, please give your name.

Here then are the questions:

1 Would you please rate the following personal problems by ticking the appropriate boxes:

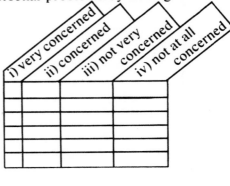

	i) very concerned	ii) concerned	iii) not very concerned	iv) not at all concerned
a) Taxes are too high				
b) Fear of crime				
c) Unhappy family/personal relations				
d) Can't afford children's higher education				
e) Standard of living is declining				
f) Not enough time for family/friends				

2 Would you say that compared with five years ago your life today is: (please circle the letter)
a) much more satisfactory/ b) more satisfactory/ c) about the same/ d) less satisfactory/ e) much less satisfactory.

3 Would you say that you get from your work: (please circle)
a) a great deal of satisfaction/ b) some satisfaction/ c) indifference/ d) relatively little satisfaction/ e) virtually no satisfaction.

4 Would you say that in your life religion is: (please circle)
a) very important/ b) somewhat important/ c) not very important/ d) not at all important.

5 Would you please rate the following common services by ticking the appropriate boxes:

	i) excellent	ii) good	iii) average	v) very bad	iv) fairly poor
a) Police protection					
b) Mail delivery					
c) Public transportation					
d) Telephone service					
e) Television					

6 In recent years a great deal of discussion has centred on racial equality. What are your feelings on this topic: (please circle)
a) far too little has been done/ b) not enough has been done/ c) about the right amount has been done/ d) too much has been done/ e) far too much has been done.

7 In recent years a great deal of discussion has centered on sexual equality at work and elsewhere. What are your feelings on this topic: (please circle)
a) far too little has been done/ b) not enough has been done/ c) about the right amount has been done/ d) too much has been done/ e) far too much has been done.

8 Under what conditions do you feel your country should intervene in a war: (please circle)
a) to assist a threatened ally/ b) to assist an invaded ally/ c) only if my country is directly threatened/ d) only if my country is invaded/ e) never.

9 Which of the following do you think is the biggest threat confronting the world today: (please circle)
a) terrorism/ b) inflation/ c) unemployment/ d) collapse of world monetary system/ e) population explosion/ f) increased size of government/ g) holocaust/ h) other please specify) ....
....

10 With respect to couples living together before marriage do you feel it's all right if: (please circle)
a) the couple are engaged/ b) the couple are in love but not yet engaged/ c) the couple know one another well/ d) they both simply want to/ e) unacceptable under any conditions.

11 Would you please list below in order the three well known living figures whom you respect most:

A. ....

B. ....

C. ....

12 Would you please list below in order the three well known living figures whom you respect least:

A. ....

B. ....

C. ....

13 Listed below are a variety of professions. Would you please circle the number from 1-10 in terms of your respect for and trust in each. Thus, if you think lawyers command 100% of your respect and trust you would give them a '10', whereas if you only think they command 30% of your respect and trust you would give them a '3'.

a) Politician	1 2 3 4 5 6 7 8 9 10
b) Professional soldier	1 2 3 4 5 6 7 8 9 10
c) Advertising executive	1 2 3 4 5 6 7 8 9 10
d) Businessman	1 2 3 4 5 6 7 8 9 10
e) Lawyer	1 2 3 4 5 6 7 8 9 10
f) Government Official	1 2 3 4 5 6 7 8 9 10
g) Doctor	1 2 3 4 5 6 7 8 9 10
h) Teacher	1 2 3 4 5 6 7 8 9 10
i) Minister	1 2 3 4 5 6 7 8 9 10

14 Do you feel abortion should be: (please circle)
a) permitted under all circumstances/ b) permitted when both partners agree/ c) permitted when woman wants/ d) no opinion/ e) not permitted save under exceptional circumstances/ f) totally outlawed/ g) other (please specify) ....

15 In your judgement what is the biggest loss your country has suffered in the last ten years: (please circle)
a) too much welfare state/ b) government corruption /c) lack of economic stability/ d) life far less personalized/ e) problem occasioned by drugs and crime/ f) erosion of traditional values/ g) disintegration of family values/ h) other (please specify) ....

16 Law and Order is a major issue today. Do you feel: (please circle)
a) the police need far more power/ b) the current laws are adequate/ c) we need to do more to protect suspects/ d) other (please specify)...............
....

17 Do you feel unions have: (please circle)
a) far too much power/ b) too much power/ c) the right amount of power/ d) too little power/ e) far too little power.

18 What do you feel is our biggest central problem in the educational system today: (please circle)
a) too discriminatory/ b) pupil to teacher ratio declining/ c) children's reliance on drugs and alcohol/ d) peer group pressure towards lack of discipline/ e) other (please specify)...
....

19 Tobacco is a controversial issue today. Do you feel that tobacco smoking should be: (please circle)
a) unrestricted/ b) subject to less controls than it is today/ c) allowed to remain as it is/ d) subject to more controls than it is today/ e) prohibited.

20 Some people strongly believe that a family should have no more than two children to keep population at its present level or (given infant mortality) to make it slightly decline. Do you: (please circle)
a) strongly agree/ b) agree/ c) no opinion/ d) disagree/ e) strongly disagree.

21 Unfaithfulness in marriage is wrong:
a) agree /b) agree when the marriage is patently unhappy /c) think its natural /d) disagree under all but exceptional circumstances /e) totally disagree.

22 Some people believe that the world is undergoing an "energy crisis". Do you: (please circle)
a) strongly agree/ b) agree/ c) no opinion/ d) disagree/ e) strongly disagree.

23 Would you be prepared to pay a higher price for petrol and/or have a less comfortable car to save energy: (please circle)
a) pay a higher price for petrol /b) have a less comfortable car /c) pay a higher price for petrol and have a less comfortable car /d) neither /e) no opinion.

24 Energy generated by a nuclear power plant is a current issue. What do you think about nuclear power plants: (please circle)
a) approve/ b) disapprove.

25 Illegal aliens are a current issue. Which of the opinions below most accurately summarises your feelings on this topic: (please circle)
a) they should all be sent home/ b) those who might be undesirable should be sent home/ c) those presently in country should be allowed the opportunity to stay if they have no criminal record/ d) tighter controls should be instituted for the future but existing illegal aliens and their family members should be allowed to stay.

26 Some people feel protecting our environment is important – others feel that our environment should suffer if it would create jobs. Given this stark choice, which do you favour: (please circle)
a) jobs/ b) environment.

27 Do you feel there is too much violence and sex on television and in films: (please circle)
a) strongly agree/ b) agree/ c) no opinion/ d) disagree/ e) strongly disagree.

28 Do you feel in today's world the United Nations is: (please circle)
a) very important/ b) moderately important/ c) no opinion/ d) moderately unimportant/ e) very unimportant.

29 Do you think that we should reduce our standard of trying to help that of the third world: (please circle)
a) strongly agree/ b) agree/ c) no opinion/ d) disagree/ e) strongly disagree.

30 Pollution standards are a major issue today. There are many ways of improving pollution standards. Would you please circle below your opinion on these different ways:
a) willing to pay significantly more for petrol /b) willing to pay more for my car to make it "cleaner" /c) willing to pay to insulate my home to make it more efficient /d) willing to pay increased taxes to improve waste disposal /e) willing to pay more for products to eliminate industrial pollution.

31 Do you believe advertising is: (please circle)
a) misleading and should be eliminated/ b) an area which should be under much tighter control/ c) not important/ d) important to the economy/ e) normally informative and truthful.

32 Do you believe most people cheat on their taxes: (please circle)
a) yes/ b) no.

33 Do you believe that homosexuality/lesbianism amongst consenting adults should remain legal: (please circle)
a) strongly agree/ b) agree/ c) no opinion/ d) disagree/ e) strongly disagree.

34 Do you believe that we need: (please circle)
a) more capitalism/ b) to remain at our current level of capitalism versus socialism/ c) more socialism.

35 Do you believe that emphasising human rights in the Eastern block countries is: (please circle)
a) the only proper thing to do/ b) no opinion/ c) not our business.

36 Soccer hooliganism is a major problem today. Which of the following

do you agree with: (please circle)
a) soccer hooligans should be imprisoned/b)soccer hooligans should be banned from future matches/c) no opinion/d) soccer hooliganism is no worse in Britain than elsewhere/e) soccer hooliganism is merely "youthful high spirits".

37 Do you feel the House of Lords should be abolished: (please circle)
a) strongly agree/ b) agree/ c) no opinion/ d) disagree/ e) strongly disagree.

38 Northern Ireland has been much in the news. Which of the following statements most clearly reflects your own point of view: (please circle) a) we should get troops out of Northern Ireland/b) this is Britain's Vietnam – there appears no way out/c) no opinion/d) we should continue to pursue present policies/e) we should crack down more on militants on both sides/f) other (please specify) . . . .
. . . .

39 Do you feel that in joining the Common Market we made the right move: (please circle)
a) strongly agree/ b) agree/ c) no opinion/ d) disagree/ e) strongly disagree.

40 Do you feel that Scotland deserves more independence: (please circle)
a) strongly agree/ b) agree/ c) no opinion/ d) disagree/ e) strongly disagree.

41 Do you feel that Wales deserves more independence: (please circle)
a) strongly agree/ b) agree/ c) no opinion/ d) disagree/ e) strongly disagree.

42 How important do you believe the British monarchy is to our system: (please circle)
a) vital/b) very important/c) important/d) no opinion/e) not very important/f) extremely unimportant/g) hurts our system.

43 Do you feel nationalised industries are: (please circle)
a) markedly more efficient than private industry/b) modestly more efficient than private industry/c) no opinion/d) moderately less efficient than private industry /e) markedly less efficient than private industry.

44 Do you feel Concorde was worth the money: (please circle)
a) strongly agree/ b) agree/ c) no opinion/ d) disagree/ e) strongly disagree.

45 Some people say that class still plays a major role in where one gets in life. Do you: (please circle)
a) strongly agree/b) agree/c) no opinion/d) modestly disagree/e) strongly disagree.

46 Some people think British pub hours should be extended. Do you: (please circle)
a) strongly agree/b) agree/c) no opinion/d) disagree/e) strongly disagree.

47 Some people think children should be allowed into pubs as a matter of course. Do you: (please circle)

a) strongly agree/b) agree/c) no opinion/d) disagree/e) strongly disagree.

48 Do you agree that the Government should introduce a wealth tax: (please circle)
a) strongly agree/ b) agree/ c) no opinion/ d) disagree/ e) strongly disagree.

49 Do you agree that we should have capital punishment: (please circle)
a) strongly agree/ b) agree/ c) no opinion/ d) disagree/ e) strongly disagree.

In order to accurately report responses to the questionnaire by different types of people, we would appreciate receiving the following personal information. Dependent upon the degree to which you are willing to tell us these facts, our research will be more accurate: (please tick where appropriate)

1 Male ☐   Female ☐
2 Age
3 Income:
    £3,000 or less ☐
    £3,000 – £4,000 ☐   £7,500 – £10,000 ☐
    £4,000 – £5,000 ☐   £10,000 – £15,000 ☐
    £5,000 – £7,500 ☐   Over £15,000 ☐

4 Married ☐ Single ☐ Divorced ☐ Widowed. ☐
5 Education
        Didn't complete secondary school
        Completed secondary school
        Some further education
        Completed degree or diploma
        Advanced degree
6 Religion:
Protestant ☐   Catholic ☐
other (please specify)

7 Name of town and county in which you live:

....

8 Your name if you wish to be included in the credits for the next work.

          Please send your completed questionnaire to:
          The Research Director
          The Book of Numbers
          Heron House
          Chiswick Mall
          London W4 2PR
          England